**Welcome to *My Math*** – your very own math book! You can write in it – in fact, you are encouraged to write, draw, circle, explain, and color as you explore the exciting world of mathematics. Let's get started. Grab a pencil and finish each sentence.

My name is ______________________________.

My favorite color is ______________________________.

My favorite hobby or sport is ______________________________.

My favorite TV program or video game is

______________________________.

My favorite class is ______________________________.

**mhmymath.com**

**STEM** McGraw-Hill is committed to providing instructional materials in Science, Technology, Engineering, and Mathematics (STEM) that give all students a solid foundation, one that prepares them for college and careers in the 21st century.

Send all inquiries to:
McGraw-Hill Education
8787 Orion Place
Columbus, OH 43240

ISBN: 978-0-07-905767-9 (***Volume 1***)
MHID: 0-07-905767-5

Printed in the United States of America.

6 7 8 9 LWI 23 22 21 20

Grade K • Volume 1

**Authors:**

Carter • Cuevas • Day • Malloy
Altieri • Balka • Gonsalves • Grace • Krulik • Molix-Bailey
Moseley • Mowry • Myren • Price • Reynosa • Santa Cruz
Silbey • Vielhaber

# GO digital ▶ connectED.mcgraw-hill.com

## ▶ Log In

1 Go to **connectED.mcgraw-hill.com.**

2 Log in using your username and password.

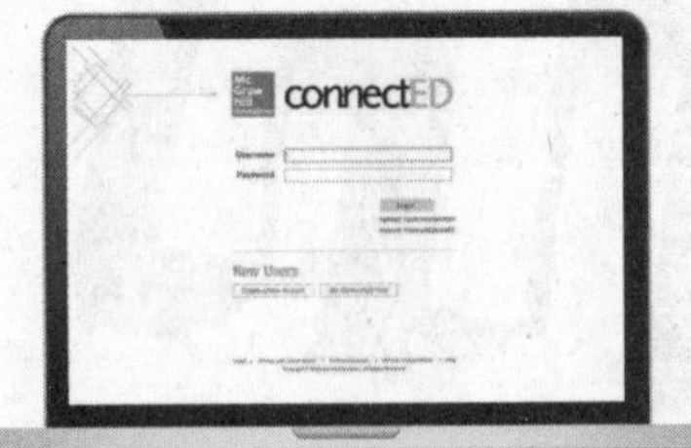


3 Click on the Student Edition icon to open the Student Center.

## ▶ Go to the Student Center

4 Click on Menu, then click on the **Resources** tab to see all of your online resources arranged by chapter and lesson.

5 Click on the **eToolkit** in the Lesson Resources section to open a library of eTools and virtual manipulatives.

6 Look here to find any assignments or messages from your teacher.

7 Click on the **eBook** to open your online Student Edition.

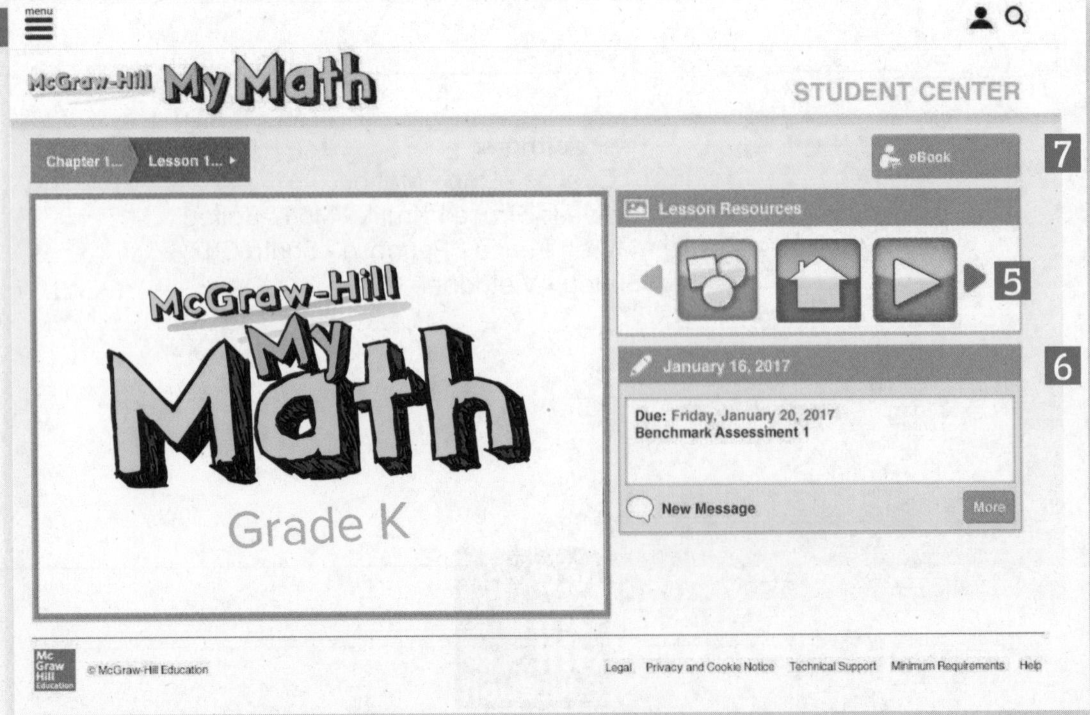

Username:

Password:

# Explore the eBook!

8 Click the **speaker icon** at the top of the eBook pages to hear the page read aloud to you.

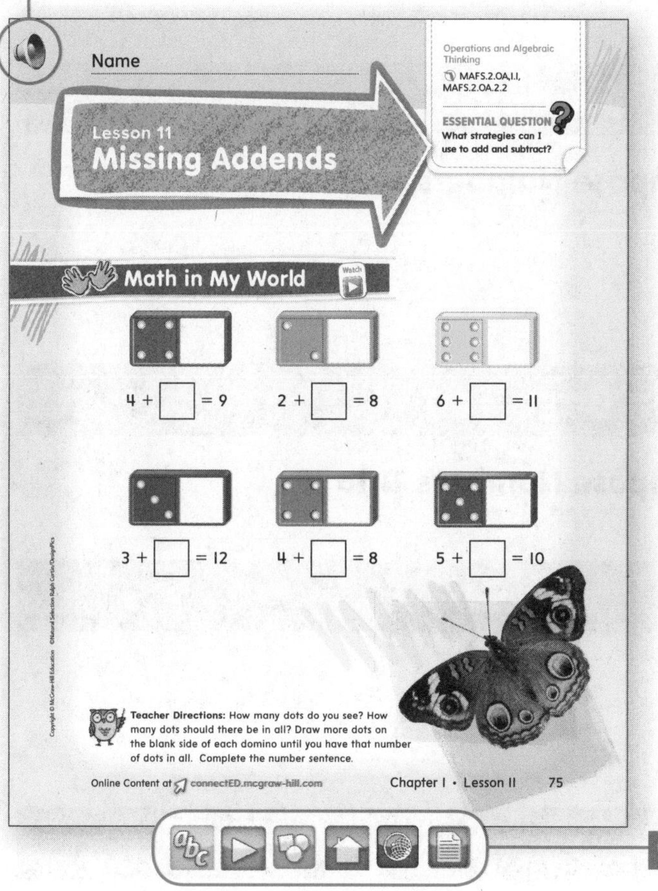


More resources can be found by clicking the icons at the bottom of the eBook pages.

 Practice and review your Vocabulary.

 Animations and videos allow you to explore mathematical topics.

 Explore concepts with eTools and virtual manipulatives.

 eHelp helps you complete your homework.

 Explore these fun digital activities to practice what you learned in the classroom.

 Worksheets are PDFs for Math at Home, Problem of the Day, and Fluency Practice.

# Contents in Brief

## Organized by Domain

Chapter

# Numbers 0 to 5

Counting and Cardinality

ESSENTIAL QUESTION

How do we show how many?

## Getting Started



## Lessons and Homework



## Wrap Up

There are Brain Builders problems in every lesson.

Chapter

# Numbers to 10

Counting and Cardinality

ESSENTIAL QUESTION

What do numbers tell me?

## Getting Started



## Lessons and Homework



## Wrap Up



We like healthful food!

Chapter

# Numbers Beyond 10

Counting and Cardinality

ESSENTIAL QUESTION
How can I show numbers beyond 10?

## Getting Started



## Lessons and Homework



## Wrap Up



Look for this!
eHelp
Click online and you can get more help while doing your homework.

Operations and Algebraic Thinking

ESSENTIAL QUESTION
How can we show a number in other ways?

# Chapter 4 Compose and Decompose Numbers to 10

## Getting Started



## Lessons and Homework



## Wrap Up

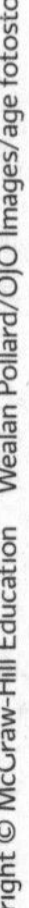

# Chapter 5 Addition

ESSENTIAL QUESTION

How can I use objects to add?

## Getting Started



## Lessons and Homework



## Wrap Up



Party time!

Look for this! Click online and you can find tools that will help you explore concepts.

Operations and Algebraic Thinking

Chapter 6

# Subtraction

ESSENTIAL QUESTION

How can I use objects to subtract?

## Getting Started



## Lessons and Homework



## Wrap Up

Chapter

# Compose and Decompose Numbers 11 to 19

ESSENTIAL QUESTION

How do we show numbers 11 to 19 in another way?

## Getting Started



## Lessons and Homework



## Wrap Up



Vocab abc

**Look for this!** Click online and you can find activities to help build your vocabulary.

Let it snow!

Chapter

# Measurement

Measurement and Data

ESSENTIAL QUESTION

How do I describe and compare objects by length, height, and weight?

## Getting Started



## Lessons and Homework



## Wrap Up

Chapter

# Classify Objects

Measurement and Data

ESSENTIAL QUESTION

How do I sort objects?

## Getting Started



## Lessons and Homework



## Wrap Up



Bright idea!

Chapter

# 10 Position

Geometry

**ESSENTIAL QUESTION**
How do I identify positions?

## Getting Started



## Lessons and Homework



## Wrap Up

Chapter

# 11 Two-Dimensional Shapes

Geometry

ESSENTIAL QUESTION

How can I compare shapes?

## Getting Started



## Lessons and Homework



## Wrap Up



Let's learn shapes!

Chapter 12

# Three-Dimensional Shapes

Geometry

**ESSENTIAL QUESTION**

How do I identify and compare three-dimensional shapes?

## Getting Started



## Lessons and Homework



## Wrap Up

Shapes are fun!

Chapter 1

# Numbers 0 to 5

**ESSENTIAL QUESTION**

How do we show how many?

## Let's Go to the Farm!

Watch a video!

Name ________________

# Chapter 1 Project

## Create a Mural

1. What objects would your group like to draw?
2. Write the number of objects your group will be drawing in the box below.
3. Plan how your group will draw the objects on the mural. Create your own drawing of it in the box below.

Name

# Am I Ready?

1

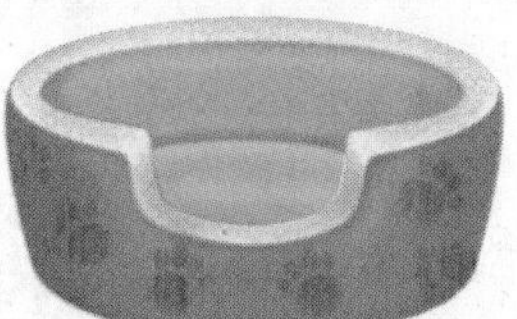

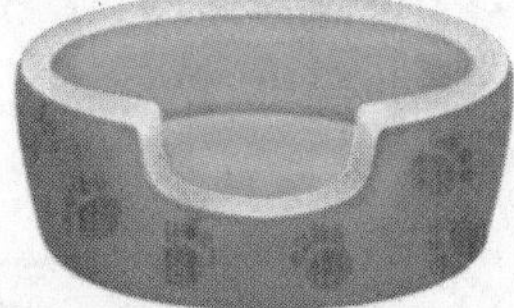

2

3

4

**Directions: 1.** Match each cat to one cat bed by drawing a line from each cat to a bed. **2.** Color three apples. **3.** Look at the picture. Draw that many flowers. **4.** Look at the picture. Draw that many baseballs.

Name

# My Math Words

## Review Vocabulary

more

less

**Directions:** Say each word. Trace each word. Tell how the groups of corn show the meaning of each word.

# My Vocabulary Cards

count

equal to

five 5

four 4

greater than

less than

**Teacher Directions:**
**Ideas for Use**

- Have students choose a number card from 1 to 4. Tell them to work with a classmate to find the card that shows one more.
- Ask each student to choose a card and draw a picture to show its meaning. Have them work with a classmate to guess which word they chose.

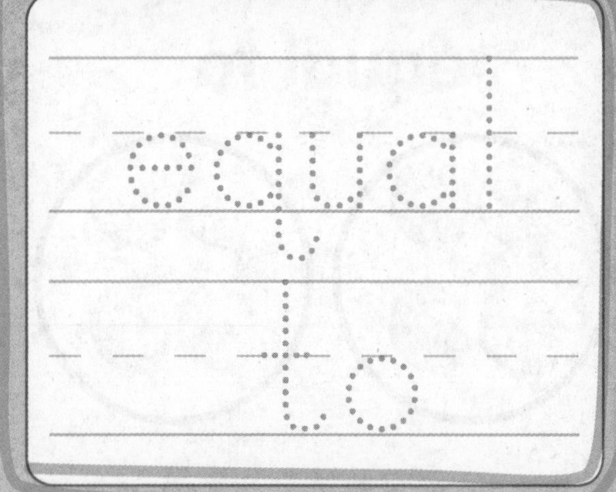

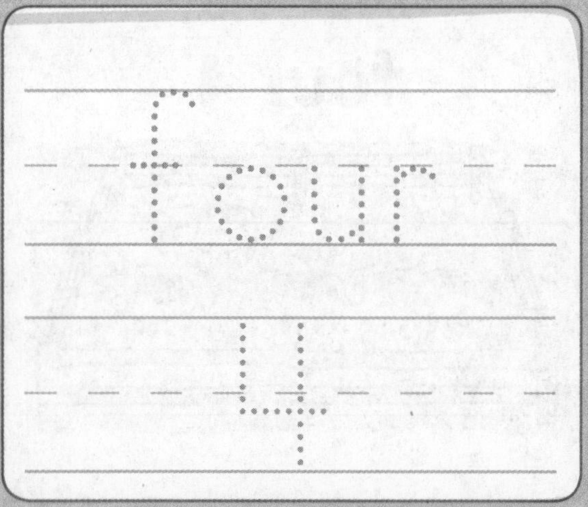

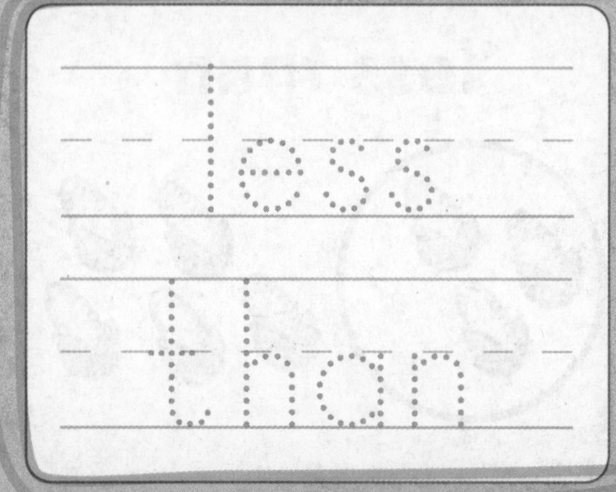

# My Vocabulary Cards

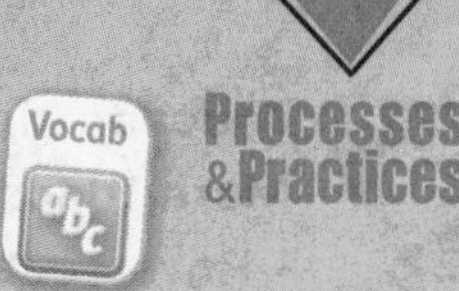

number

3 2 5 0

one 1

three 3

two 2

zero 0

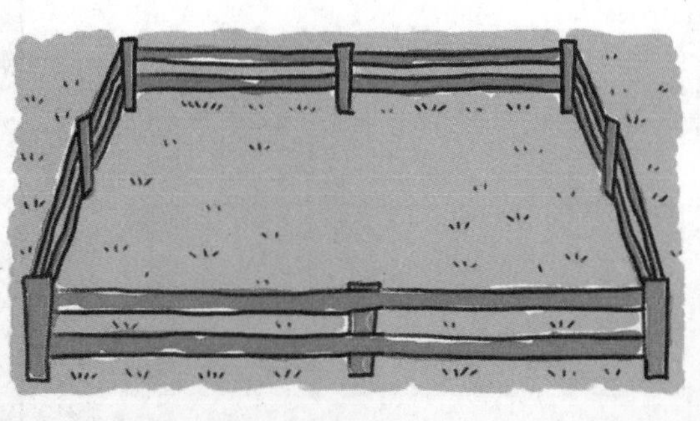

**Teacher Directions:**
**More Ideas for Use**

- Have each student choose a number card and a piece of paper. Guide students to draw other objects in a group to show how many.
- Have students use the blank card to create a vocabulary card of their own.

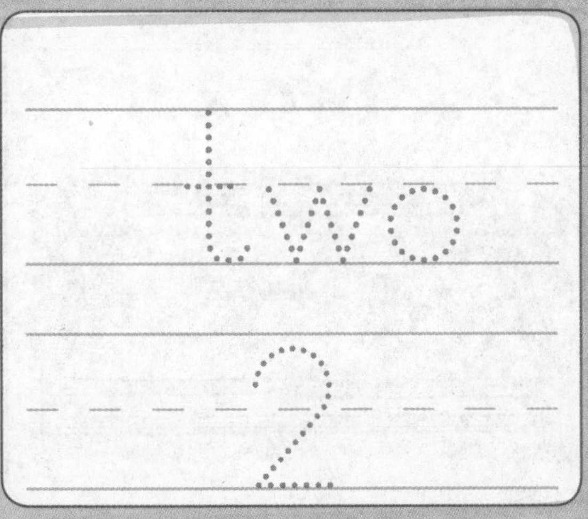


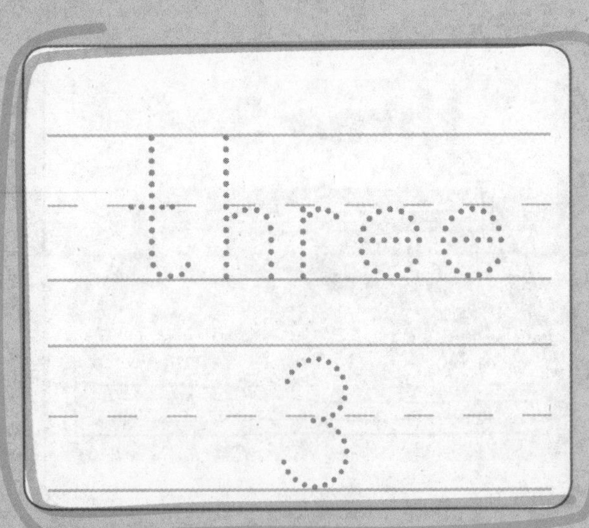

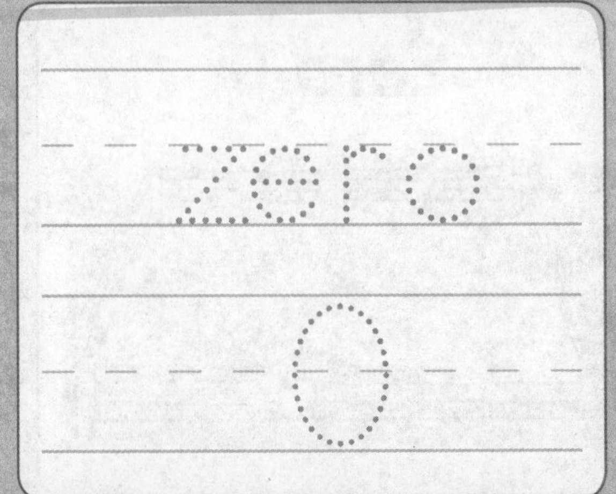

# My Foldable

FOLDABLES® Follow the steps on the back to make your Foldable.

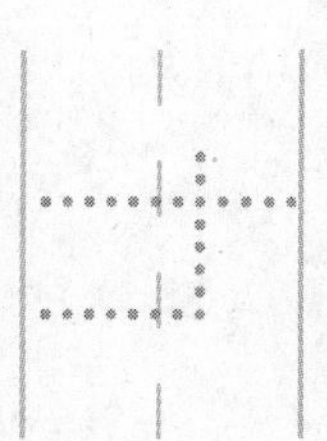

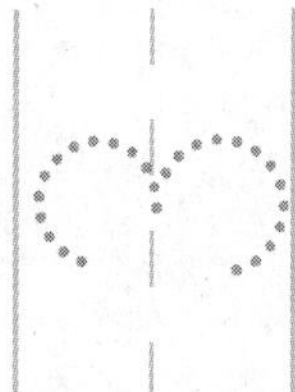

FOLDABLES
Study Organizer

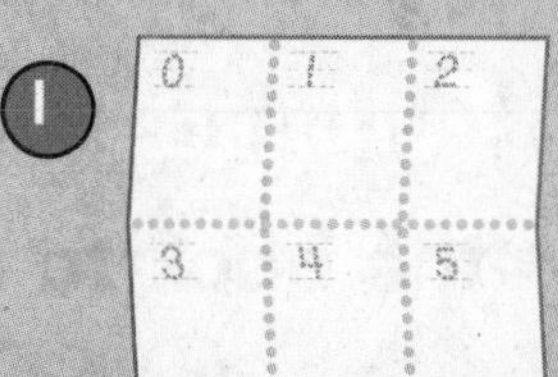
1
0
1
2
3
4
5

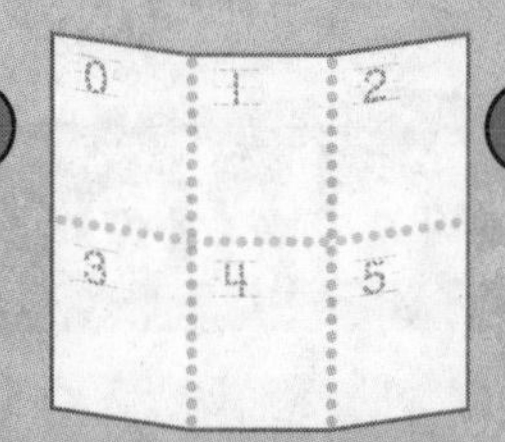
2
0
1
2
3
4
5

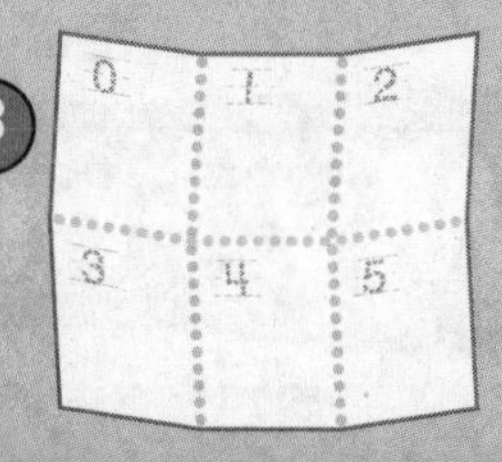
3
0
1
2
3
4
5

Name

Lesson 1

# Count 1, 2, and 3

**ESSENTIAL QUESTION**
How do we show how many?

## Math in My World

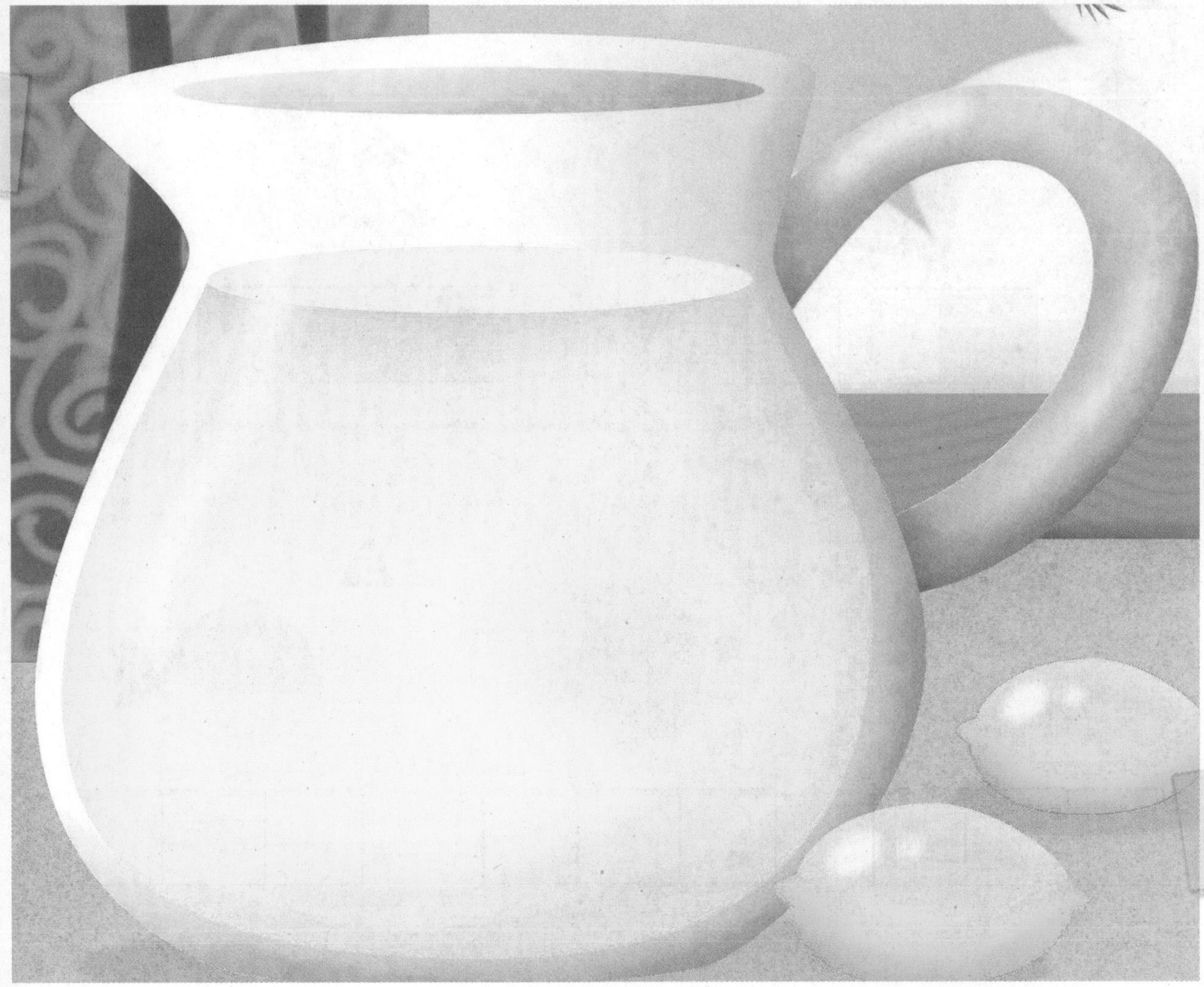

**Teacher Directions:** Use ●. Show a group of 1, 2, and 3 counters. Touch, count, and say how many in each group. Draw one lemon in the lemonade. Count how many lemons are on the page. Say how many in all.

Processes
&Practices

# Guided Practice

count

1

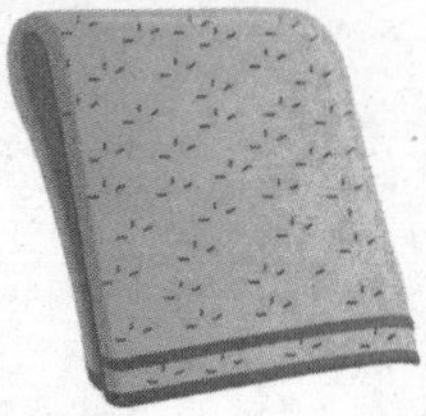

2

3

4

5

**Directions: 1–5.** Identify the object. Use the picture to count how many of that object. Use color tile(s) to show how many. Color one box for each object counted. Say how many.

Name

## Independent Practice

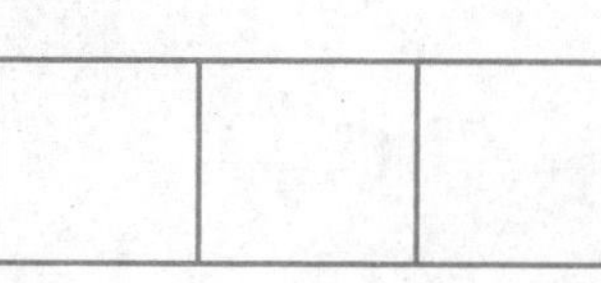

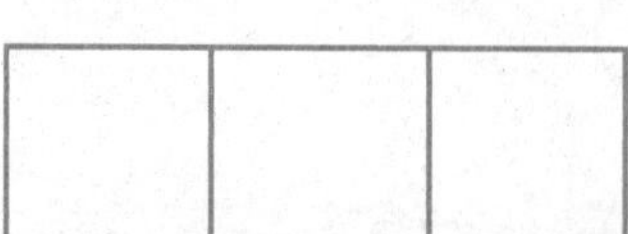

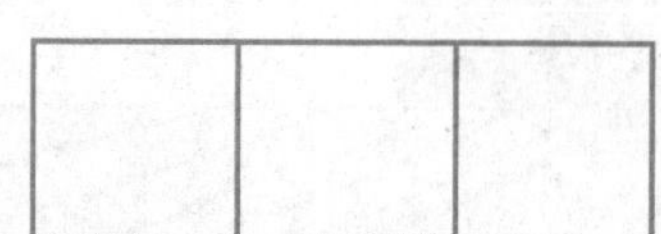

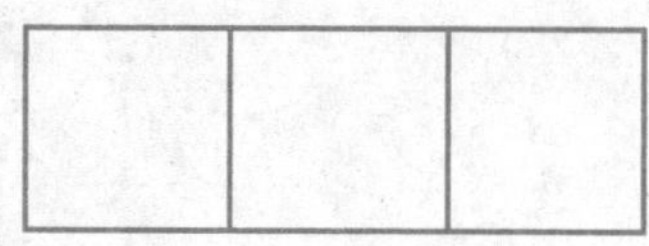

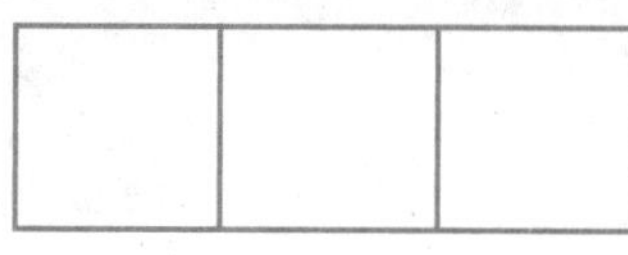

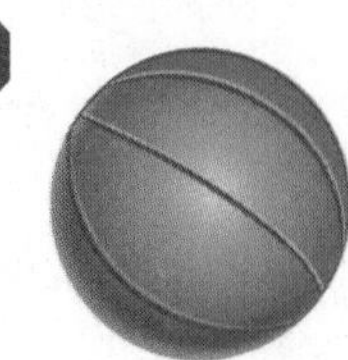

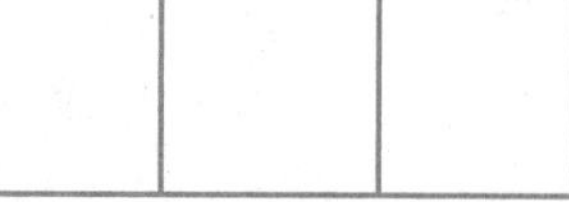

**Directions: 6–11.** Identify the object. Use the picture to count how many of that object. Use color tile(s) to show how many. Color one box for each object counted. Say how many.

# Brain Builders

12

**Directions: 12.** Count the boxes that are colored on each basket. Draw one orange in the basket that shows one box colored. Draw two apples in the basket that shows two boxes colored. Draw three grapes in the basket that shows three boxes colored. Draw another orange. Color in one more box to show the number of oranges. Tell a friend how many oranges there are now.

Name

# My Homework

Lesson 1

Count 1, 2, and 3

## Homework Helper

Need help? connectED.mcgraw-hill.com

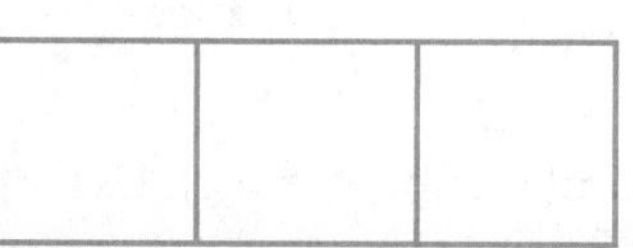

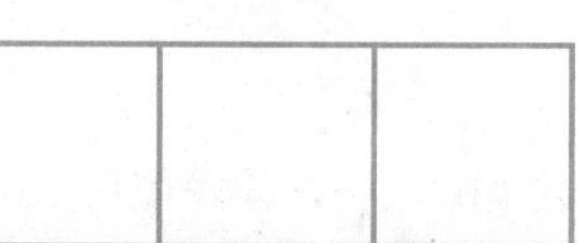

**Directions: 1–4.** Identify the object. Use the picture to count how many of that object. Use dry cereal to show how many. Color one box for each object counted. Say how many.

5

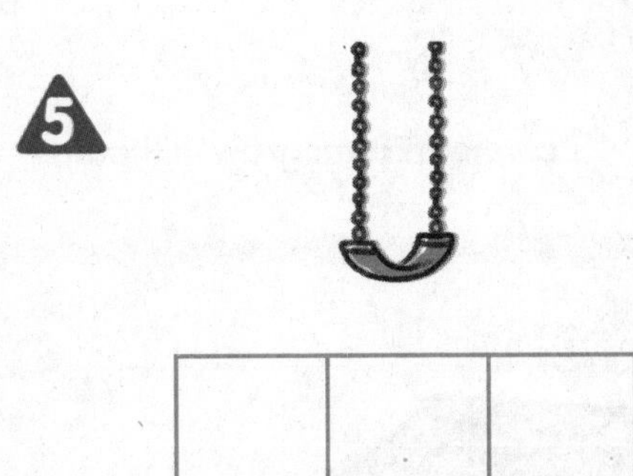

6

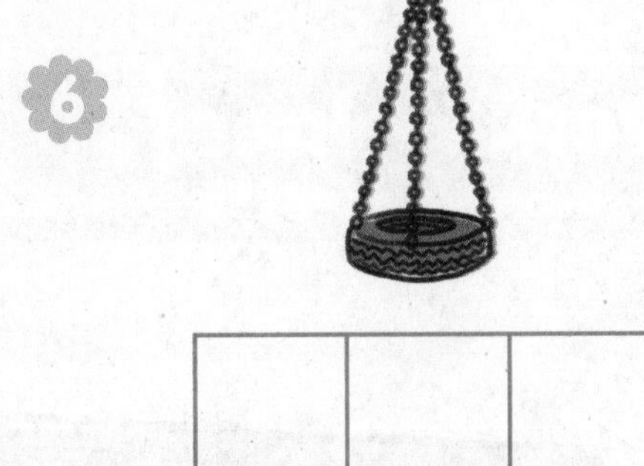

## Vocabulary Check

7 count

**Directions: 5–6.** Identify the object. Use the picture to count how many. Use dry cereal to show how many. Color one box for each object counted. Say how many. **7.** Draw three objects. Count and say how many.

**Math at Home** Gather items such as paper clips or rubber bands. Ask your child to put them in groups of one, two, and three and count the items in each group.

Name

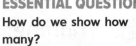

# Lesson 2
# Read and Write 1, 2, and 3

**ESSENTIAL QUESTION** 
How do we show how many?

## Math in My World

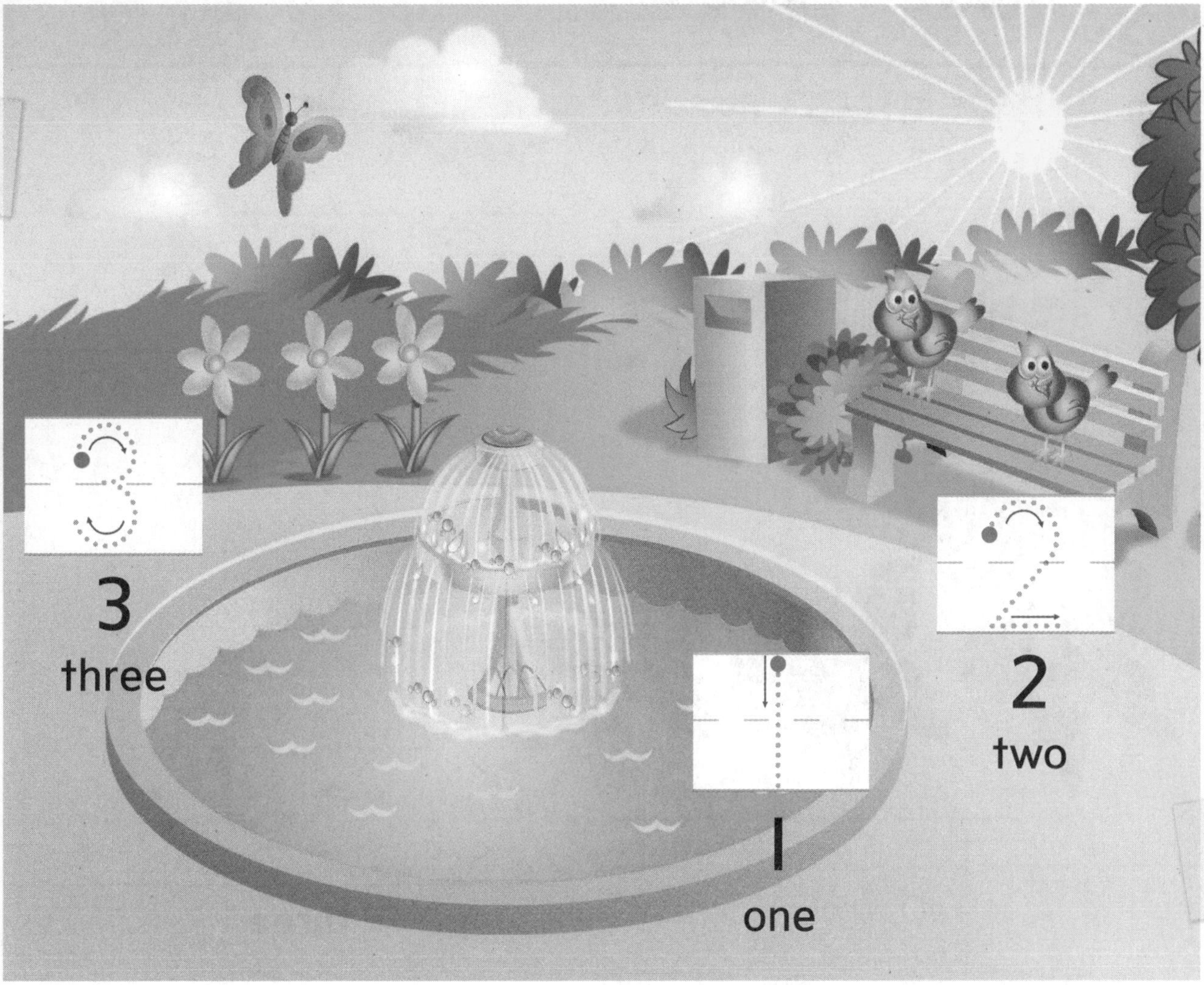

3
three

2
two

1
one

**Teacher Directions:** Look at the picture. Count the objects. Use ● to show that number. Say how many. Trace the number.

# Guided Practice

Processes & Practices

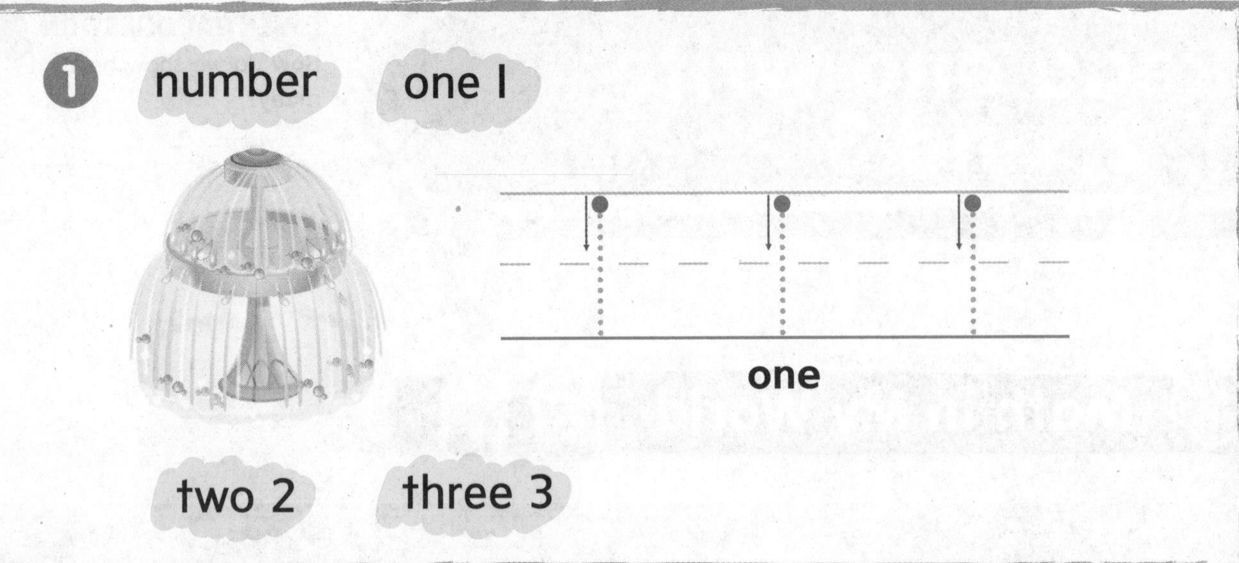


2

two

3

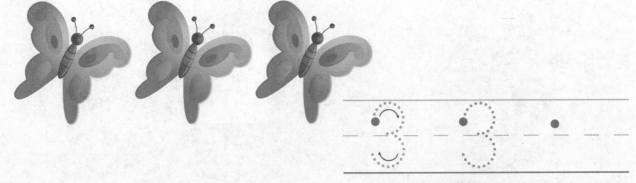

three

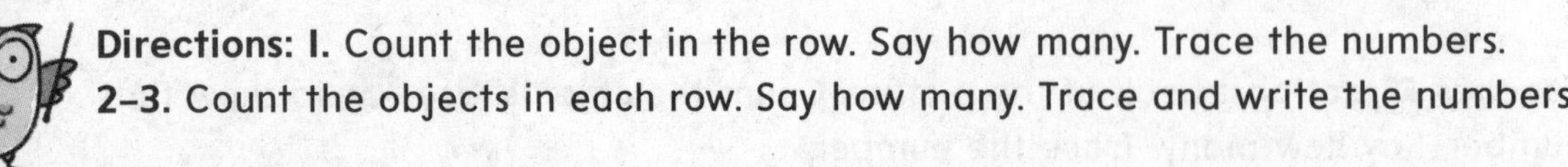

**Directions: 1.** Count the object in the row. Say how many. Trace the numbers.
**2–3.** Count the objects in each row. Say how many. Trace and write the numbers.

Name

## Independent Practice

   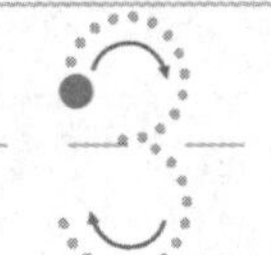

5

  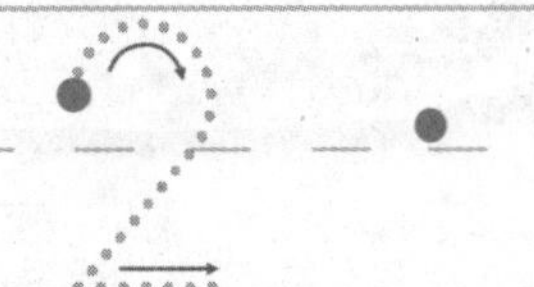

6

   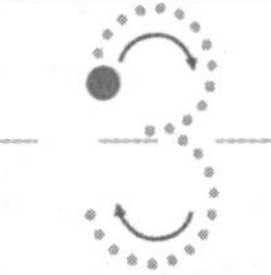

**Directions: 4–6.** Count the objects in each row. Say how many. Trace the number. Write the number two times.

Processes &Practices

## Brain Builders

7

8

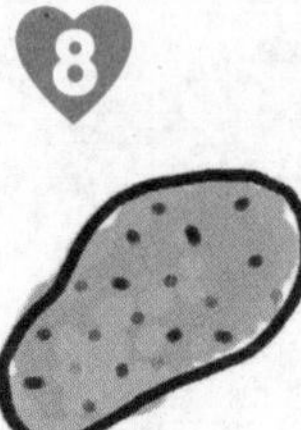

9

10

**Directions: 7.** Draw one tomato in the pot. Write the number two times. **8.** Draw one potato in the pot. Write the number two times. **9.** Draw two carrots in the pot. Write the number two times. **10.** Draw three onions in the pot. Write the number two times. Draw an X on three objects in the pot.

Name

# My Homework

**Lesson 2**

**Read and Write 1, 2, and 3**

## Homework Helper

Need help? connectED.mcgraw-hill.com

1

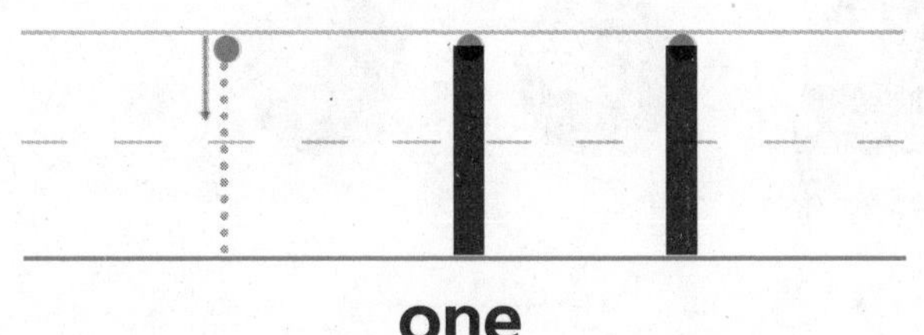

one

2

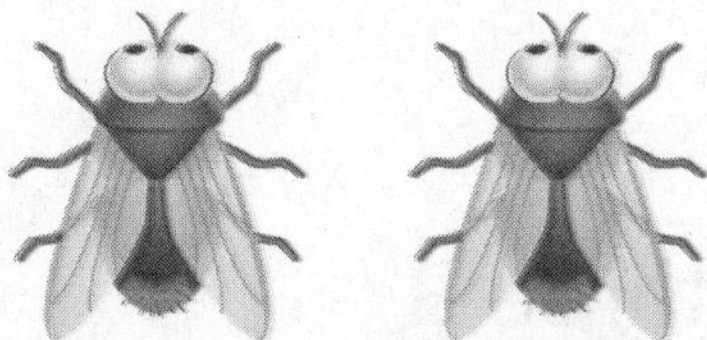

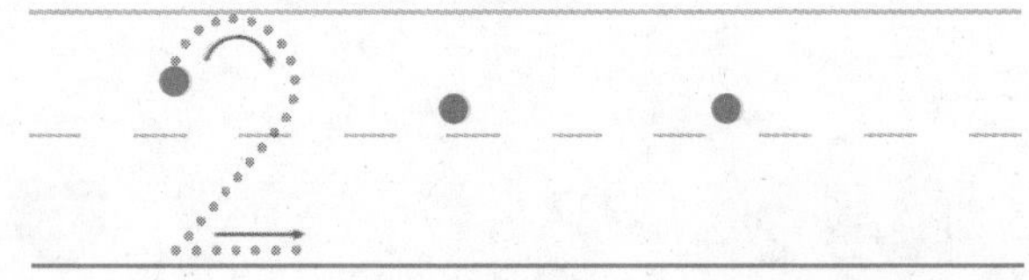

two

3

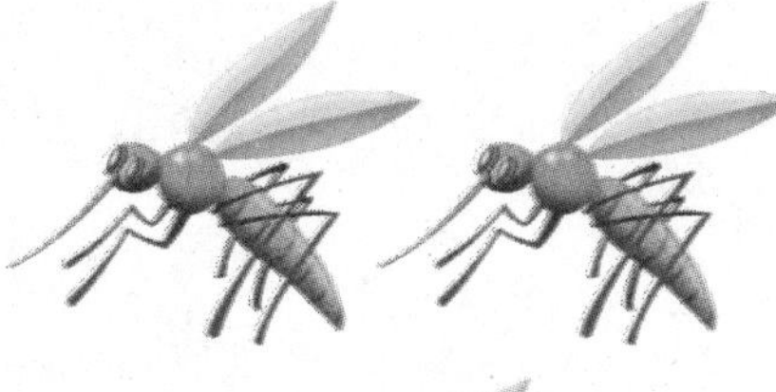

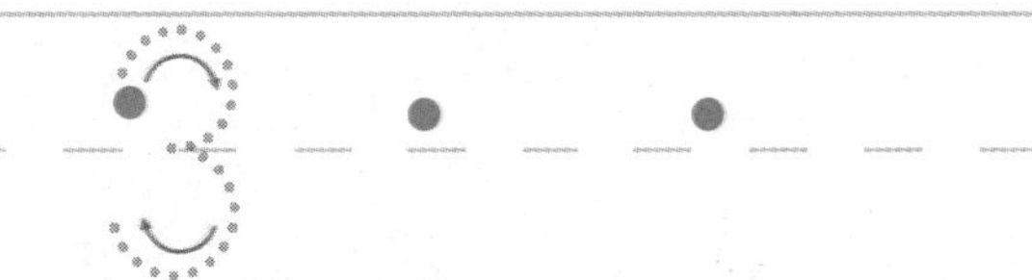

three

**Directions: 1–3.** Count the objects in each group. Say how many. Trace the number. Write the number two times.

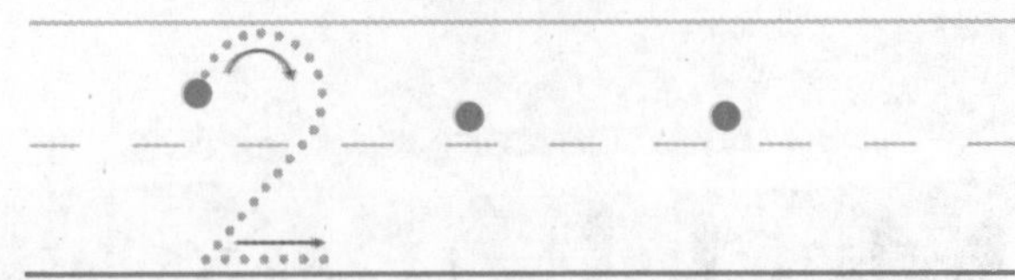

## Vocabulary Check

5 number

6 one 1

7 two 2

8 three 3

**Directions: 4.** Count the objects. Say how many. Trace the number. Write the number two times. **5.** Count the objects. Write the number. **6–8.** Say the word. Draw Xs to show how many.

**Math at Home** Show your child 3 spoons, 2 cups, and 1 bowl. Have your child count and say how many of each object. Have your child write each number.

Name

Lesson 3

# Count 4 and 5

**ESSENTIAL QUESTION**
How do we show how many?

## Math in My World

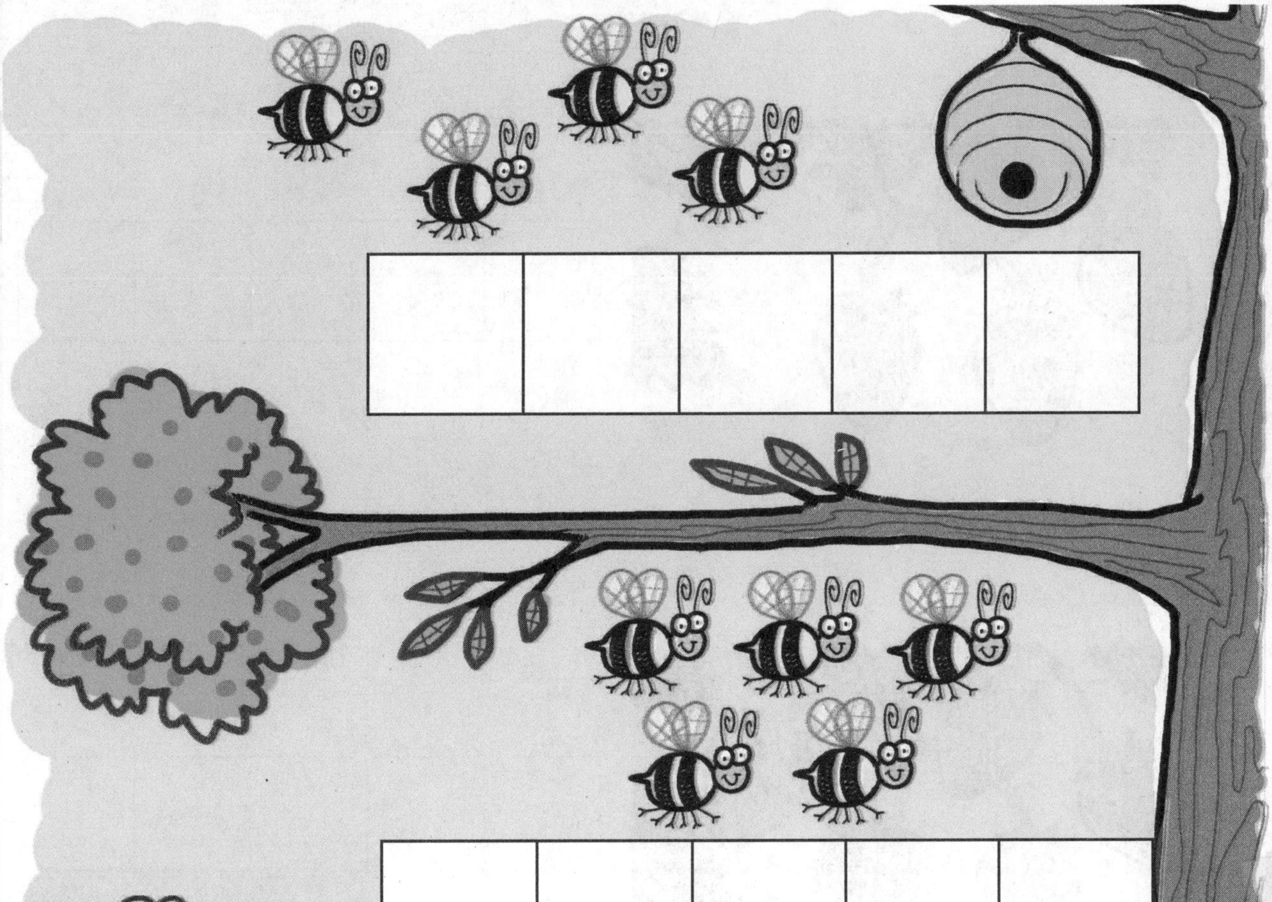

**Teacher Directions:** Count the bees in each group. Say how many. Use [cube] to show how many. Color the boxes below each group of bees to show how many.

Processes & Practices

## Guided Practice

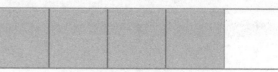

2

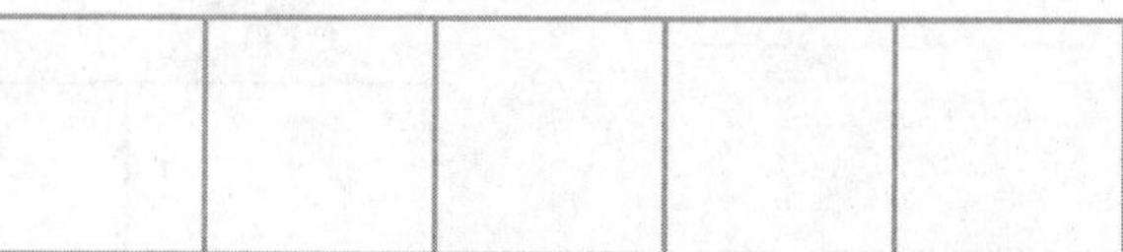

3

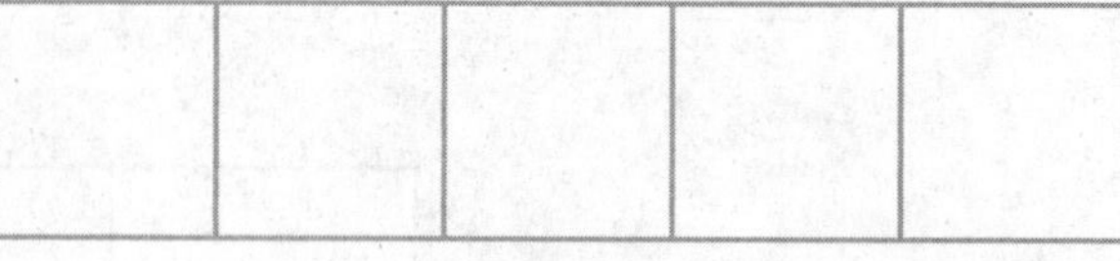

**Directions: 1.** Count the bears. Say how many. Use cubes to show how many. Color the boxes to show how many. **2–3.** Count the berries in each group. Say how many. Use cubes to show how many. Color the boxes to show how many.

Name

## Independent Practice

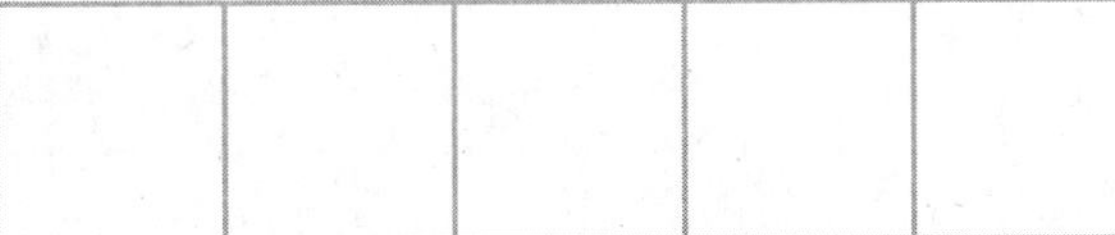

**Directions: 4–6.** Count the animals in each group. Say how many. Use cubes to show how many. Color the boxes to show how many.

# Brain Builders

7

**Directions: 7.** Count the berries in one group. Say how many. Draw that many berries in a basket and color them red. Draw the same number of nuts in the basket and color them brown. Count the berries in the other group. Say how many. Draw that many berries in the other basket and color them red. Draw the same number of nuts in the basket and color them brown. Circle the group of four berries. Draw an X on the group of five berries.

Name

# My Homework

**Lesson 3**

**Count 4 and 5**

## Homework Helper

**Need help?** connectED.mcgraw-hill.com

1

2

**Directions: 1.** Count the fish. Say how many. Use dry cereal to show how many. Color the boxes to show how many. **2.** Count the crabs. Say how many. Use dry cereal to show how many. Color the boxes to show how many.

3

4

**Directions: 3–4.** Count the frogs in each group. Say how many. Use dry cereal to show how many. Color the boxes to show how many.

**Math at Home** Help your child gather objects such as pennies or spoons. Put the objects in groups of four and groups of five. Practice counting the objects in each group.

Name

Lesson 4
# Read and Write 4 and 5

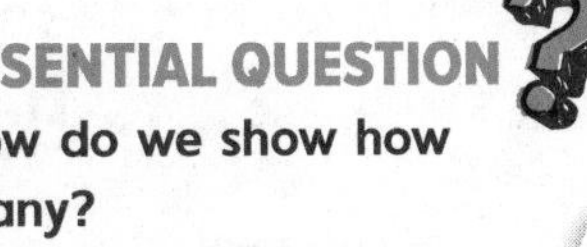

**ESSENTIAL QUESTION**
How do we show how many?

## Math in My World

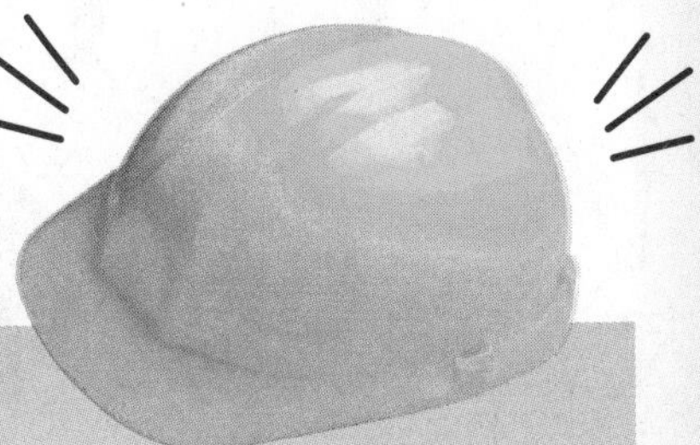

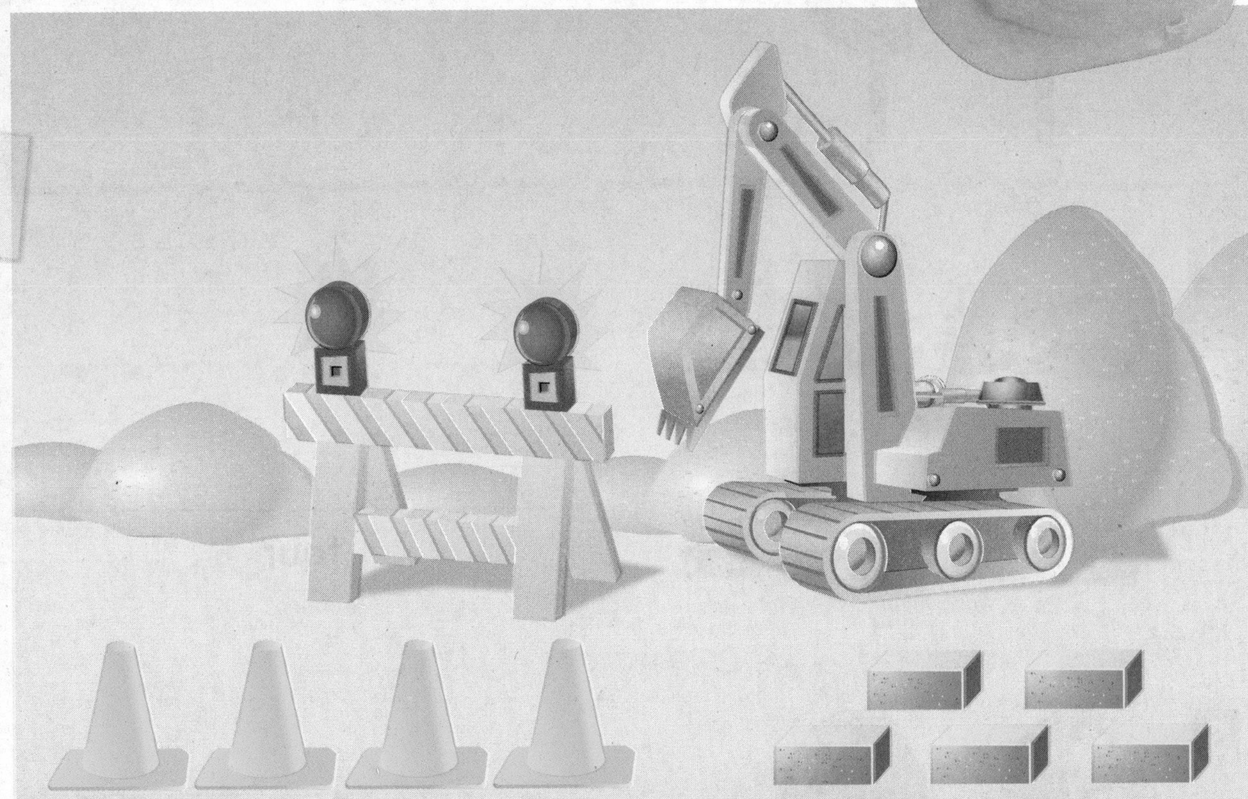

4
four

5
five

**Teacher Directions:** Count the cones. Say how many. Use ○ to show how many. Trace the number. Repeat the activity with the bricks.

# Guided Practice

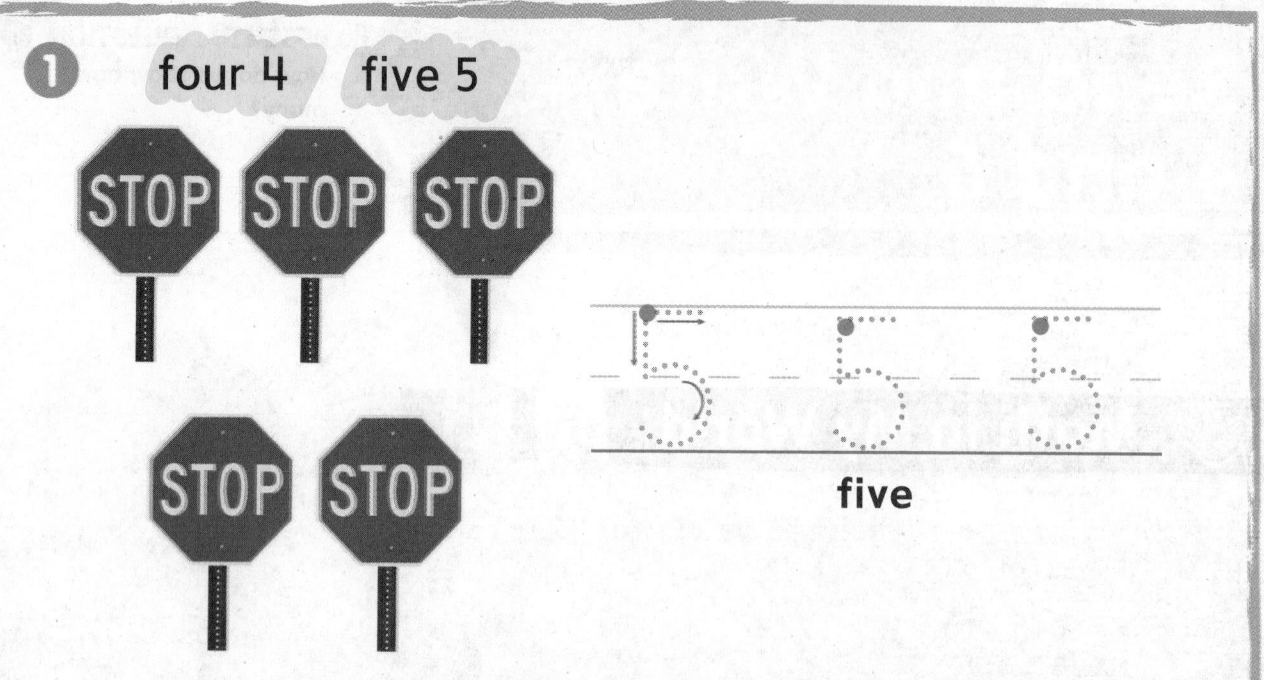

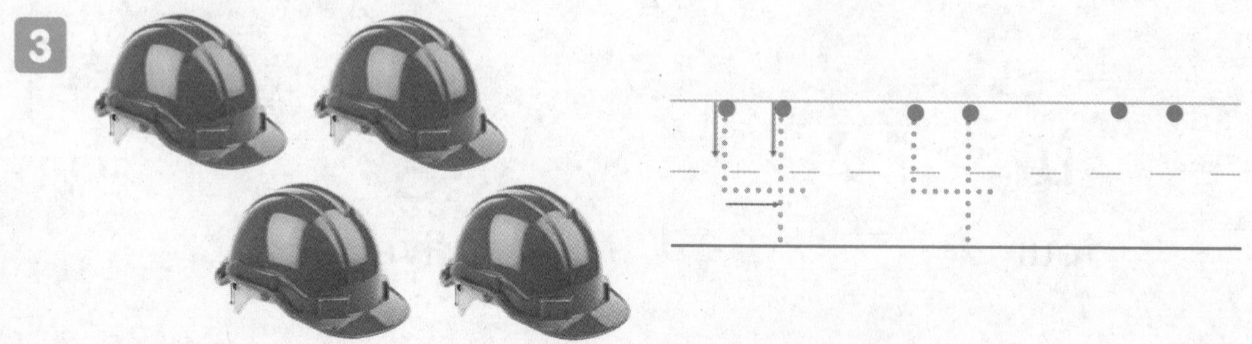

**Directions: I.** Count the objects in the group. Say how many. Trace the numbers.
**2–3.** Count the objects in each group. Say how many. Trace and write the numbers.

Name

## Independent Practice

4

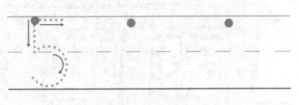

5

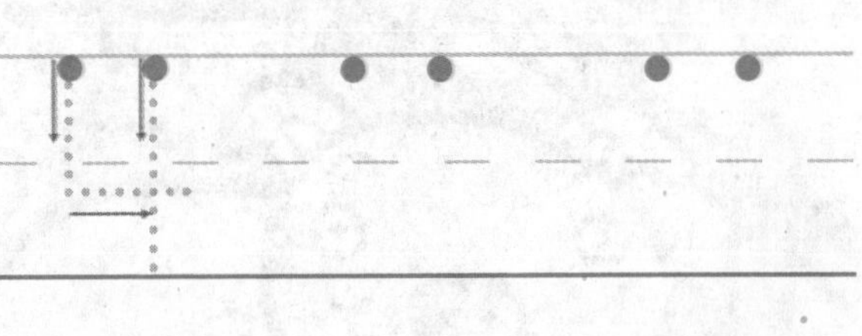

6

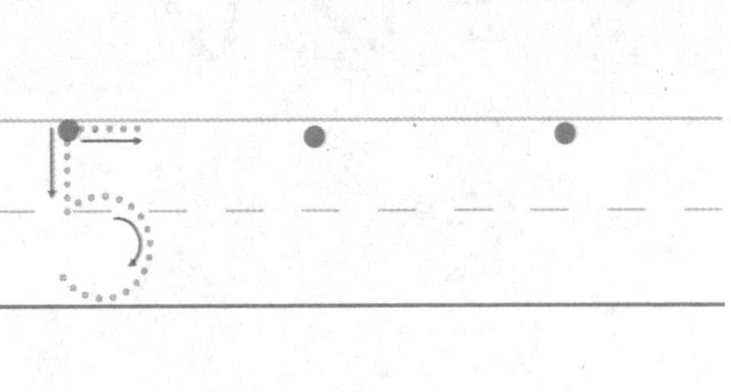

**Directions: 4–6.** Count the objects in each group. Say how many. Trace and write the numbers.

## Brain Builders

7

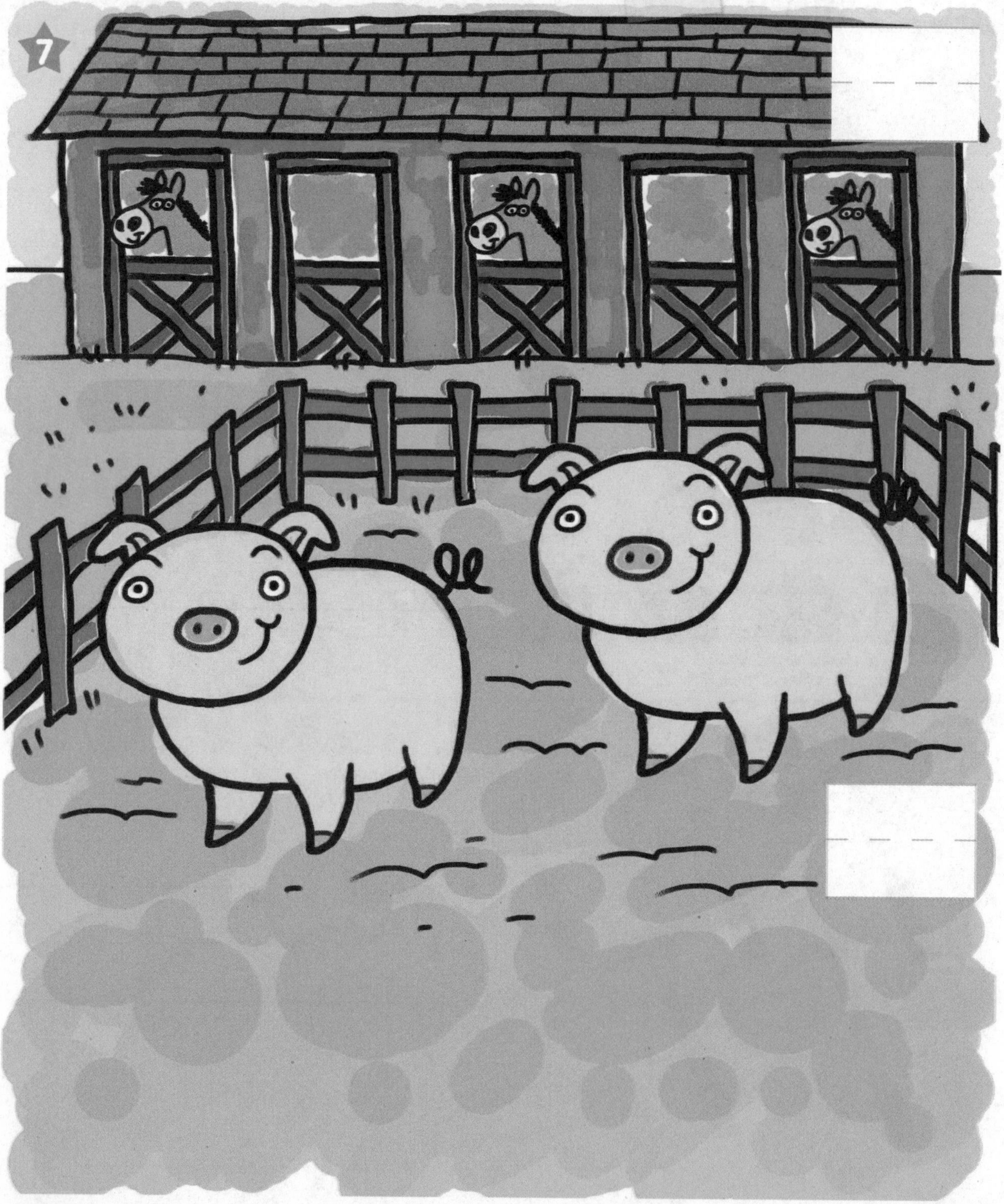

**Directions: 7.** Count the pigs. Draw more to show four. Use counters to show how many. Write the number. Count the horses. Draw more to show five. Use counters to show how many. Write the number. Put an X on four of the horses.

# My Homework

Lesson 4

**Read and Write 4 and 5**

## Homework Helper

Need help? connectED.mcgraw-hill.com

1

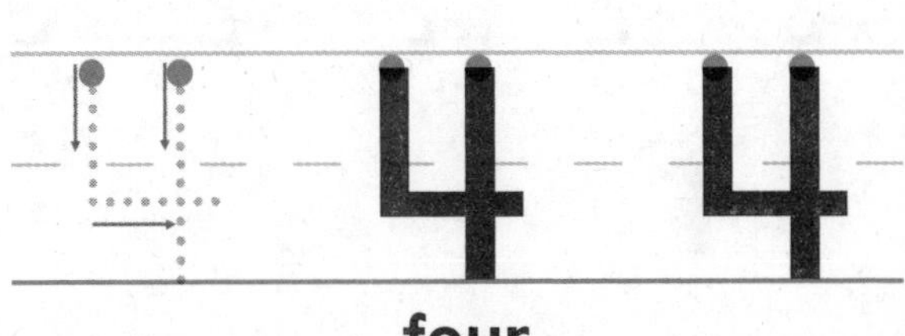


four

2

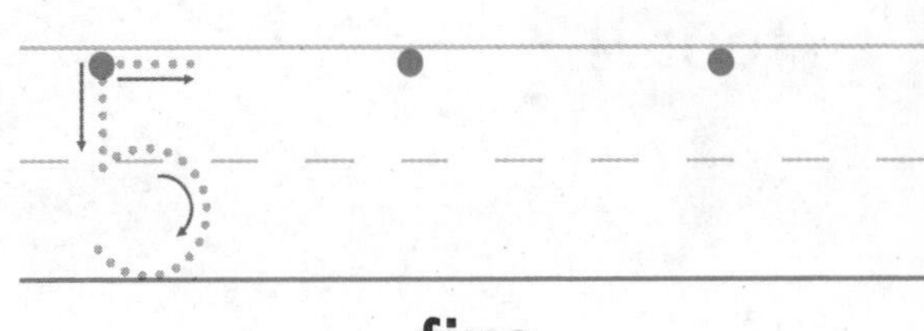

five

3

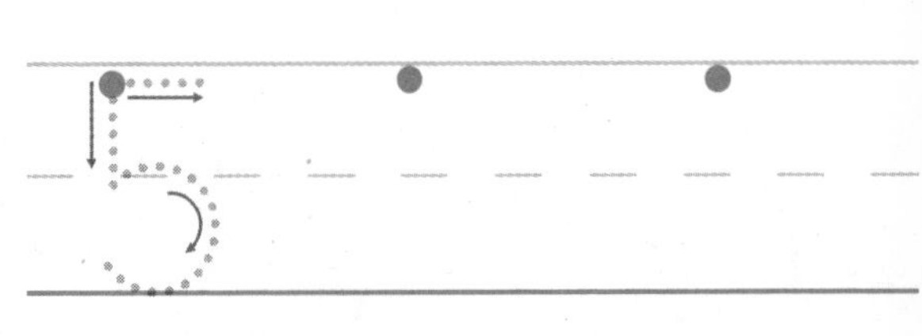

**Directions: 1–3.** Count the objects in each group. Say how many. Trace and write the numbers.

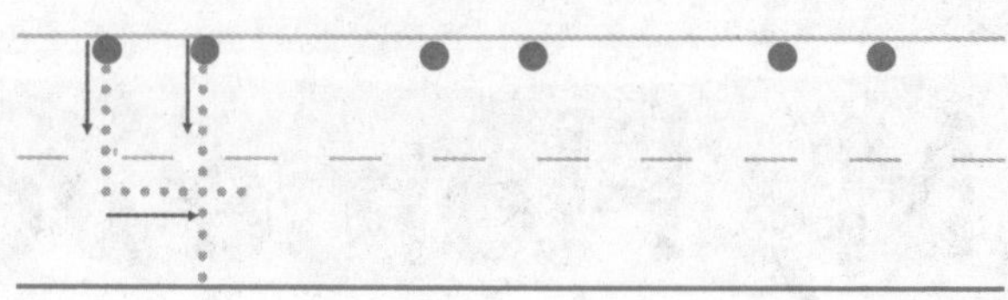

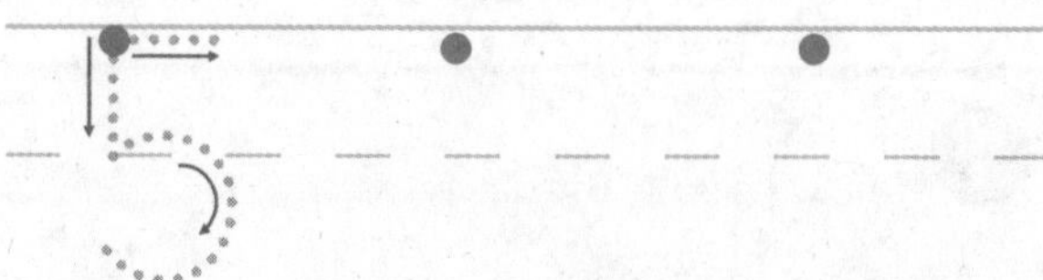


## Vocabulary Check

**Directions: 4–5.** Count the objects in each group. Say how many. Trace and write the numbers. **6–7.** Say the word. Draw objects to show how many.

**Math at Home** Make groups of four and five using pasta. Have your child count how many in each group and write the number.

Name

# Lesson 5
# Read and Write Zero

**ESSENTIAL QUESTION**
How do we show how many?

## Math in My World

Tools Watch

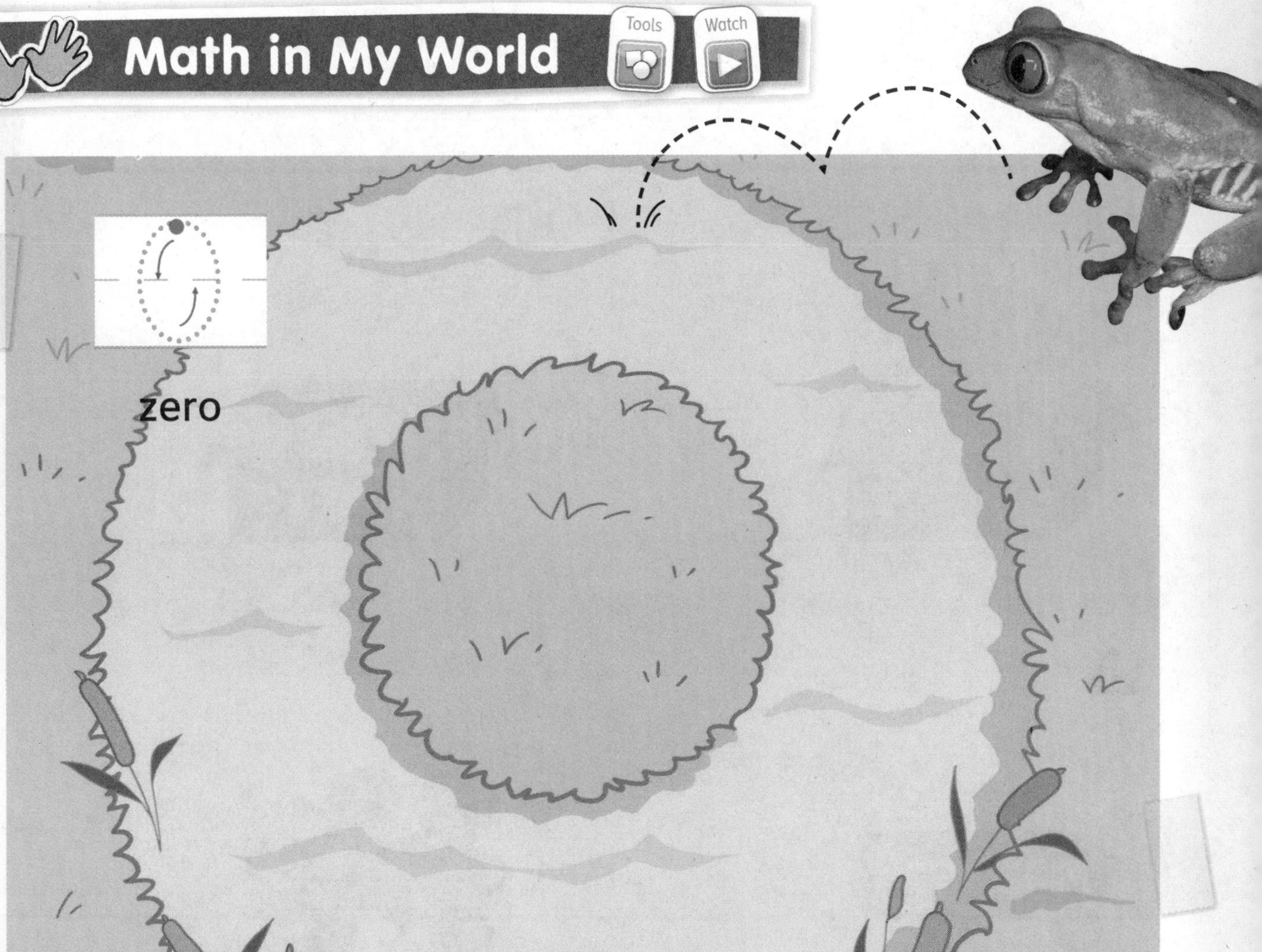

zero

Copyright © McGraw-Hill Education ©Ingram Publishing/Alamy

**Teacher Directions:** Use [cube] to act out this story. Five frogs are in the pond. Count the frogs. Five frogs jumped out. Count the frogs left in the pond. Trace the number zero.

Processes &Practices

## Guided Practice

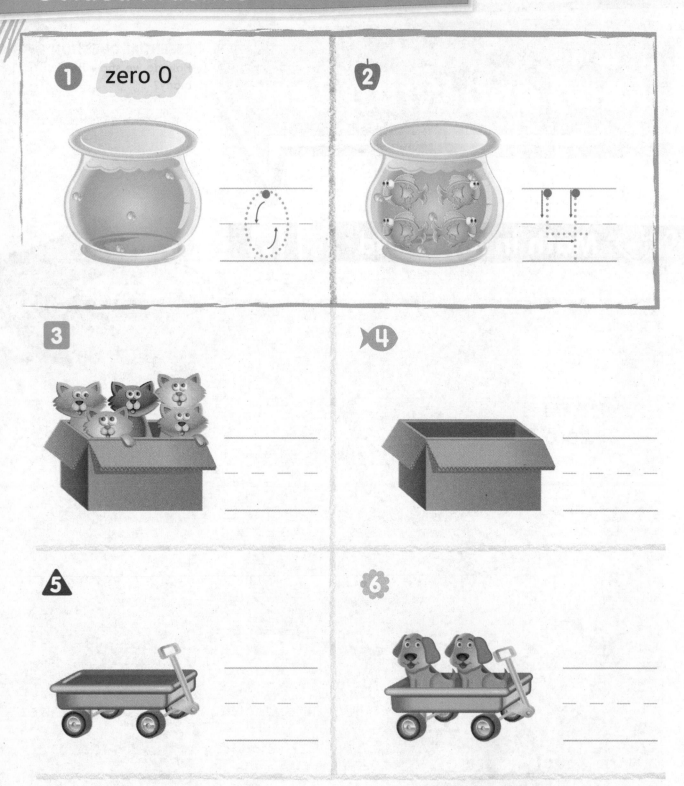

1 zero 0

2

3

4

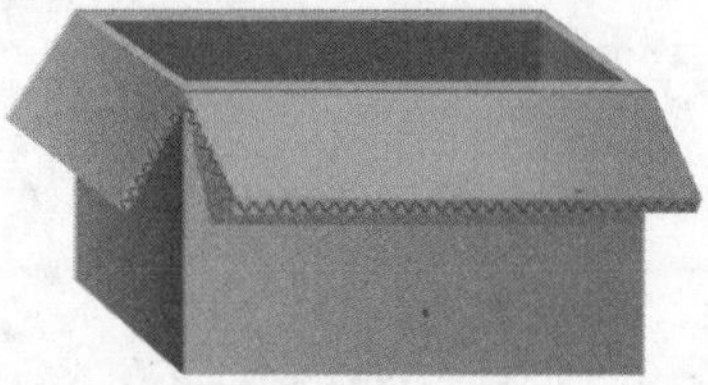

5

6

**Directions: 1–2.** Count how many fish are in the bowl. Trace the number. **3–4.** Count how many cats are in the box. Write the number. **5–6.** Count how many dogs are in the wagon. Write the number.

Name

## Independent Practice

7

8

9

10

11

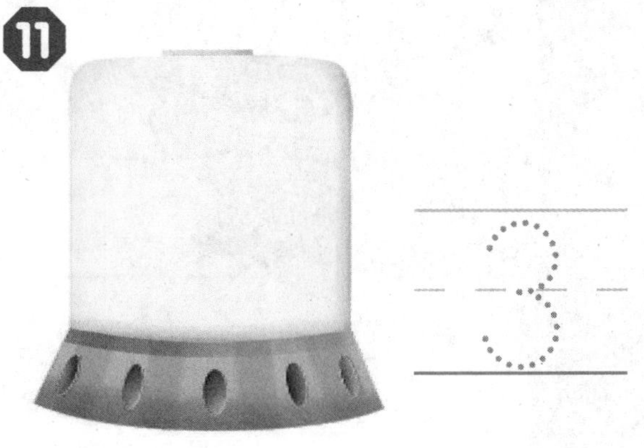

12

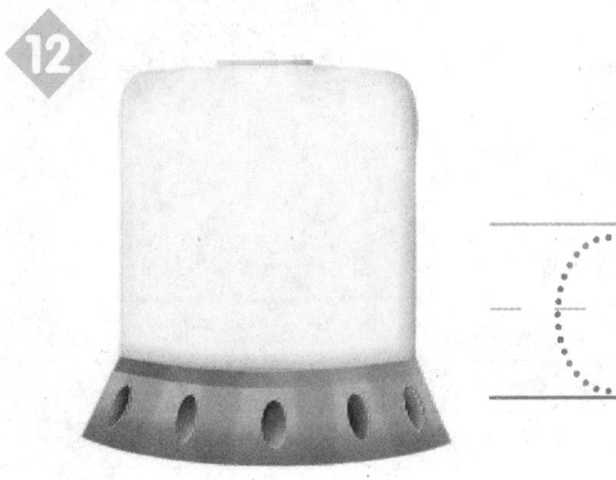

**Directions: 7–8.** Count how many birds are in the cage. Write the number. **9–10.** Count how many turtles are on the rock. Write the number. **11–12.** Trace the number. Draw bugs to show how many. Say how many.

# Brain Builders

13

**Directions: 13.** Count the birds in each nest. Say how many. Write the number. Draw an X on the nests with birds inside them. Draw a circle around the nest with zero birds. Explain to a friend how you know there are zero birds in the nest.

Name

# My Homework

Lesson 5

Read and Write Zero

## Homework Helper

Need help? connectED.mcgraw-hill.com

1
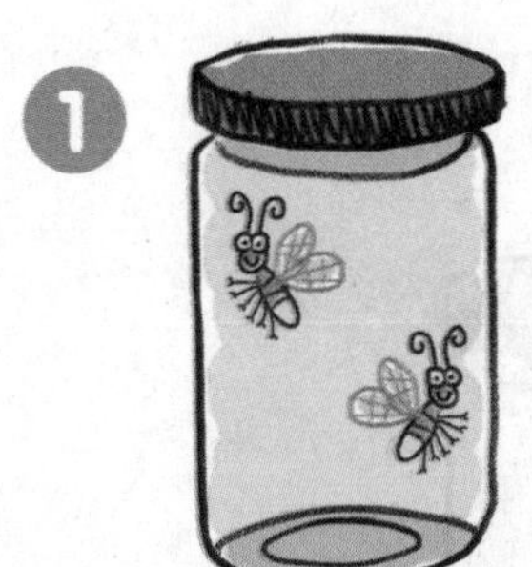

2

2

0

3
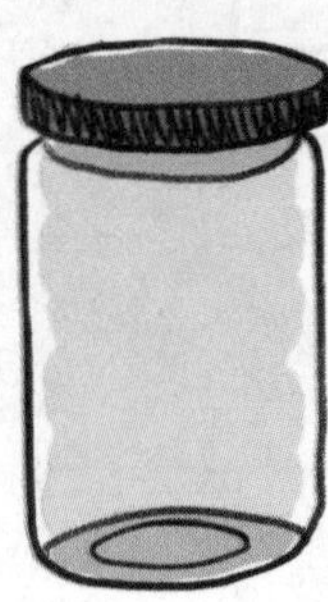

4

5

6

**Directions: 1–6.** Count the bugs in each jar. Say how many. Write the number that shows how many. Draw a circle around the jars with zero bugs.

## Vocabulary Check

**zero 0**

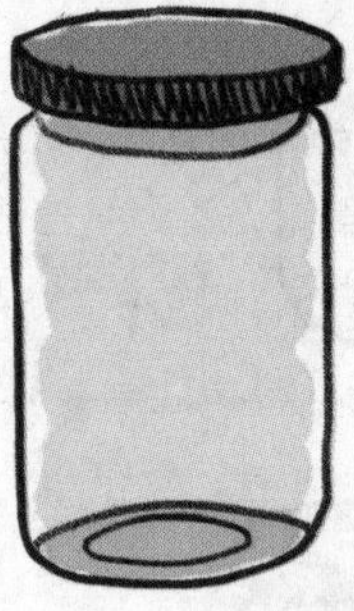

**Directions: 7–9.** Count the bugs in each jar. Say how many. Write the number. Draw a circle around the jars with zero bugs.

**Math at Home** Look at a family photo. Ask your child questions about the photo that have zero as the answer. Have your child practice writing the number zero.

Name

# Check My Progress

## Vocabulary Check 

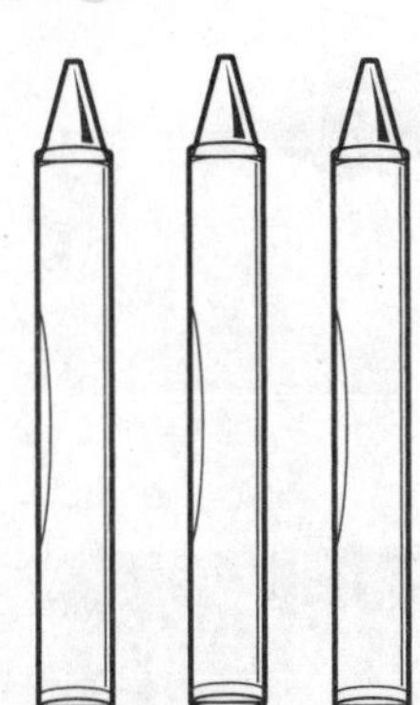

5

## Concept Check

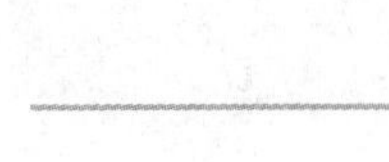

**Directions: 1.** Count the crayons. Color them. Say how many. **2.** Look at the group of objects. Circle each number. **3–4.** Count the objects in each group. Say how many. Write the number.

5

6

7

**Directions: 5–6.** Count the objects in each group. Say how many. Write the number. **7.** Draw an X on each group of four frogs. Circle each group of five frogs.

Name

Lesson 6

# Equal To

**ESSENTIAL QUESTION**
How do we show how many?

## Math in My World

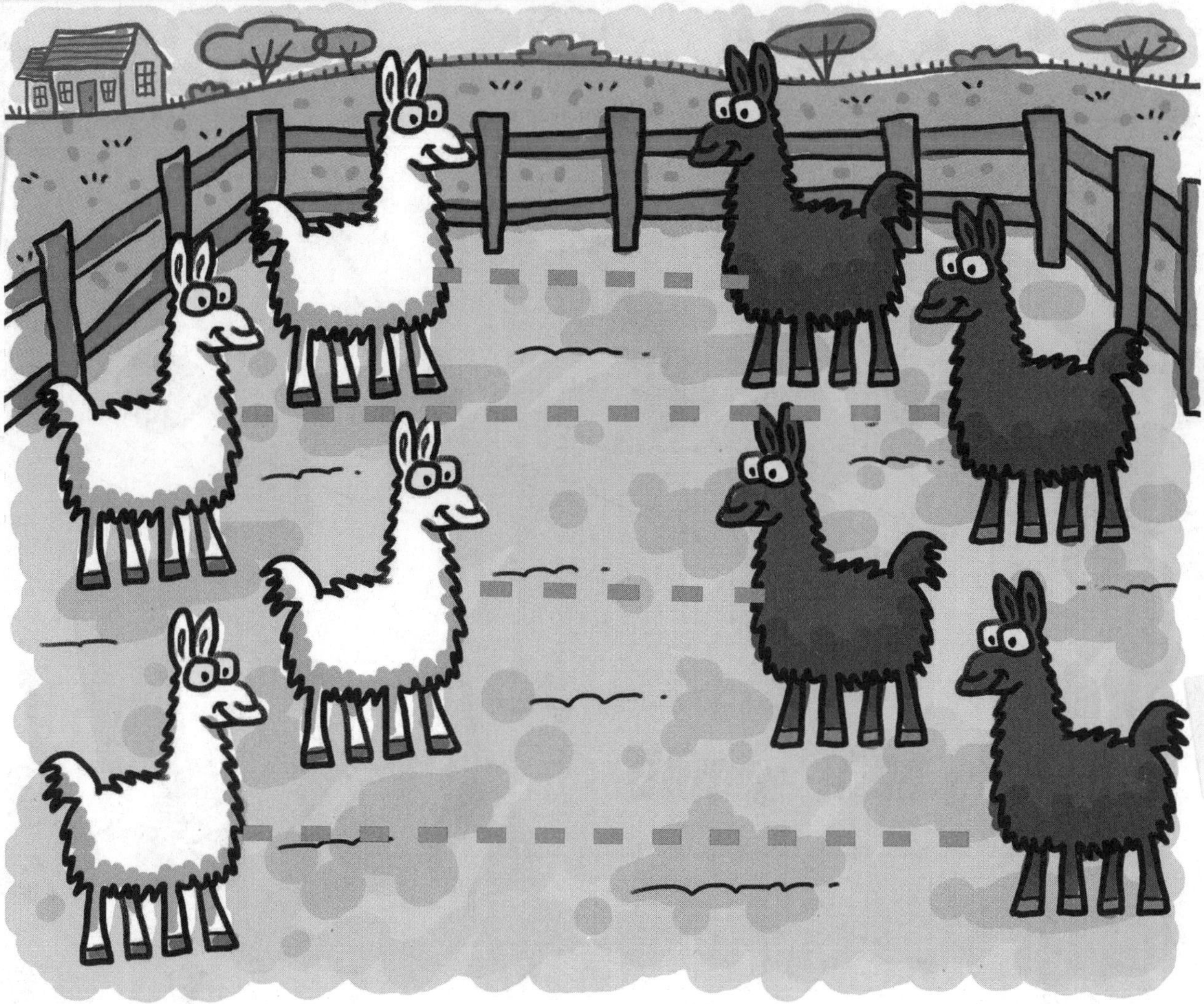

**Teacher Directions:** Place on each white llama. Place on each brown llama. Describe the sizes of the two groups. Trace the lines to match objects in the groups.

# Guided Practice

Processes & Practices

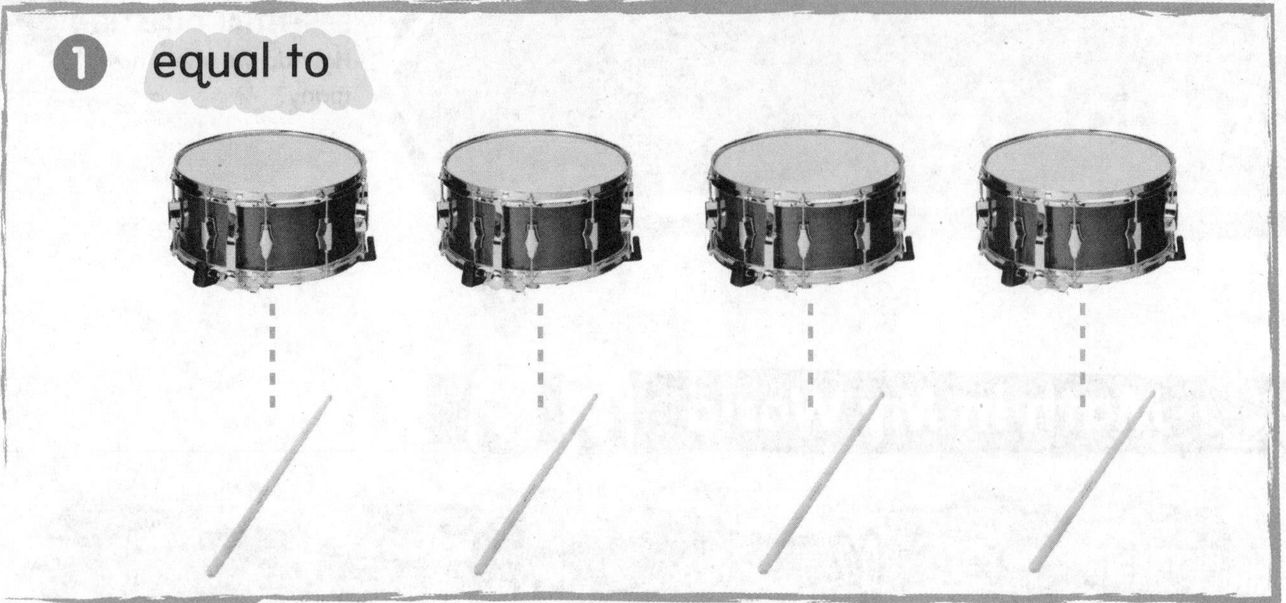

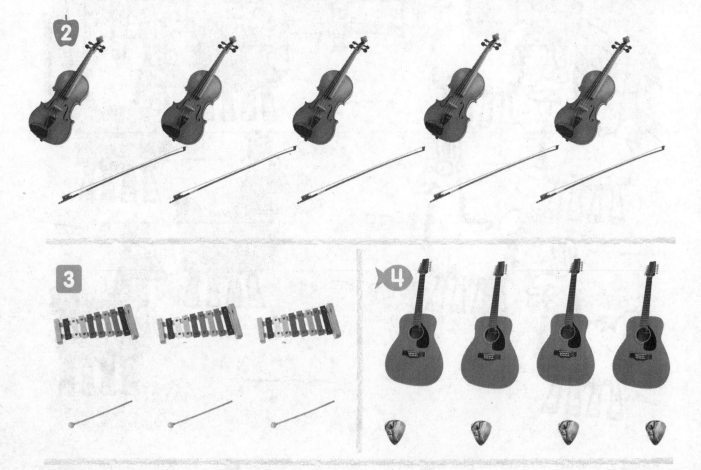

2

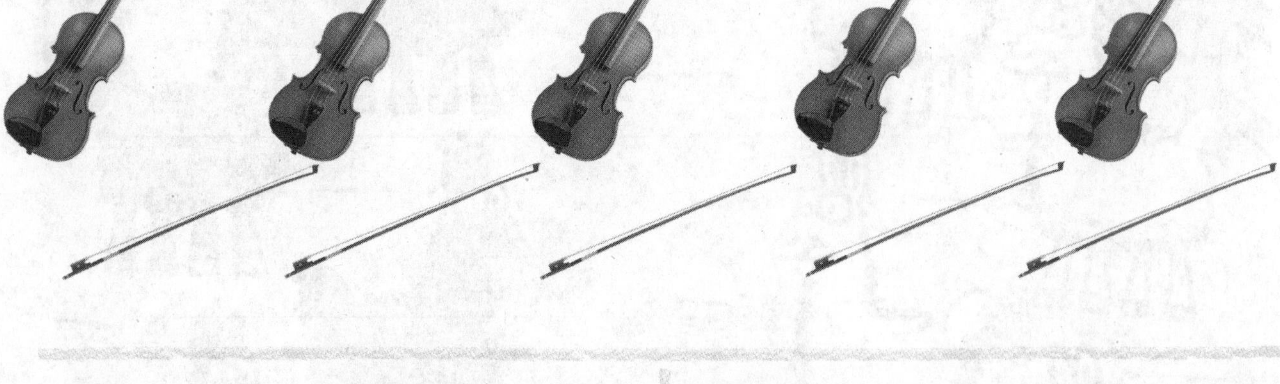

3

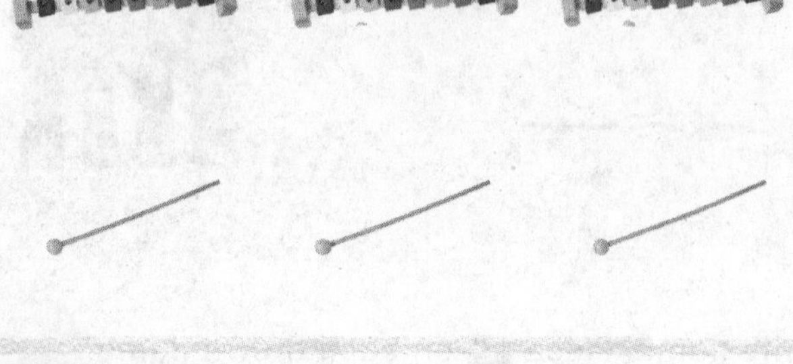

4

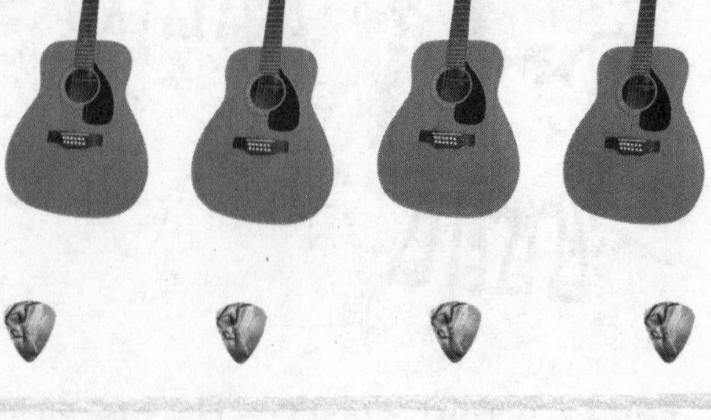

**Directions: 1–4.** Draw a line from an object in one group to match an object in the other group. Describe the sizes of the groups.

Name

## Independent Practice

5

6

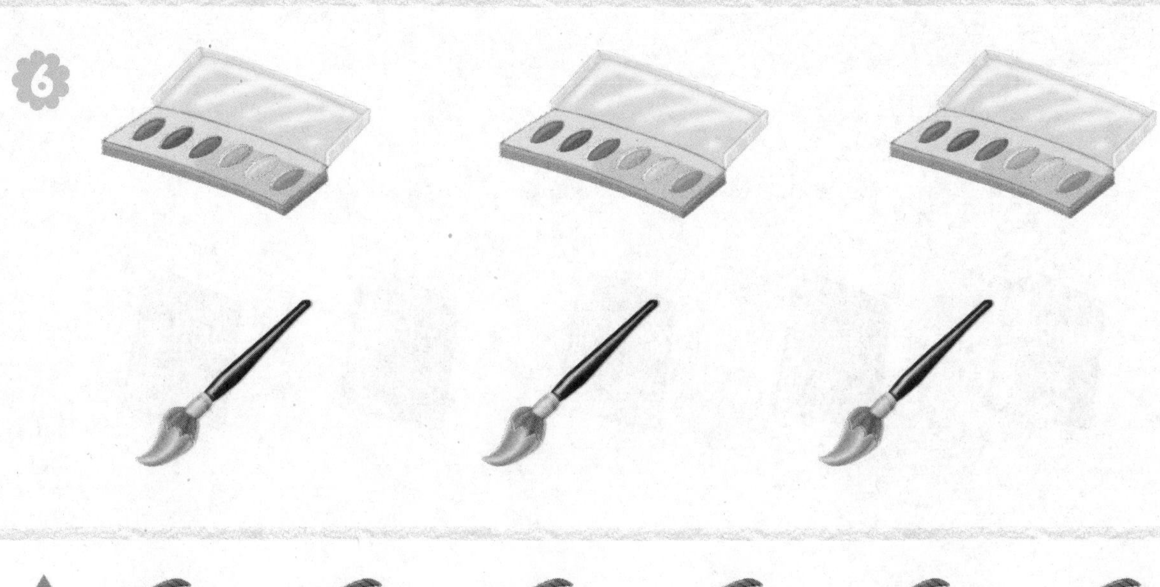

7

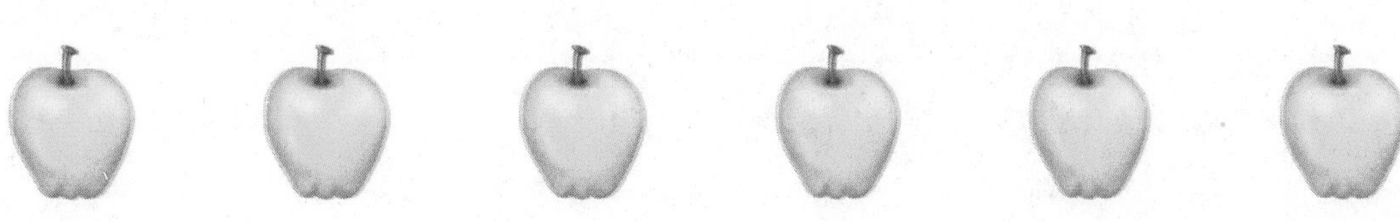

**Directions: 5–7.** Draw a line from an object in one group to match an object in the other group. Describe the sizes of the groups.

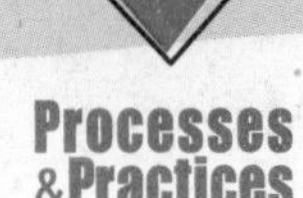

Processes
&Practices

## Brain Builders

8

**Directions: 8–9.** Use counters to show a group that is equal to the group of objects shown. Draw the counters. Draw lines to show that the number of objects in one group is equal to the number of objects in the other group. Write the number of objects in each group. Tell a friend what you notice about the numbers.

Name

Lesson 6

Equal To

# My Homework

## Homework Helper

Need help? connectED.mcgraw-hill.com

1

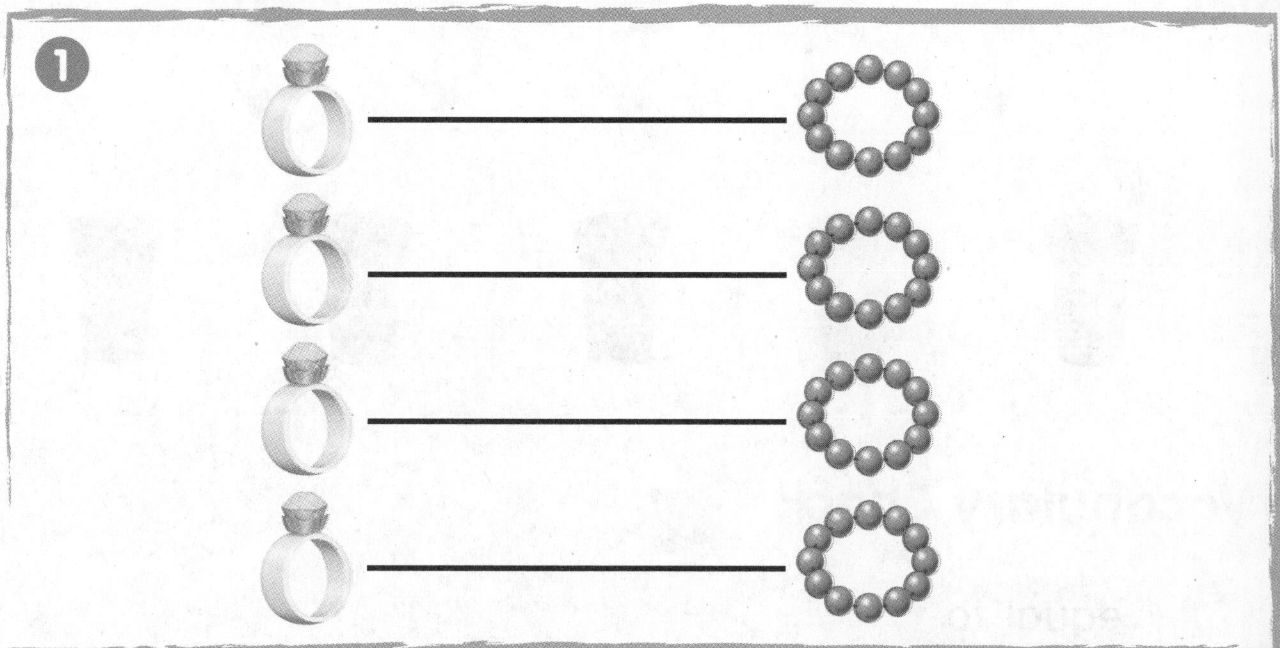

2

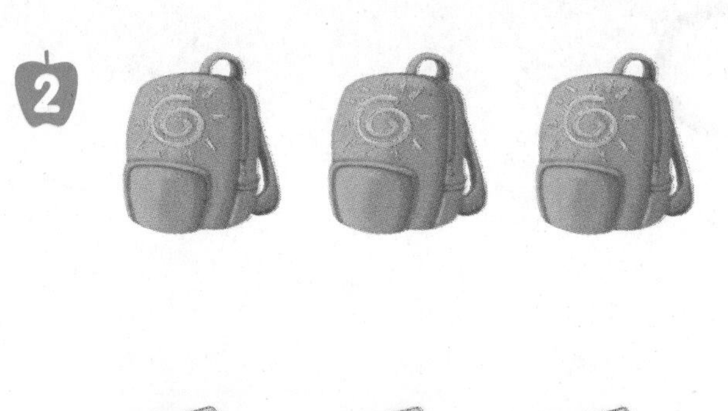

3

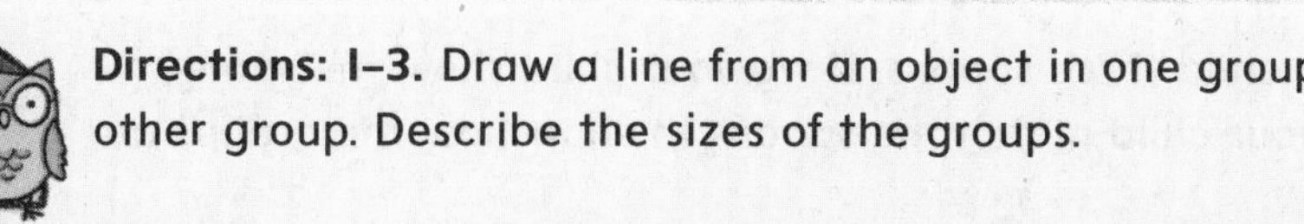

**Directions:** 1–3. Draw a line from an object in one group to match an object in the other group. Describe the sizes of the groups.

   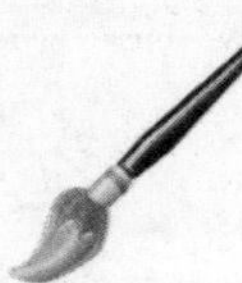

## Vocabulary Check

 equal to

**Directions: 4–5.** Draw a line from an object in one group to match an object in the other group. Describe the sizes of the groups. **6.** Draw a group of objects that is equal to the number of objects in the group shown. Draw lines to match objects in the groups.

**Math at Home** Show your child three spoons. Have your child show the same number of spoons. Have your child match the spoons to show how the groups are equal.

Name

Lesson 7
# Greater Than

**ESSENTIAL QUESTION**
How do we show how many?

## Math in My World

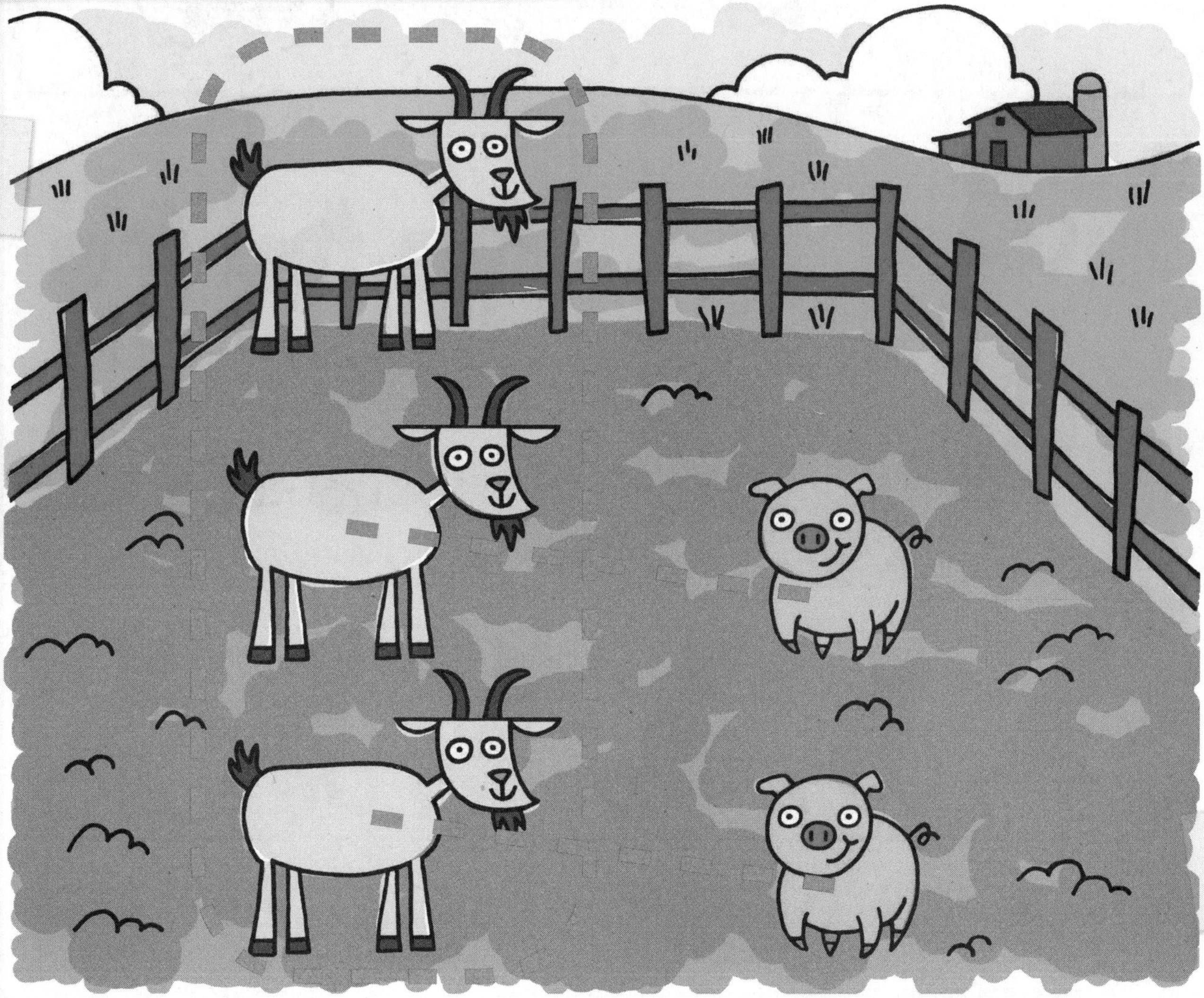

**Teacher Directions:** Place a ● on each goat. Place a ● on each pig. Trace the lines from an object in one group to match an object in the other group. Describe the sizes of the two groups. Trace a circle around the group that shows the greater number of objects.

## Guided Practice

Processes &Practices

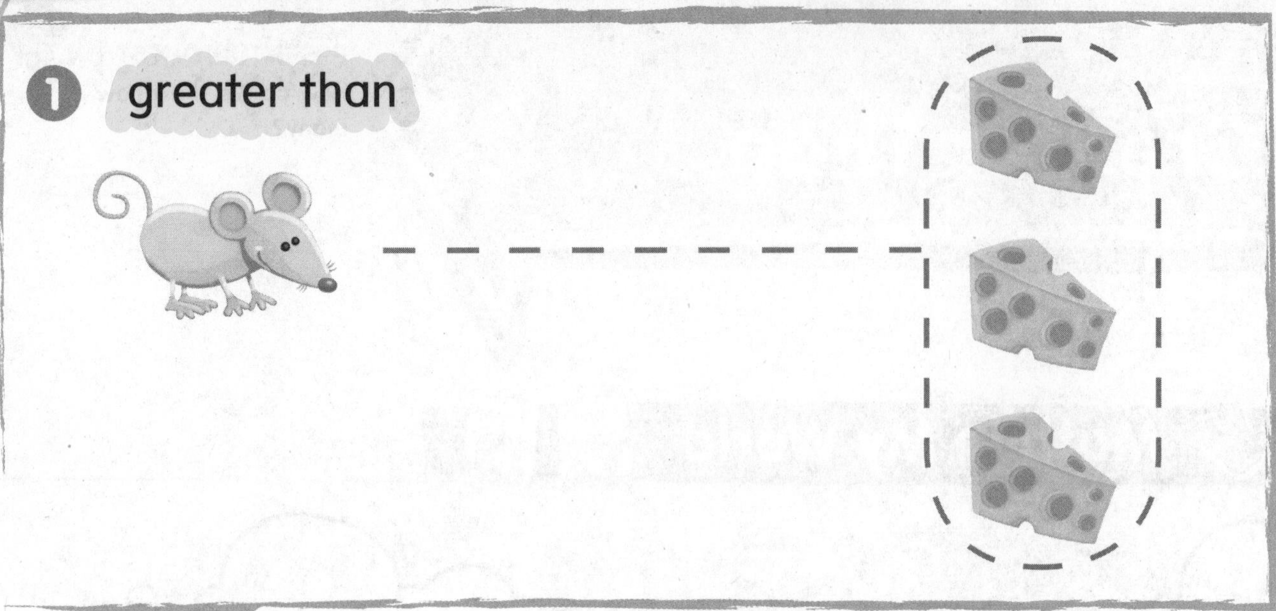


2

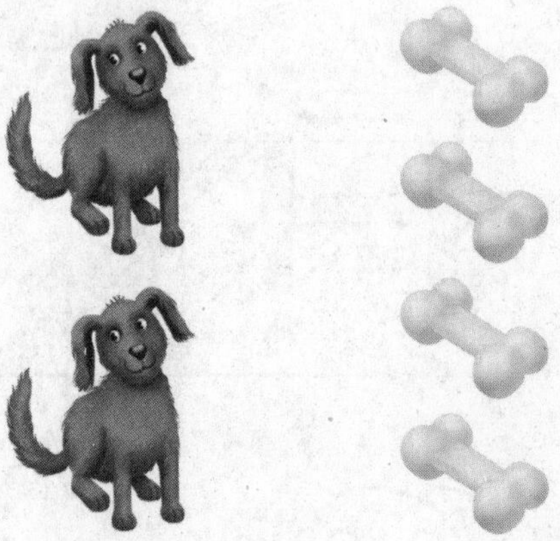

3

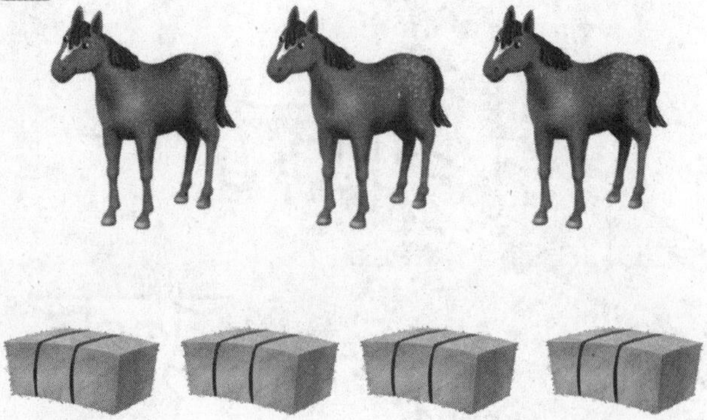

4

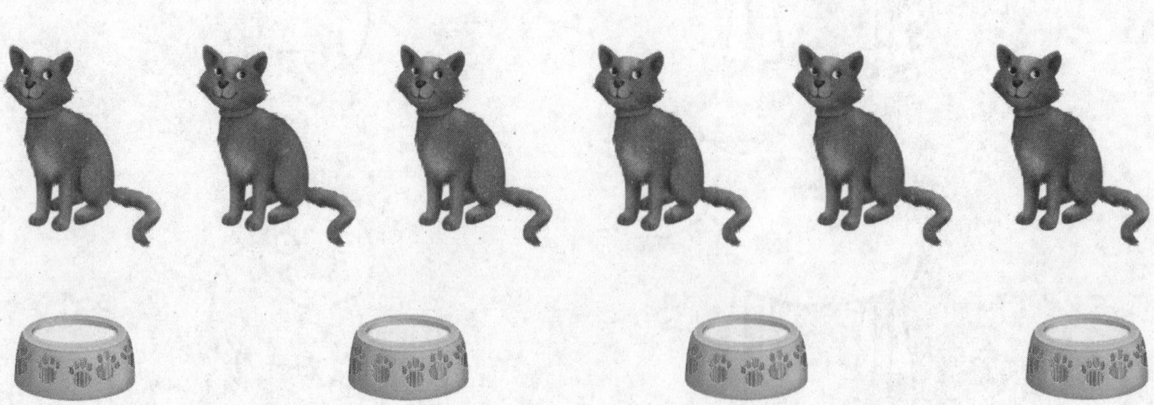

**Directions: 1–4.** Draw a line from an object in one group to match an object in the other group. Describe the sizes of the groups. Circle the group that shows the greater number of objects.

Name

## Independent Practice

5

6

7

**Directions: 5–7.** Draw a line from an object in one group to match an object in the other group. Describe the sizes of the groups. Circle the group that shows the greater number of objects.

# Brain Builders

Processes & Practices

8

9

**Directions: 8–9.** Use counters to show a group that is greater than the group of objects shown. Draw the counters. Draw lines to match the objects in one group with the counters in the other group. Tell a friend how you know which group has a greater number of objects.

Name

Lesson 7

Greater Than

# My Homework

## Homework Helper

Need help? connectED.mcgraw-hill.com

1

2

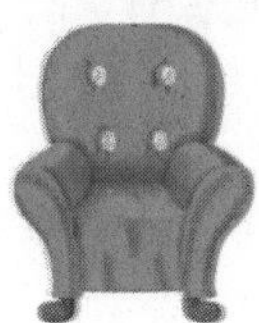 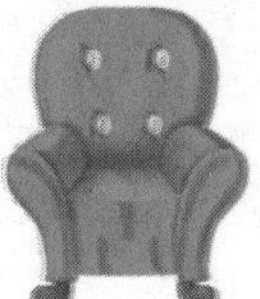 

3

**Directions: 1–3.** Draw a line from an object in one group to match an object in the other group. Describe the sizes of the groups. Circle the group that shows the greater number of objects.

## Vocabulary Check

greater than

**Directions: 4–5.** Draw a line from an object in one group to match an object in the other group. Describe the sizes of the groups. Circle the group that shows the greater number of objects. **6.** Draw a group of objects that is greater than the group of objects shown.

**Math at Home** Show five fingers. Have your child show an amount of fingers that is greater than five.

Name

Lesson 8
# Less Than

**ESSENTIAL QUESTION**
How do we show how many?

## Math in My World

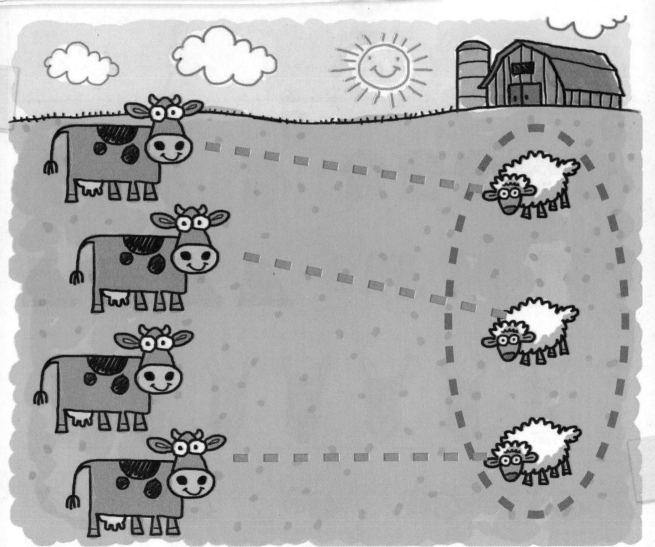

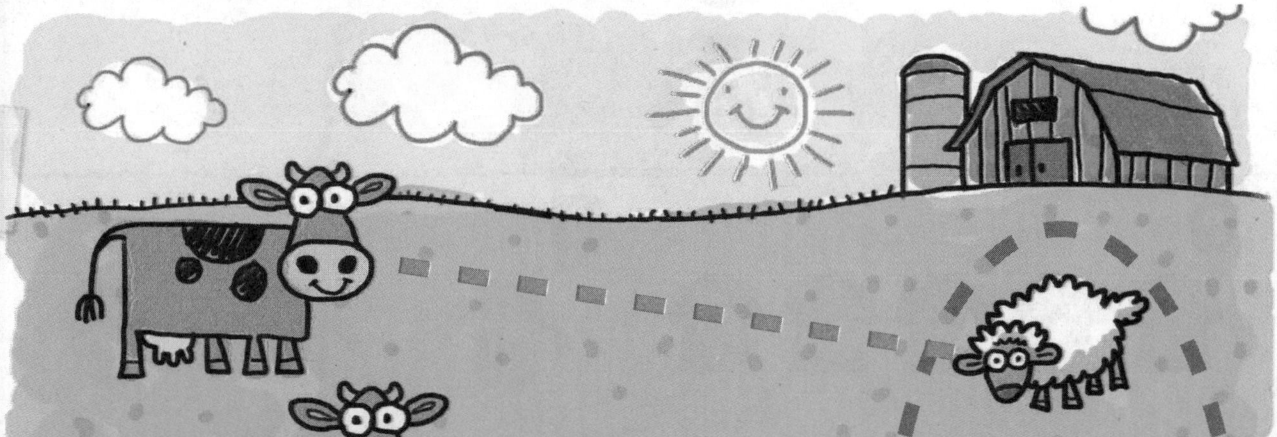

**Teacher Directions:** Place a ● on each cow. Place a ● on each sheep. Trace the lines from an object in one group to match an object in the other group. Describe the sizes of the two groups. Trace a circle around the group of objects that is less.

Processes & Practices

## Guided Practice

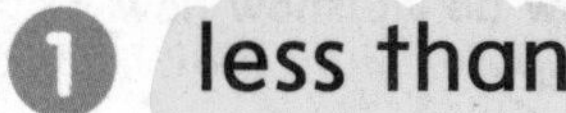

1 less than

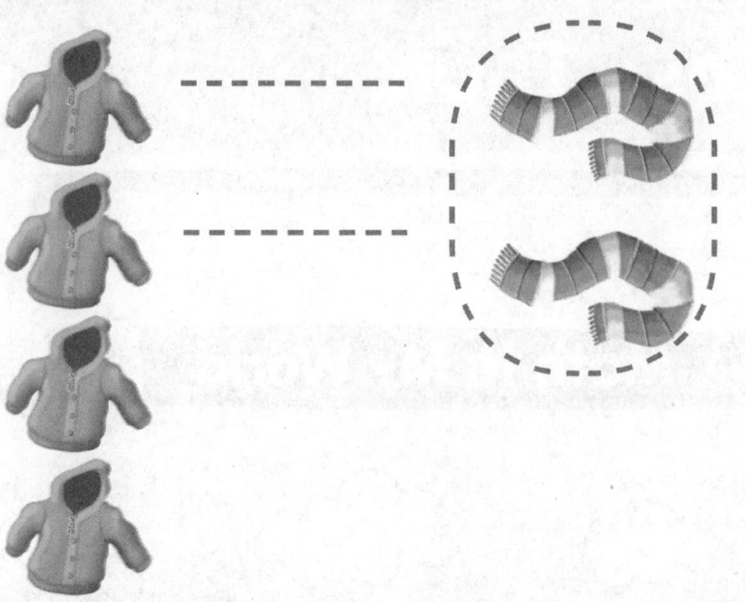

2

3

4

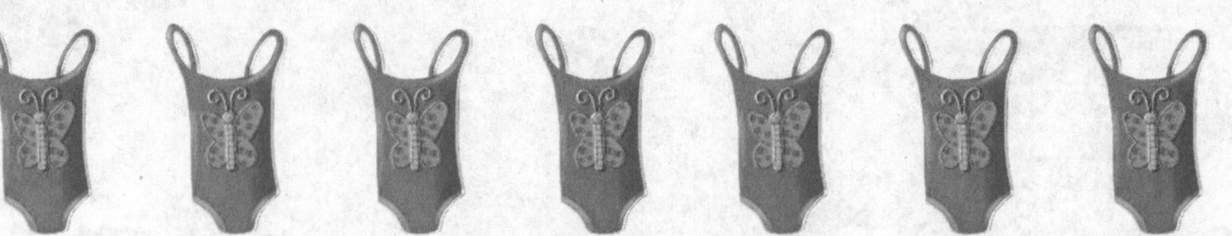

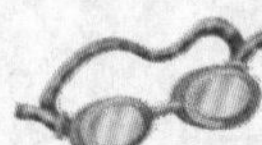

**Directions: 1–4.** Draw a line from an object in one group to match an object in the other group. Describe the sizes of the groups. Circle the group of objects that is less.

Name

## Independent Practice

5

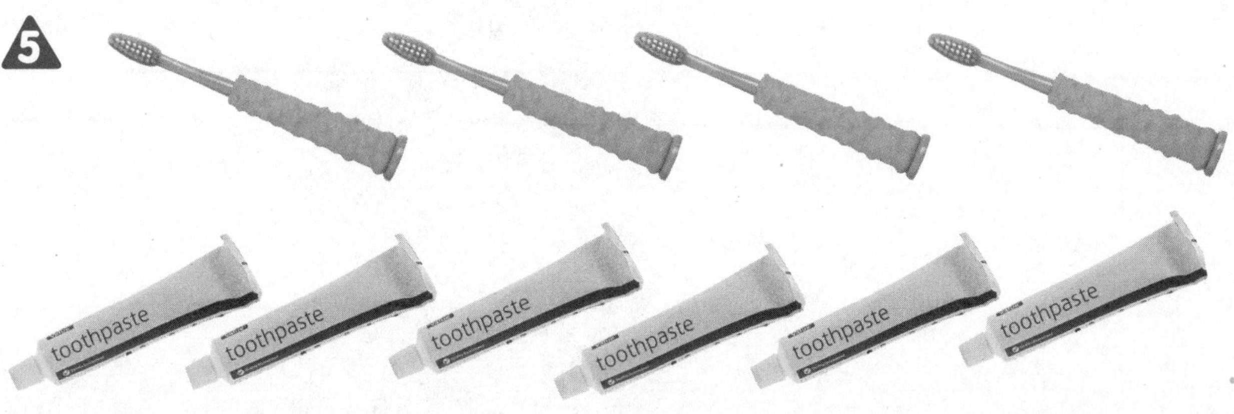


6

7

**Directions: 5–7.** Draw a line from an object in one group to match an object in the other group. Describe the sizes of the groups. Circle the group of objects that is less.

Processes
&Practices

# Brain Builders

8 

9 

**Directions: 8–9.** Count the objects. Draw a group of objects that is less than the group of objects shown. Draw a line from an object in one group to match an object in the other group. Explain to a friend why some objects do not have a match.

Name

# My Homework

**Lesson 8**

Less Than

## Homework Helper

Need help? connectED.mcgraw-hill.com

1

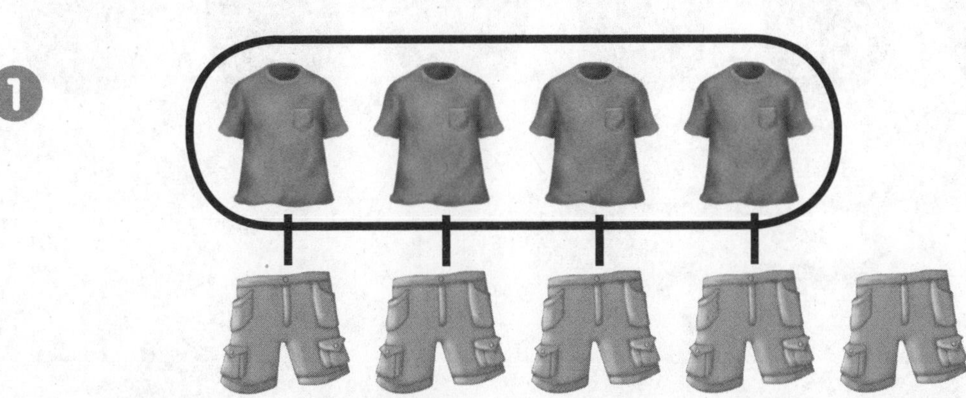

2

3

**Directions: 1–3.** Draw a line from an object in one group to match an object in the other group. Describe the sizes of the groups. Circle the group of objects that is less.

 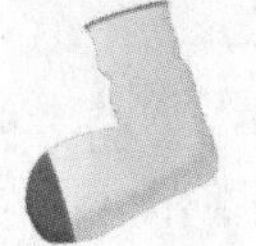   

    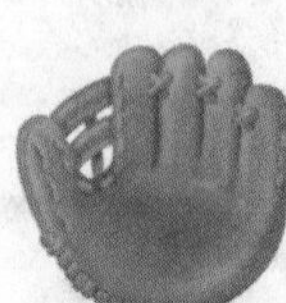 

## Vocabulary Check

less than

**Directions: 4–5.** Draw a line from an object in one group to match an object in the other group. Describe the sizes of the groups. Circle the group of objects that is less. **6.** Draw a group of objects that is less than the group of objects shown.

**Math at Home** Gather three pencils and five crayons. Compare pencils and crayons. Discuss which group is less than the other group.

Name

Lesson 9

# Compare Numbers 0 to 5

**ESSENTIAL QUESTION**
How do we show how many?

## Math in My World

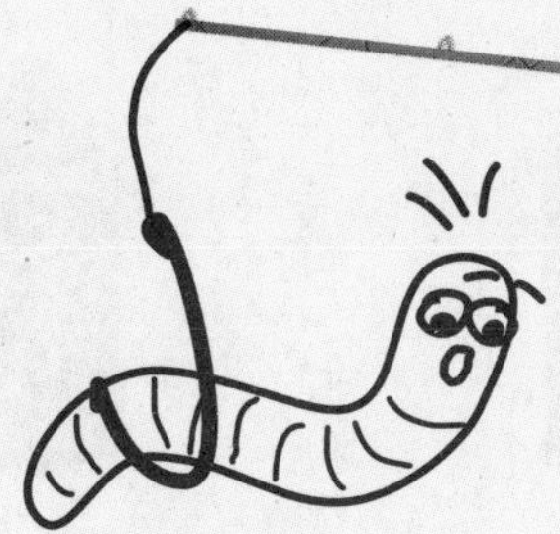

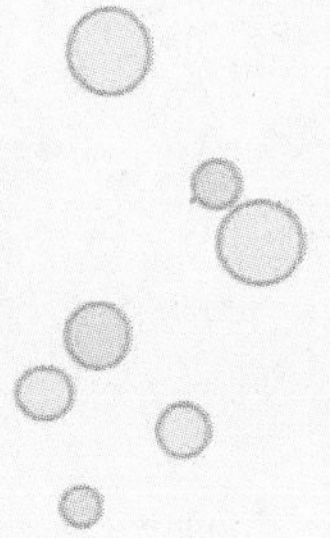

**Teacher Directions:** Draw a group of 1, 2, 3, 4, or 5 fish. Compare your drawing with another student's drawing. Tell if the other student's group is greater than, less than, or equal to your group.

## Guided Practice

Processes &Practices

1

5

2

2

3

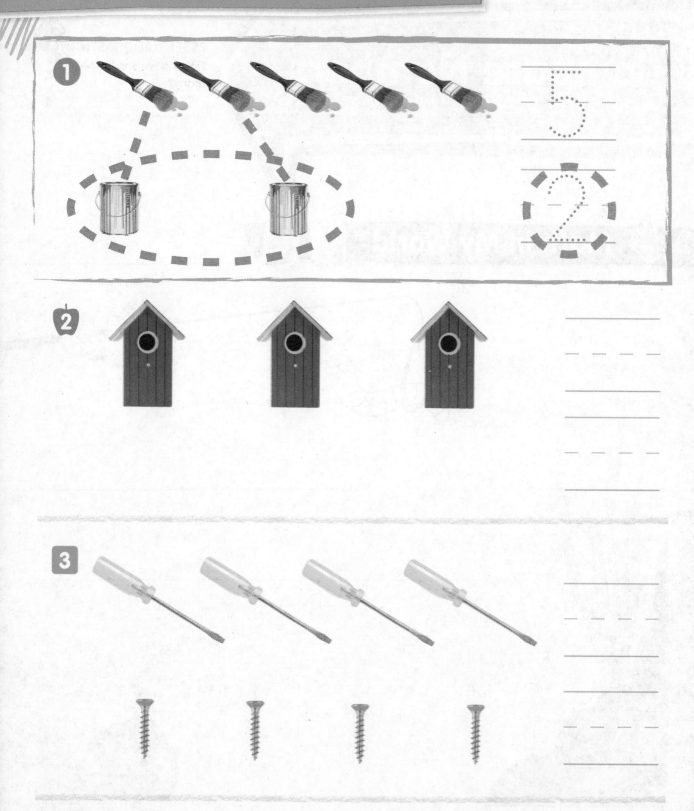

**Directions: 1–2.** Draw lines to match the objects in each group. Count the objects. Write the numbers. Circle the group and number that is less than the other group and number. **3.** Draw lines to match the objects in each group. Count the objects. Write the numbers. Draw a box around the groups and numbers to show equal to.

Name

## Independent Practice

4

5

6

**Directions: 4–5.** Draw lines to match the objects in each group. Count the objects. Write the numbers. Draw an X on the group and number that is greater than the other group and number. **6.** Draw lines to match the objects in each group. Count the objects. Write the numbers. Draw a box around the groups and numbers to show equal to.

# Brain Builders

**Directions: 7.** Draw a group of fish that is greater than the group of fish shown. Write the number. **8.** Draw a group of fish that is less than the group of fish shown. Write the number. Tell a friend if this group is more than your group from Exercise 7. **9.** Draw a group of fish that is equal to the group of fish shown above. Write the number. Tell a friend if this group is more than or less than your group from Exercise 8.

Name

# My Homework

Lesson 9

Compare Numbers 0 to 5

## Homework Helper

Need help? connectED.mcgraw-hill.com

1

**Directions: 1–2.** Draw lines to match the objects in each group. Count the objects. Write the numbers. Draw a circle around the group of objects and number that is less than the other group and number. **3.** Draw lines to match objects in each group. Count the objects. Write the numbers. Draw a box around the groups and numbers to show equal to.

4

5

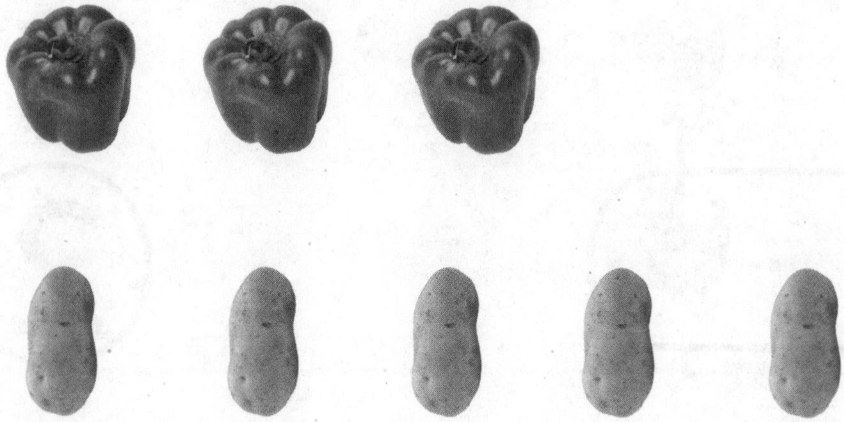

6

**Directions: 4–5.** Draw lines to match the objects in each group. Count the objects. Write the numbers. Draw an X on the group of objects and number that is greater than the other group and number. **6.** Draw lines to match the objects in each group. Count the objects. Write the numbers. Draw a box around the groups and numbers to show equal to.

**Math at Home** Show two groups of toys with up to 5 items in each group. Ask your child which group is greater than, less than, or equal to. Have your child write the numbers.

Name

# Check My Progress

## Vocabulary Check 

1 greater than

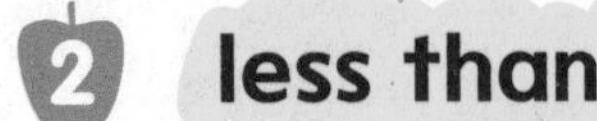
2 less than

## Concept Check

3

**Directions: 1.** Draw a group of dots that is greater than the group of dots shown. **2.** Draw a group of hearts that is less than the group of hearts shown. **3.** Draw a line from an object in one group to match an object in the other group. Draw a box around the groups if the sizes of the groups are equal to each other.

4

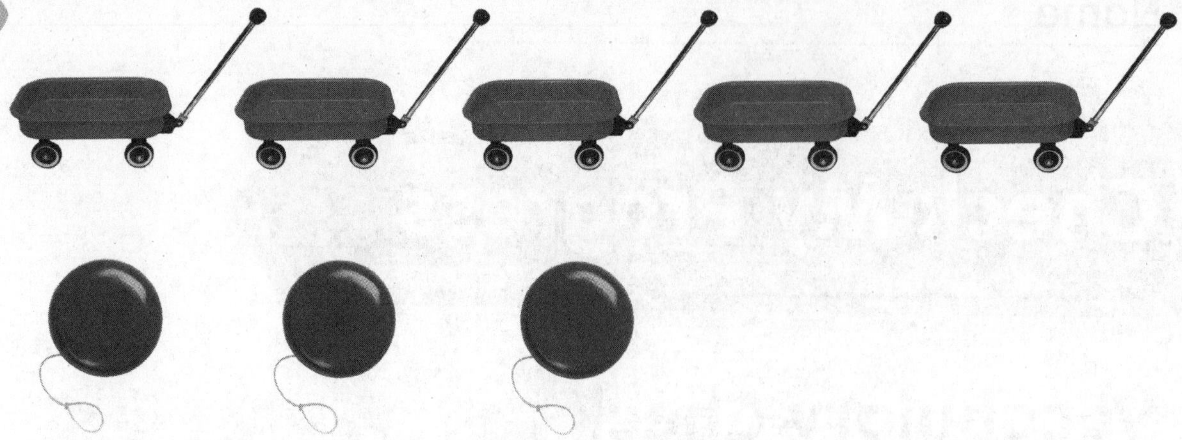

5

6

**Directions: 4.** Draw a line from an object in one group to match an object in the other group. Draw an X on the group that shows the greater number of objects. **5–6.** Draw lines to match the objects in each group. Write the numbers. Circle the group and number that is less than the other group and number.

Name

Lesson 10

# One More

**ESSENTIAL QUESTION**
How do we show how many?

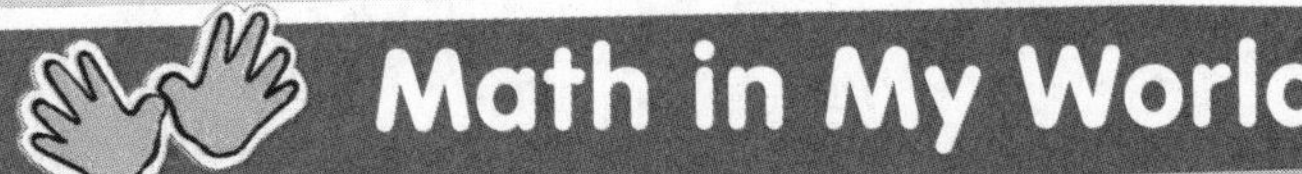

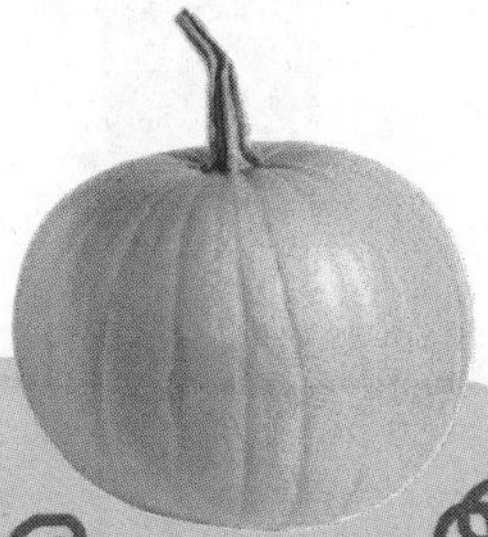

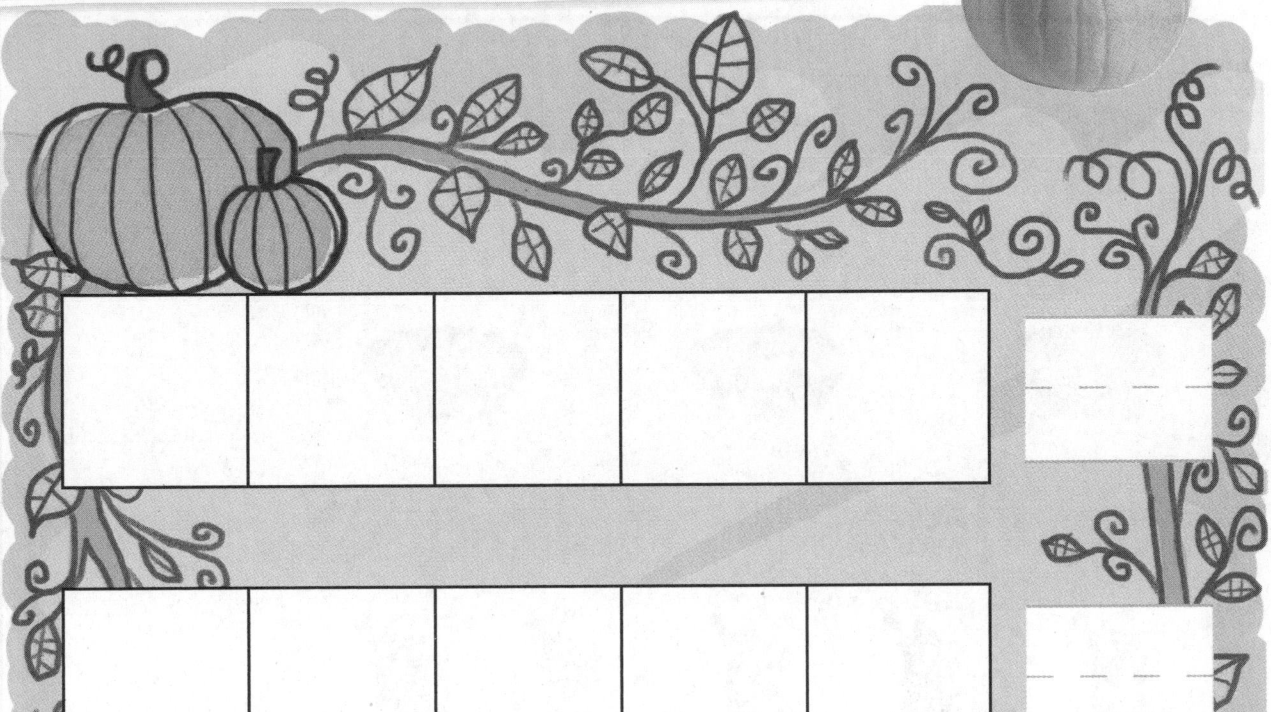

**Teacher Directions:** Use ■ in the boxes above to act out this story. Four pumpkins grew in a garden. Four and one more pumpkins grew in another garden. Color the boxes to show how many. Write the numbers. Circle the number that shows one more.

## Guided Practice

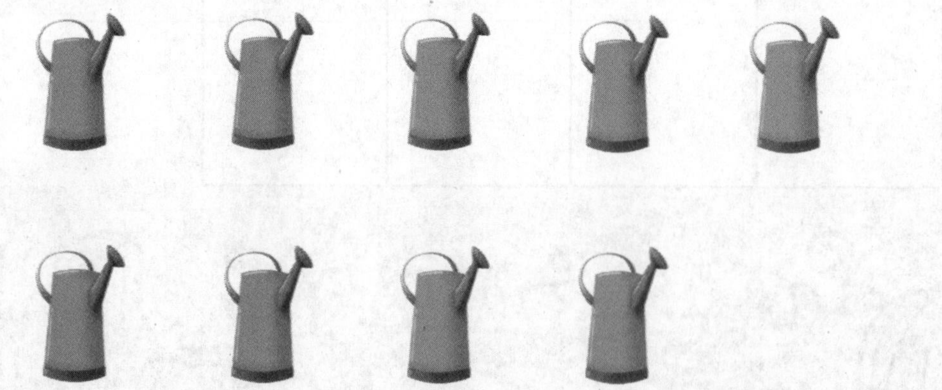

**Directions: 1–3.** Count the objects in each row. Say how many. Trace or write the numbers. Circle the number that shows one more.

Name

## Independent Practice

**Directions: 4.** Count the objects in each row. Say how many. Write the numbers. Circle the number that shows one more. **5–6.** Count the objects. Say how many. Write the number. Draw a group of objects that shows one more. Write the number.

Processes &Practices

## Brain Builders

7

**Directions: 7.** Count the corn cobs. Say how many. Use counters to show how many. Write the number. Draw a group of corn cobs that shows one more. Use counters to show how many. Write the number. Draw an X on the number that shows one more. Explain how you know it is one more to a friend.

Name

# My Homework

Lesson 10

One More

## Homework Helper

Need help? connectED.mcgraw-hill.com

1

2

2

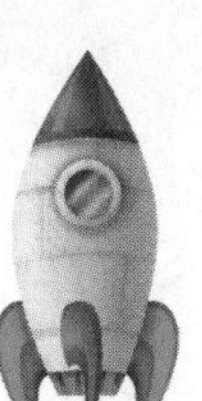

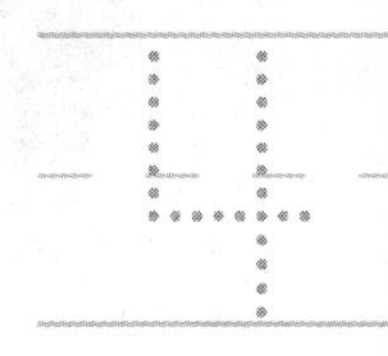

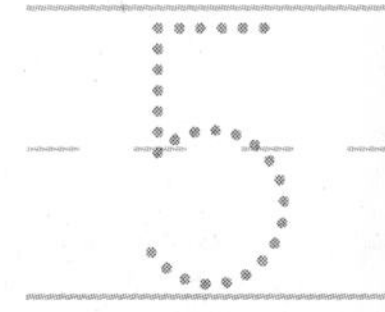

**Directions: 1–2.** Count the objects in each row. Say how many. Trace the numbers. Circle the number that shows one more.

3

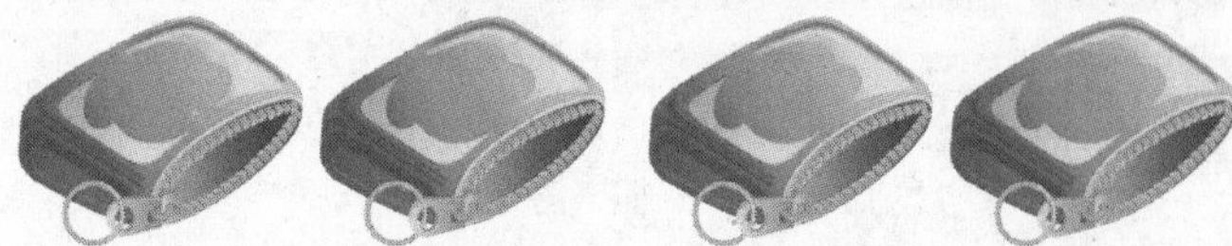

**Directions: 3.** Count the objects in each group. Say how many. Write the numbers. Circle the number that shows one more. **4–5.** Count the objects. Say how many. Write the number. Draw a group of objects that shows one more. Write the number.

**Math at Home** Show four cups. Have your child count the cups and write the number. Guide your child to draw a group of cups that shows one more and write the number.

Name

Lesson 11

# Problem Solving

**STRATEGY: Draw a Diagram**

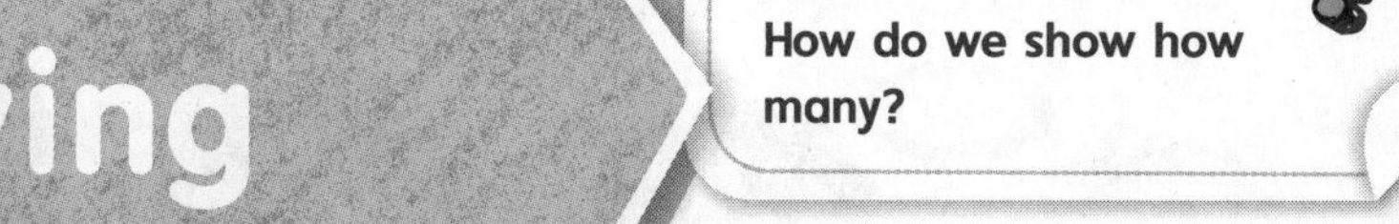

**ESSENTIAL QUESTION**

How do we show how many?

## How many nests?

## Draw a Diagram

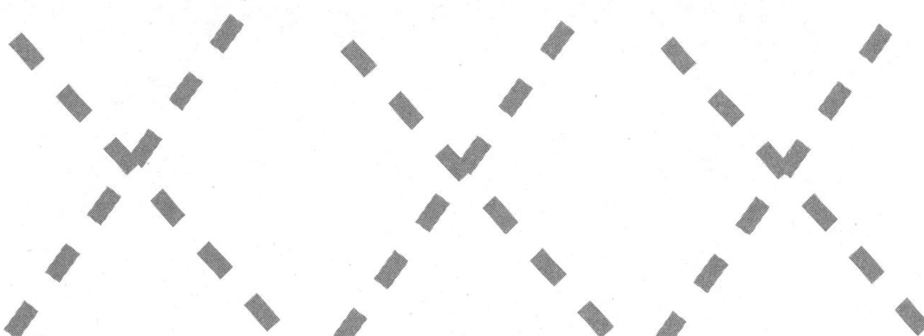

**Teacher Directions:** Trace the lines from each hen to a nest. Trace an X for each nest that now has a hen. How many nests have a hen? Write the number. Explain your answer to a friend.

## How many ducks?

## Draw a Diagram

**Directions:** Draw a line from the ring at the end of each fishing line to a duck. Draw a circle for each duck that was caught. How many ducks were caught? Write the number. Explain your answer to a friend.

Name

## How many ducks?

## Draw a Diagram

**Directions:** Each child caught one duck. Draw a line from each child's pole to the duck that was caught. Draw one circle for each duck that was not caught. How many ducks were not caught? Write the number. Explain your answer to a friend.

## How many horses?

## Draw a Diagram

**Directions:** Count the horses. Draw boxes to show how many horses if there was one more horse. Write the number. Explain your answer to a friend.

Name

# My Homework

Lesson 11

Problem Solving: Draw a Diagram

## How many butterflies?

## Draw a Diagram

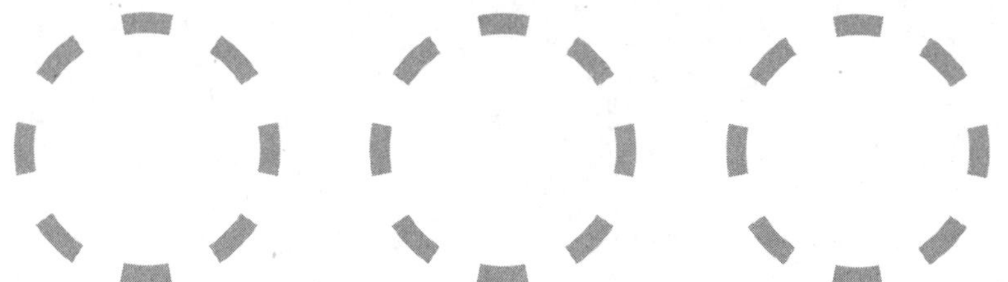

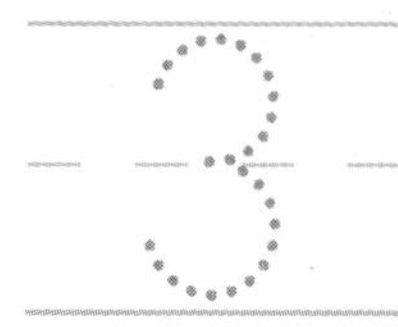

**Directions:** Trace the lines from each net to a butterfly. Trace a circle for each butterfly that was caught. How many butterflies were caught? Trace the number. Explain your answer to a family member.

## How many horses?

## Draw a Diagram

**Directions:** Draw a line from each lasso to a horse. Draw a circle for each horse that was caught. Write the number. Explain your answer to a family member.

**Math at Home** Take advantage of problem-solving opportunities during daily routines such as going to the grocery store. Have your child help you make a grocery list by drawing pictures of the grocery items needed and the number of each item needed.

Name

# My Review

Chapter 1

Numbers 0 to 5

## Vocabulary Check

two

three

five

one

four

**Directions: 1.** Use a blue crayon to color the group of five animals. **2.** Use a red crayon to color the group of three animals. **3.** Use a yellow crayon to color the group of four animals. **4.** Use an orange crayon to color the group with one animal. **5.** Use a brown crayon to color the group of two animals.

1

2

3

**Directions: 1.** Say how many fish are in the bowl. Write the number. **2.** Draw lines to match objects in each group. Count the objects. Write the numbers. Draw an X on the group and number that is greater than the other group and number. **3.** Count the tractors in each row. Write the numbers. Circle the number that is one more than three.

Name

## Brain Builders

4 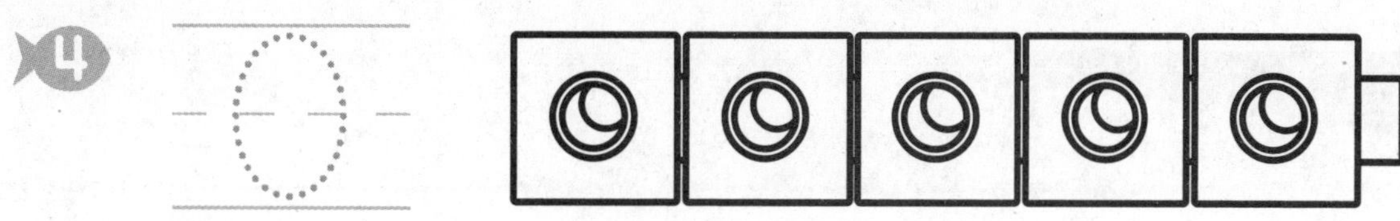

5 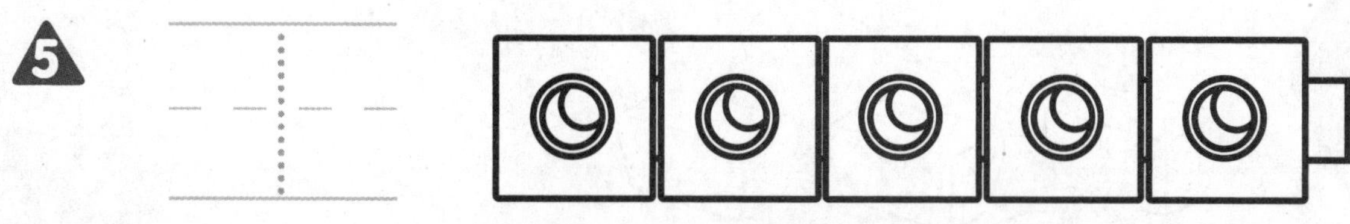

6 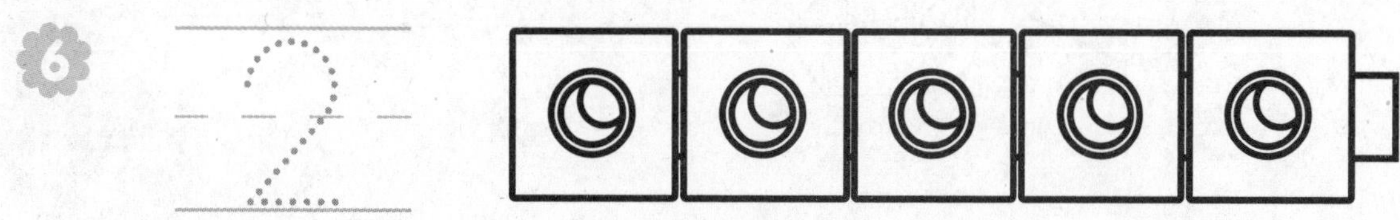

7 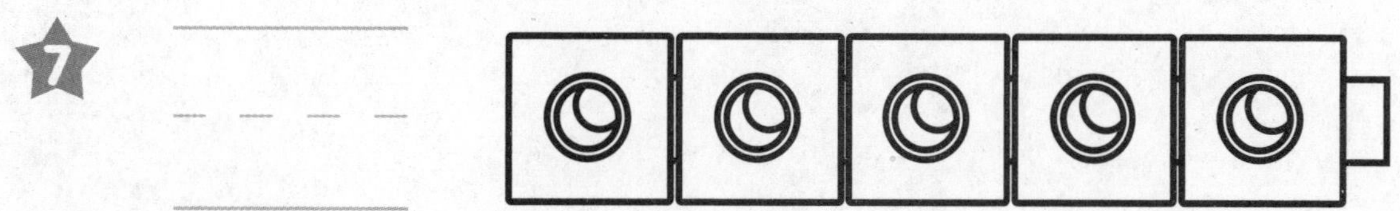

8 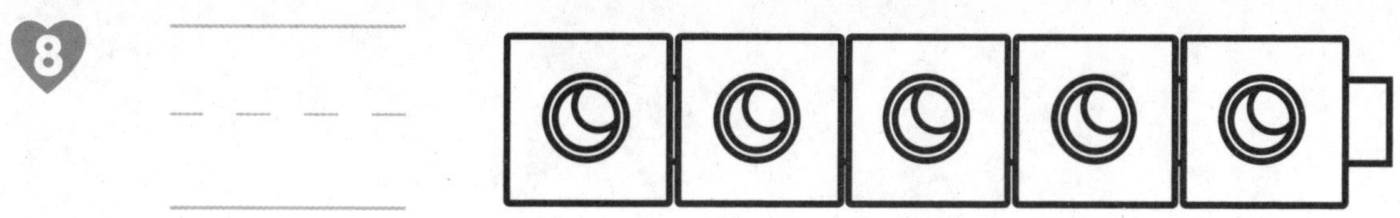

9 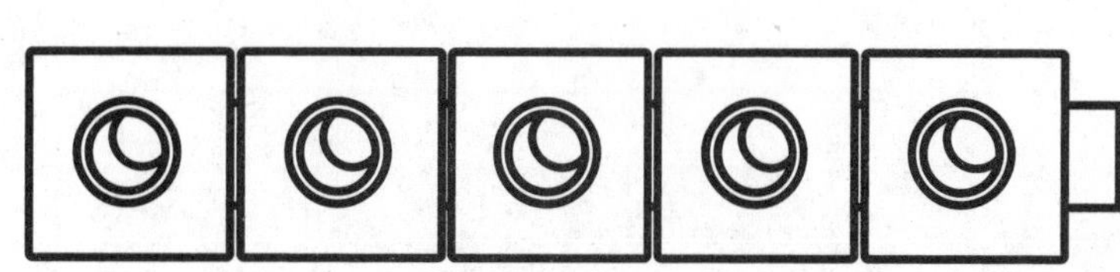

**Directions: 4–6.** Trace the number. Color the cubes to show how many. **7.** Write the number that is one more than four. Color the cubes to show how many. **8.** Write the number that is one less than four. Color the cubes to show how many. **9.** Write the number that is one more than three. Color the cubes to show how many.

# Reflect

**Chapter 1**

ESSENTIAL QUESTION

**How do we show how many?**

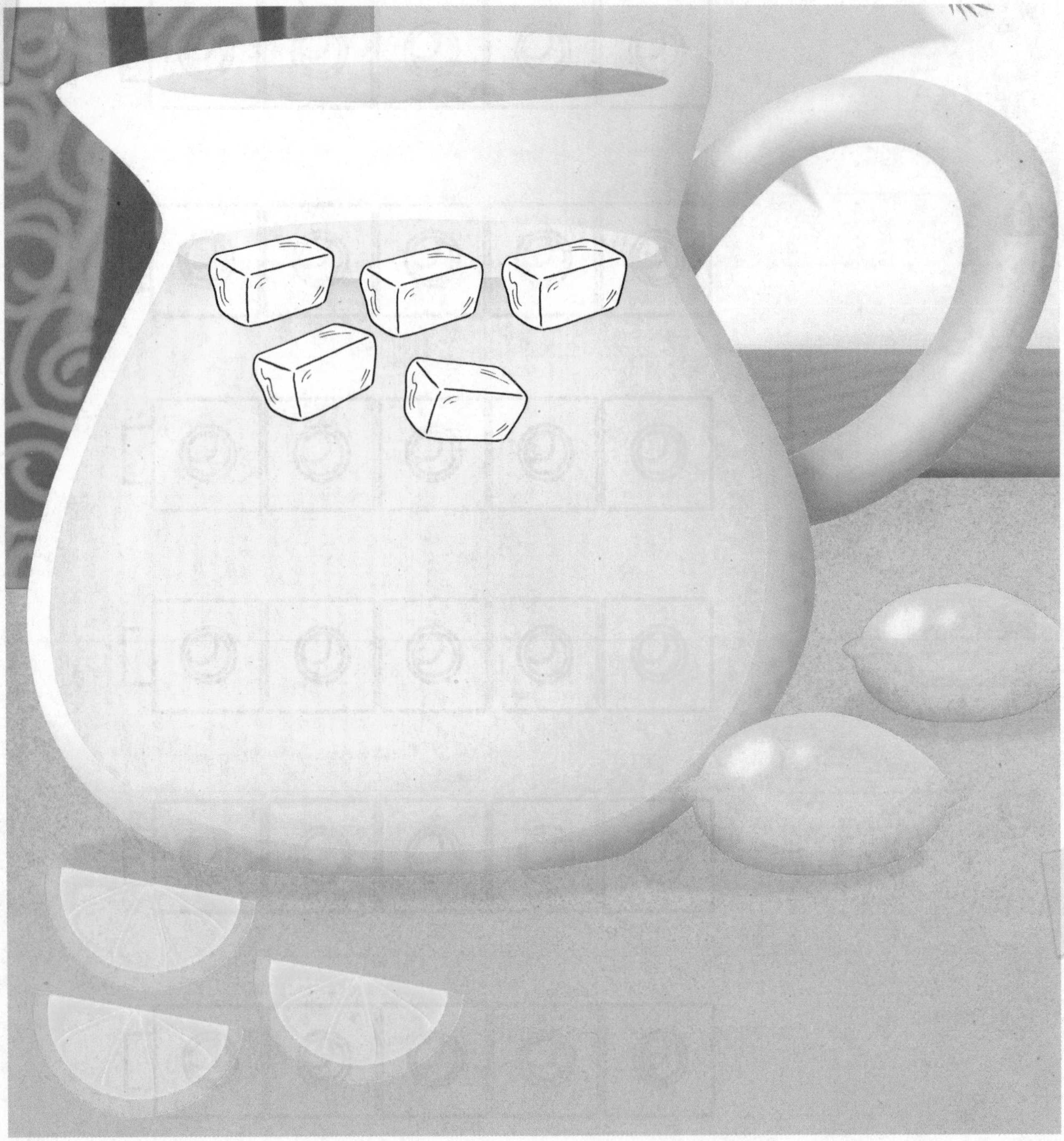

**Directions:** Draw one spoon on the table. Draw an X on the group of two objects. Draw a box around the group of three objects. Color in four of the ice cubes. Draw five lemon seeds in the lemonade. Tell a friend the number of cups on the page.

Name

Date

Score

# Performance Task

## The Pet Store

Mr. Smith's class visited the Pet Store.

**Show all your work to receive full credit.**

### Part A

Circle the fish bowl that has zero fish.

### Part B

Count the number of fish. Write the number.

## Part C

Count the number of crabs. Circle the number that is one more than that number.

3 4 5

## Part D

Circle the nest with 5 birds.

  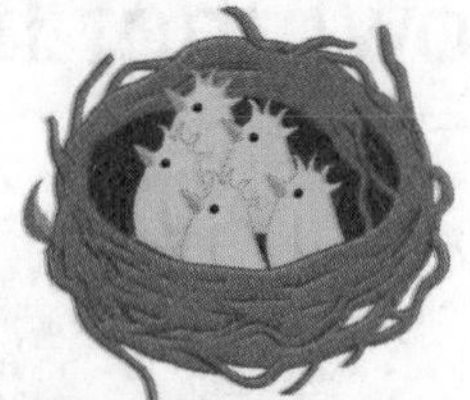

## Part E

Circle the group that has more chickens.

Chapter

# Numbers to 10

ESSENTIAL QUESTION
What do numbers tell me?

I Choose Healthful Foods!

Watch a video!

Watch

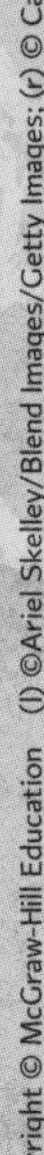

Name ..............................................................

# Chapter 2 Project

## My Number Book 0–10

1. Design your cover in the space below.
2. Use drawings and numbers to design the cover.
3. Once your book is complete, share it with the class.

Name

# Am I Ready?

1

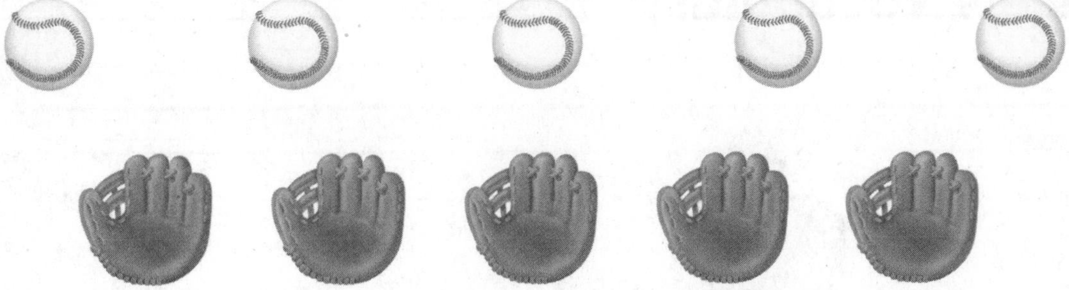

2

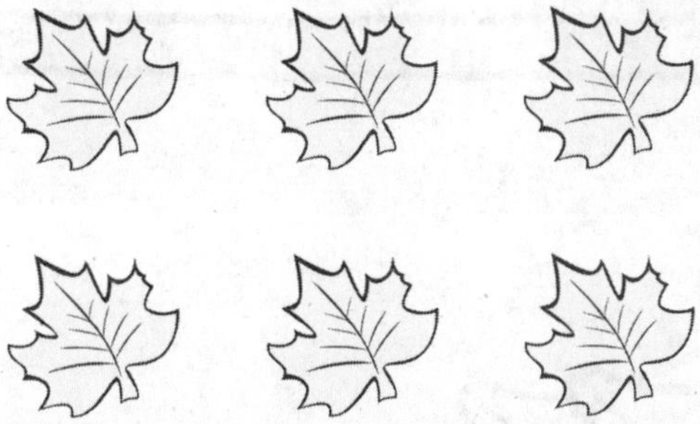

3

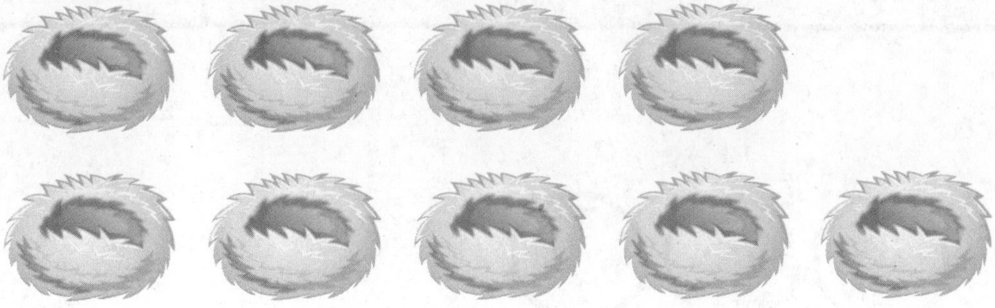

4

**Directions:** **1.** Draw a line from each baseball to a baseball mitt. **2.** Color five leaves green. **3.** Circle the row with four nests. **4.** Count the objects in each group. Write the numbers. Draw a circle around the group that has more.

Name

# My Math Words

## Review Vocabulary

one two three

**Directions:** Count the fruit on each shelf. Write the numbers. Trace each number word.

# My Vocabulary Cards

Processes &Practices

**Teacher Directions:**
**Ideas for Use**

- Have students name and copy the letters in each word. Guide students to choose two cards and compare the letters in each word.
- Have students choose a partner. Direct each partner to choose one number word. Have a partner clap the number as the other partner counts and tells how many.

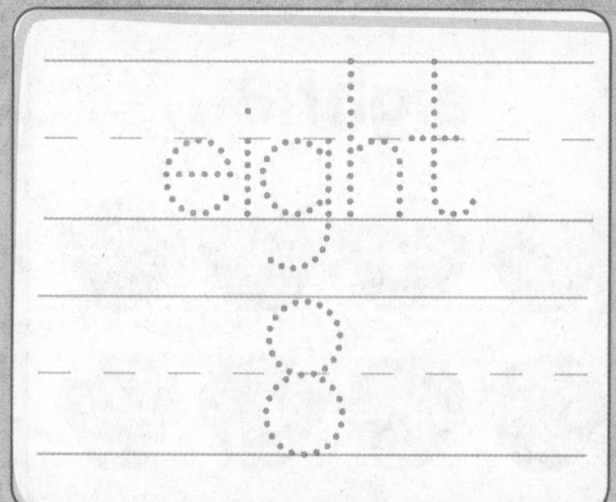

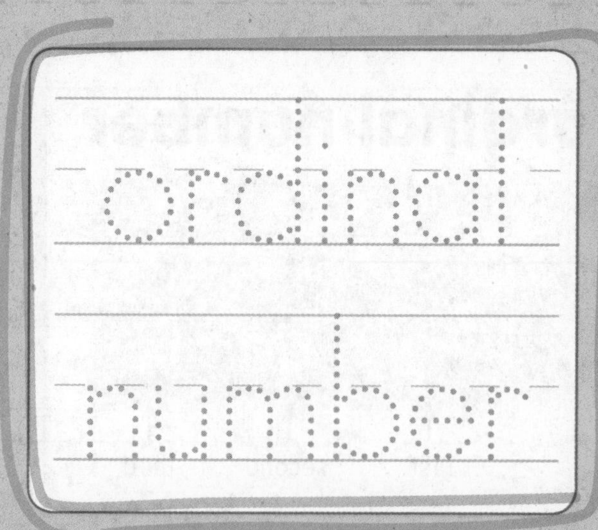


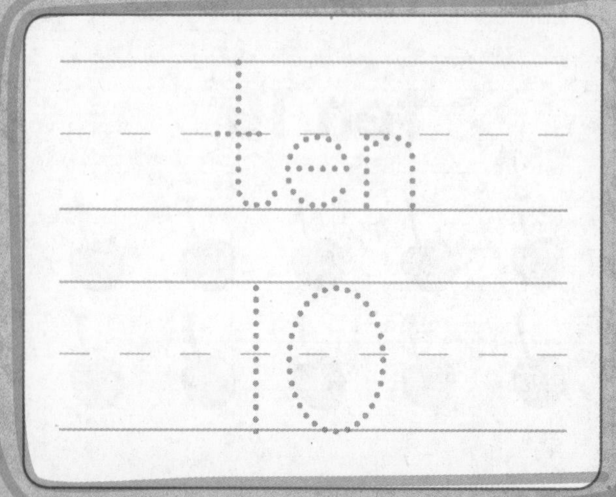

# My Foldable

FOLDABLES® Follow the steps on the back to make your Foldable.

1

2

3

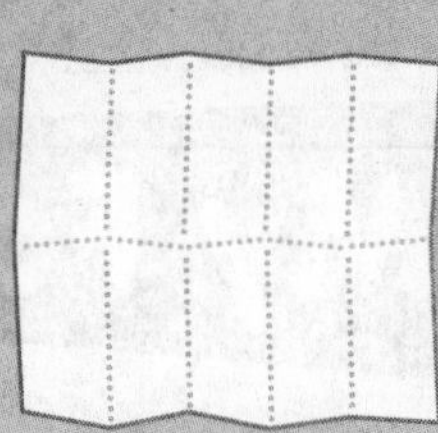

Name

# Lesson 1
# Numbers 6 and 7

**ESSENTIAL QUESTION**
What do numbers tell me?

##  Math in My World

| | | | | |
|---|---|---|---|---|
| | | | | |
| | | | | |

**Teacher Directions:** Use [red tile] and [blue tile]. Show six in the ten-frame using five red and the other blue. Draw six small tiles on a plate. Repeat with seven in the ten-frame using five red and the others blue. Draw seven small tiles on a plate. Color the boxes to show seven.

## Guided Practice

Processes & Practices

1

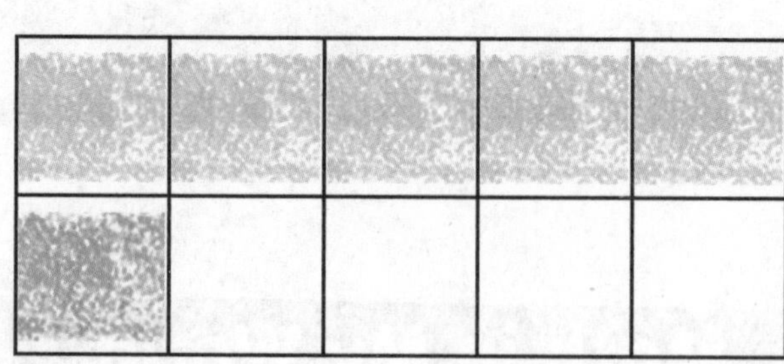

2

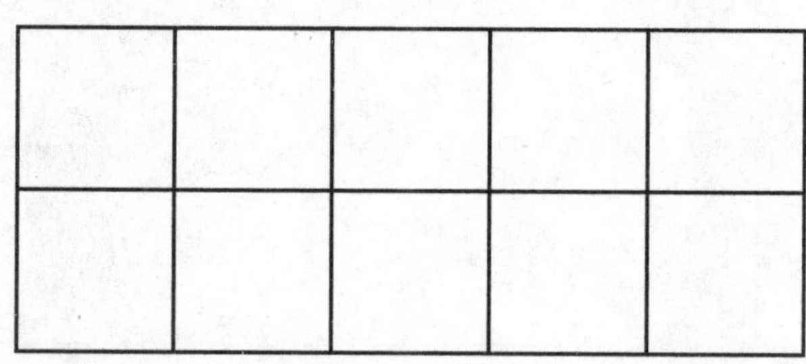

3

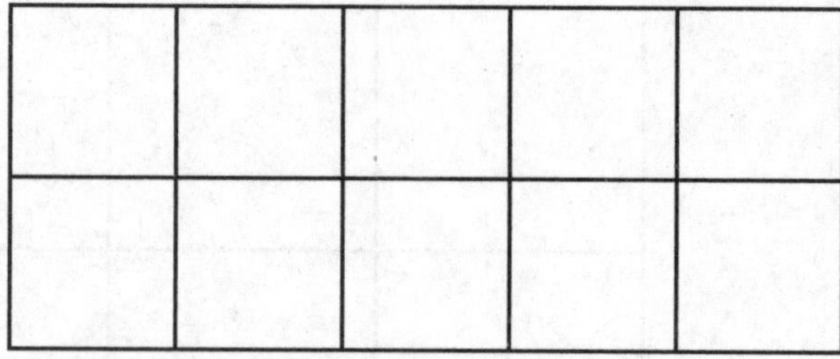

**Directions: 1–3.** Count the objects. Say how many. Use red color tiles to show five objects counted. Color one box red for each red color tile. Use blue color tiles to show the rest of the objects counted. Color one box blue for each blue color tile.

Name

## Independent Practice

4

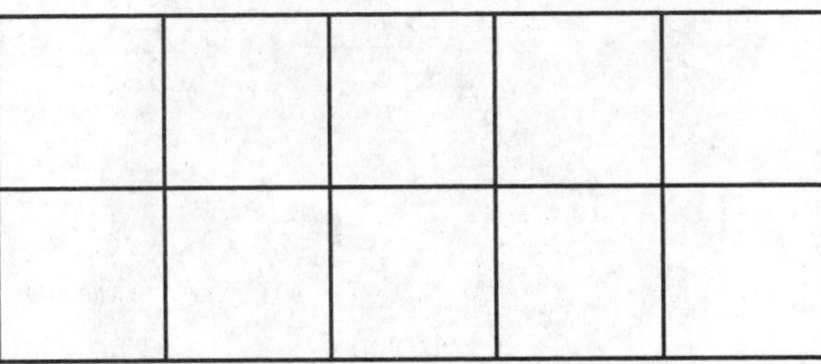

5

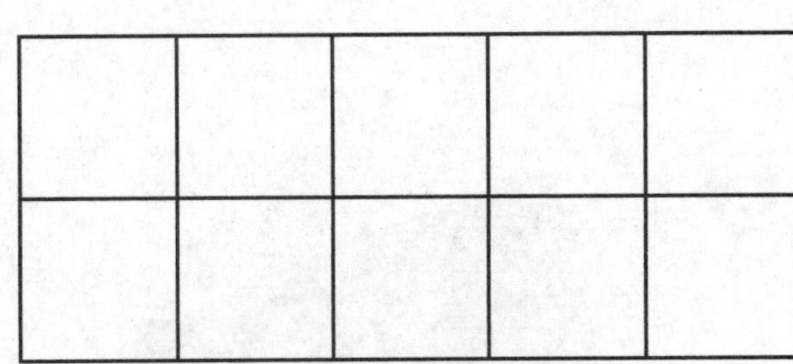

6

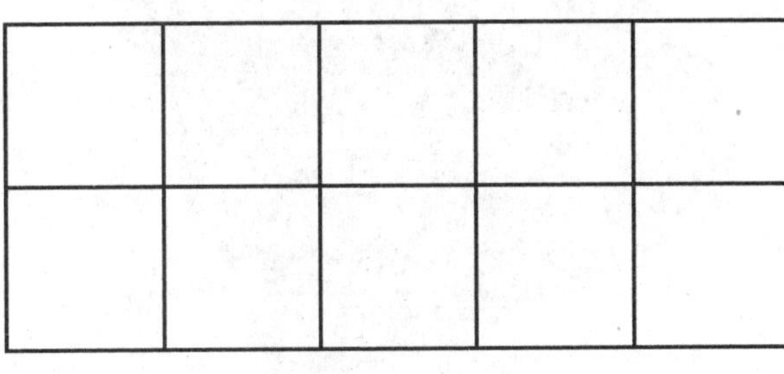

**Directions: 4–6.** Count the objects. Say how many. Use red color tiles to show five objects counted. Color one box red for each red tile. Use blue color tiles to show the rest of the objects counted. Color one box blue for each blue color tile.

Processes & Practices

# Brain Builders

7

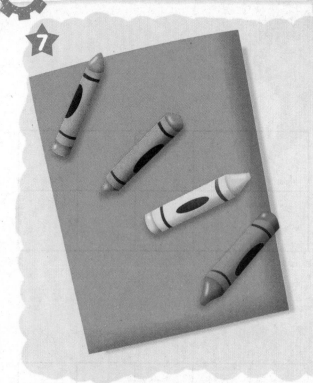

8

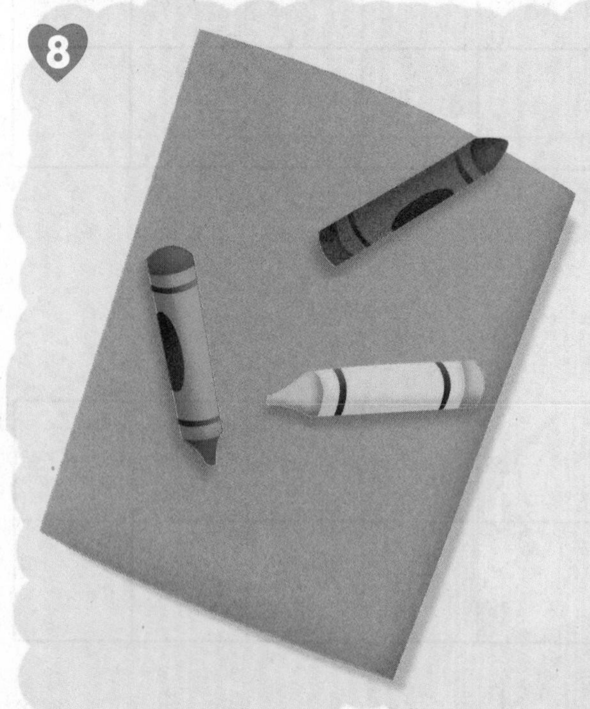

**Directions: 7.** Count the crayons. Say how many. Draw more crayons to show six. Use Work Mat 3 and color tiles to show how many. Explain to a friend how you know there are six. **8.** Count the crayons. Say how many. Draw more crayons to show seven. Use Work Mat 3 and color tiles to show how many. Explain to a friend how you know there are seven.

Name

# My Homework

Lesson 1

Numbers 6 and 7

## Homework Helper

Need help? connectED.mcgraw-hill.com

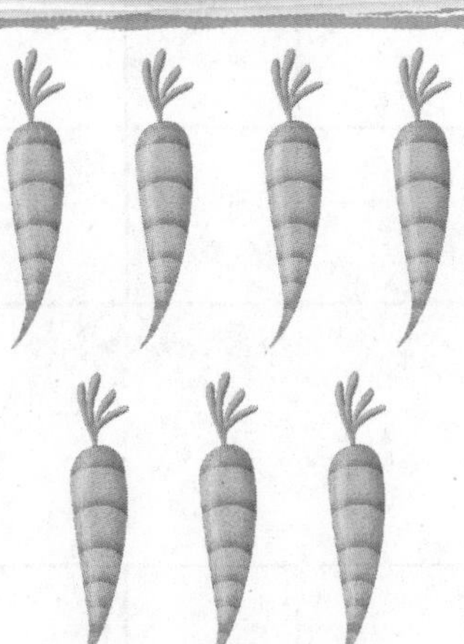

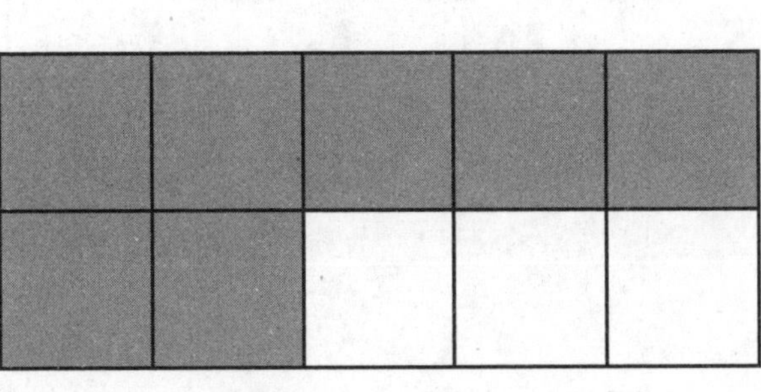

2

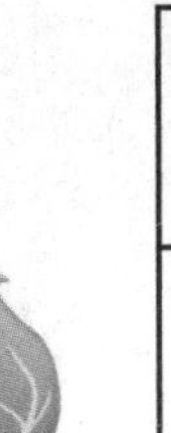

3

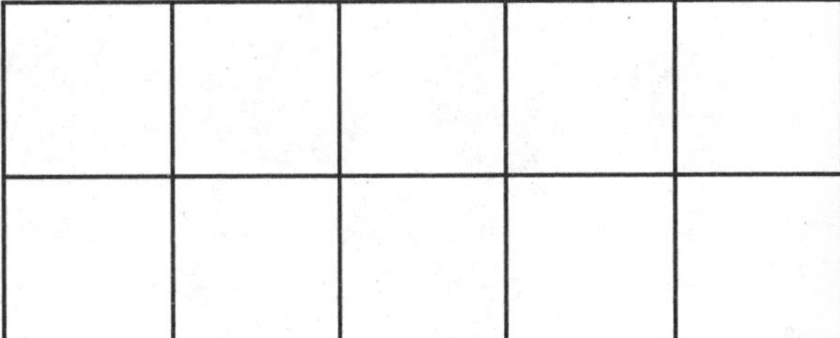

**Directions: 1–3.** Count the objects. Say how many. Use pennies to show how many. Color one box for each object counted.

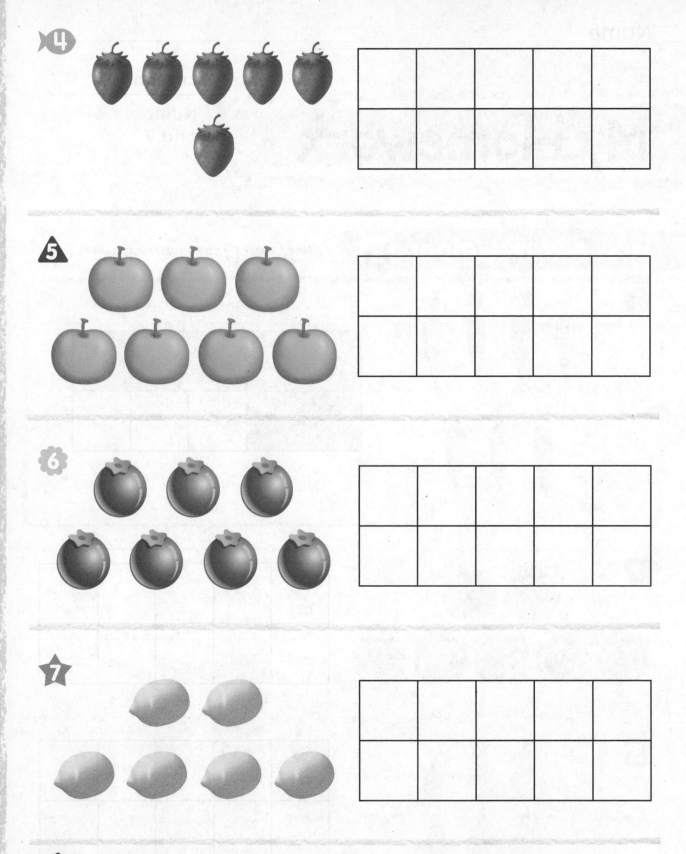

4

5

6

7

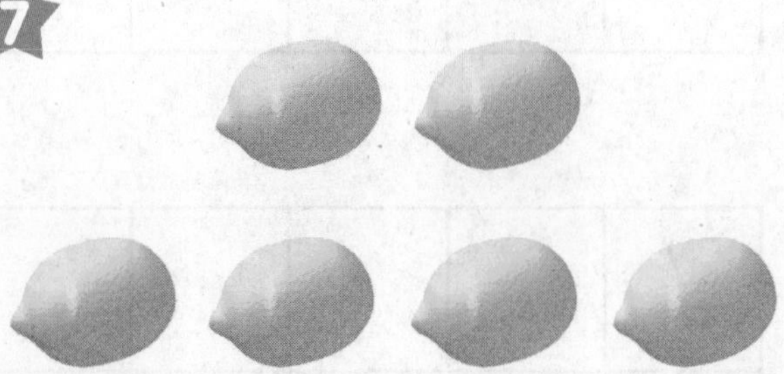

**Directions: 4–7.** Count the objects. Say how many. Use pennies to show how many. Color one box for each object counted.

**Math at Home** Take a walk with your child. Find six and seven objects such as mailboxes or houses. Have your child count the objects.

Name

# Lesson 2
# Number 8

**ESSENTIAL QUESTION**
What do numbers tell me?

## Math in My World

**Teacher Directions:** Count the legs on the large spider. Count the small spiders. Draw one line from each small spider to a leg on the large spider. Use [cube] to show five spiders on the ten-frame. Use [cube] to show the others. Color the boxes to show how many.

## Guided Practice

Processes &Practices

1

2

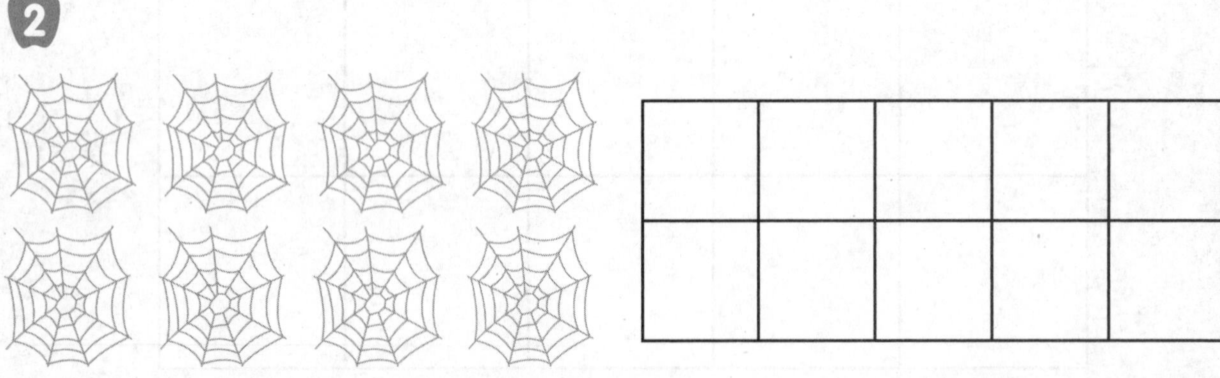

3

**Directions: 1–3.** Count the objects. Say how many. Use green connecting cubes to show five objects counted. Color one box green for each green cube. Use purple connecting cubes to show the rest of the objects counted. Color one box purple for each purple cube.

Name

## Independent Practice

4

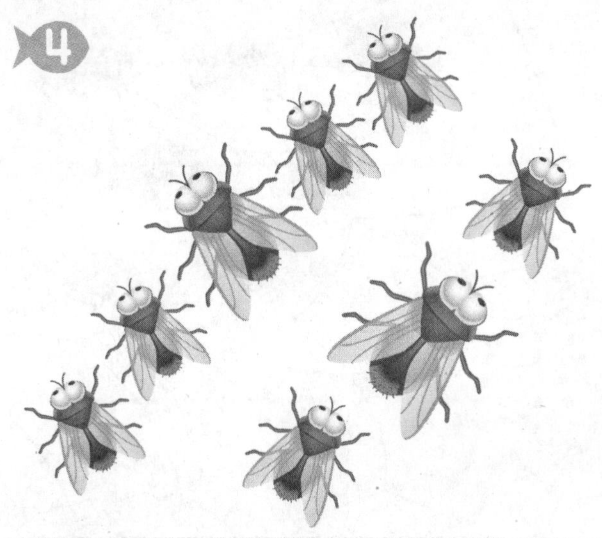

5

6

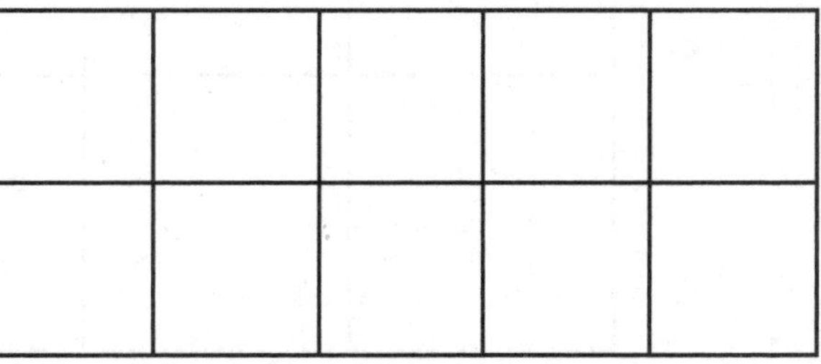

**Directions: 4–6.** Count the objects. Say how many. Use green connecting cubes to show five objects counted. Color one box green for each green cube. Use purple connecting cubes to show the rest of the objects counted. Color one box purple for each purple cube.

Processes &Practices

## Brain Builders

7

|  |  |  |  |  |
|---|---|---|---|---|
|  |  |  |  |  |
|  |  |  |  |  |

**Directions: 7.** Count the insects in each group. Say how many. Draw a circle around each group of eight. For each group that does not have eight, draw more insects to show eight. Explain to a friend how you decided how many more to draw. Color the boxes to show eight.

Name



# My Homework

## Homework Helper

Need help? connectED.mcgraw-hill.com

1

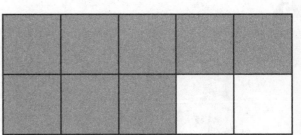

2

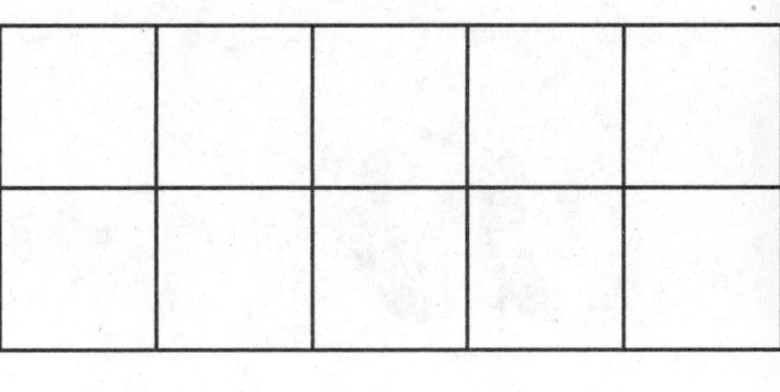

3

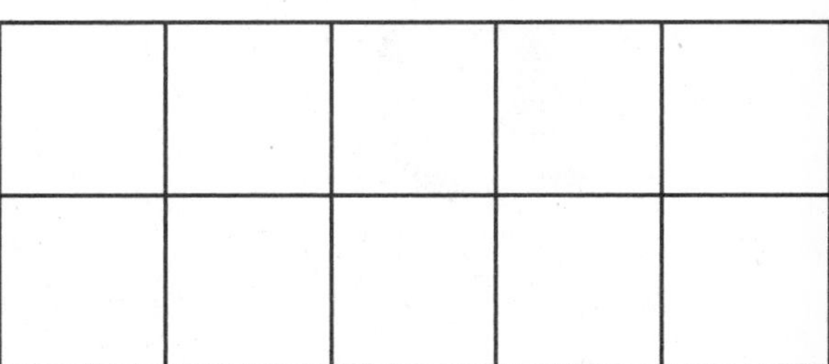

**Directions: 1–3.** Count the objects. Say how many. Use pennies to show how many. Color one box for each object counted.

4

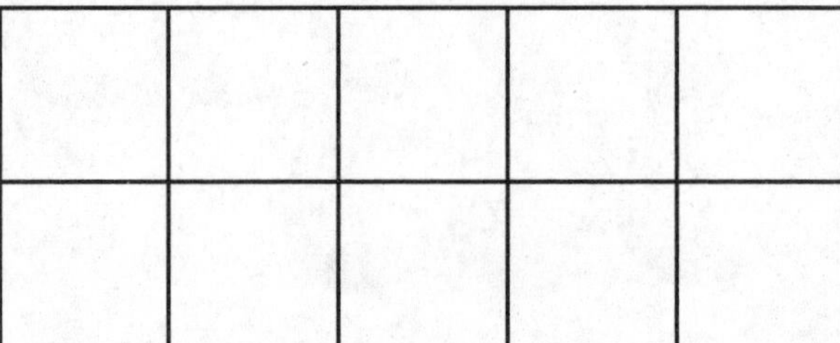

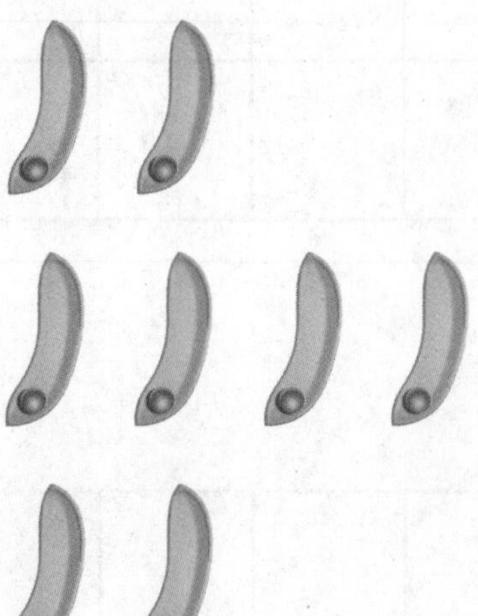

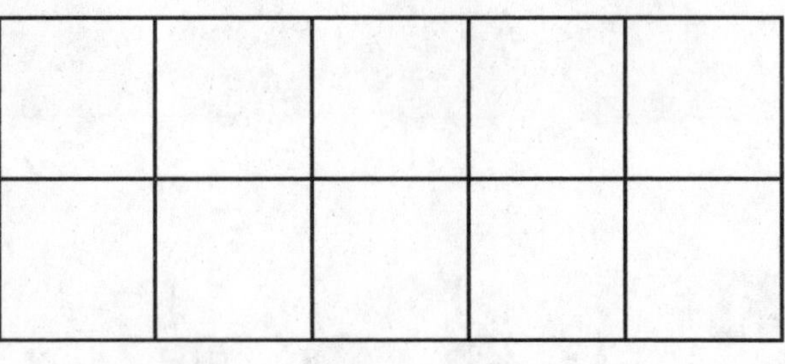

**Directions: 4–6.** Count the objects. Say how many. Use pennies to show how many. Color one box for each object counted.

**Math at Home** Have your child trace eight pennies on a piece of paper. Guide your child to color the pennies and cut them out. Draw a jar. Have your child glue the paper pennies on the jar. Count the paper pennies together.

Name

Lesson 3

# Read and Write 6, 7, and 8

**ESSENTIAL QUESTION**
What do numbers tell me?

## Math in My World

6
six

7
seven

8
eight

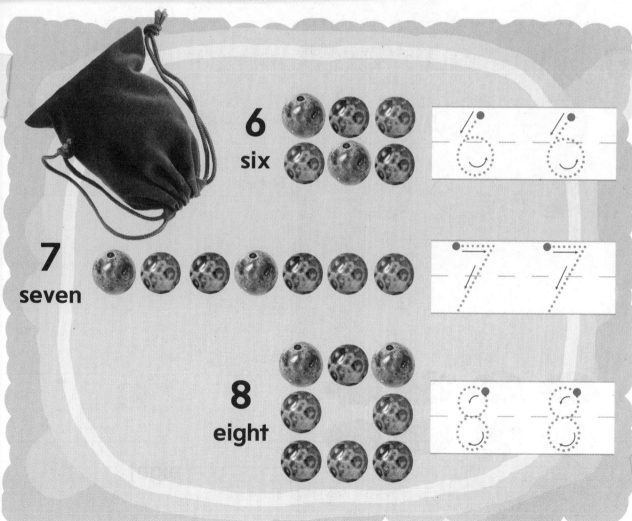

**Teacher Directions:** Count the marbles in each group. Say how many. Use [cube] to show how many. Trace the numbers.

## Guided Practice

Processes &Practices

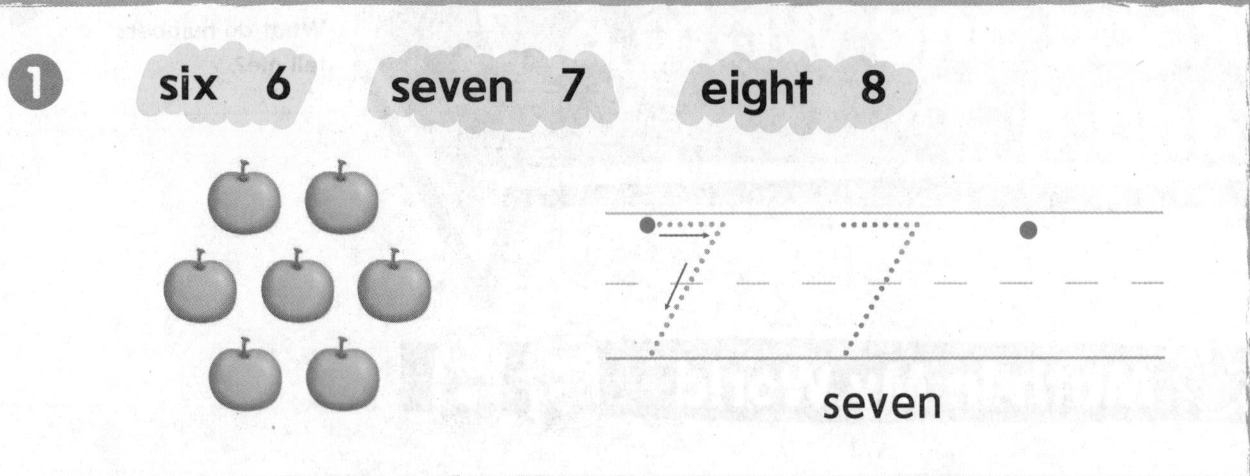


2

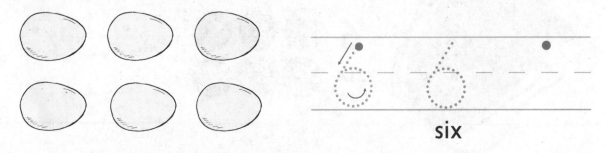

six

3

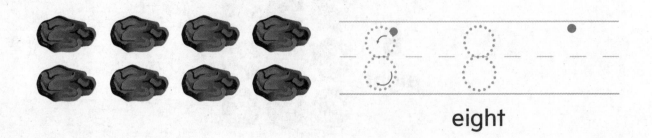

eight

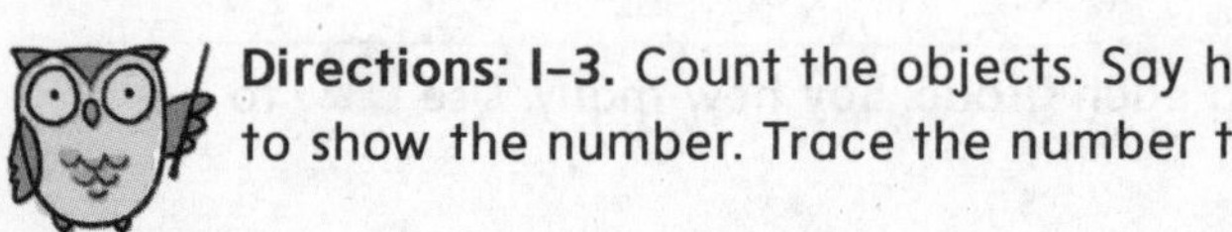

**Directions: 1–3.** Count the objects. Say how many. Use Work Mat 3 and connecting cubes to show the number. Trace the number two times. Write the number.

Name

## Independent Practice

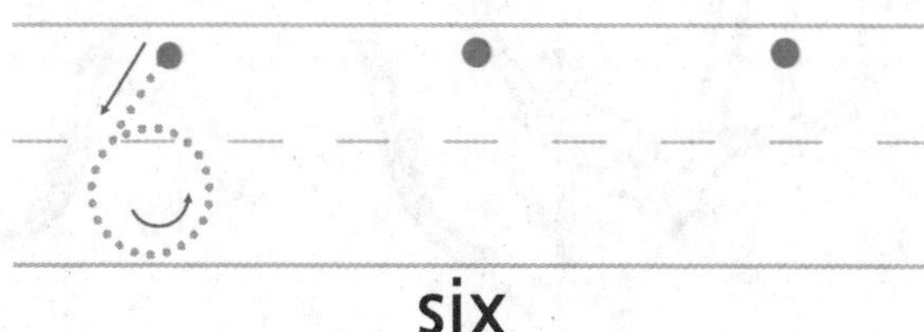

six

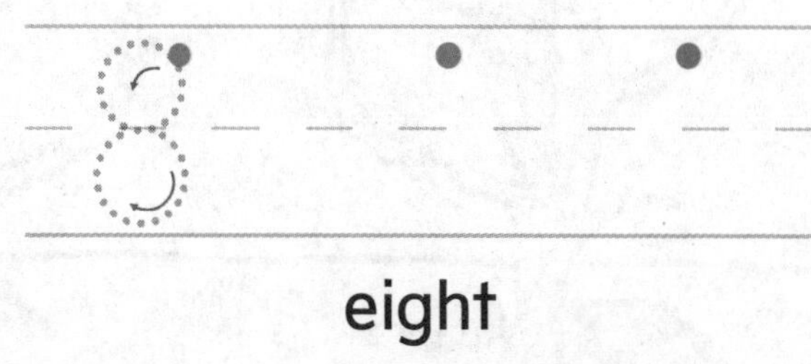

eight

6

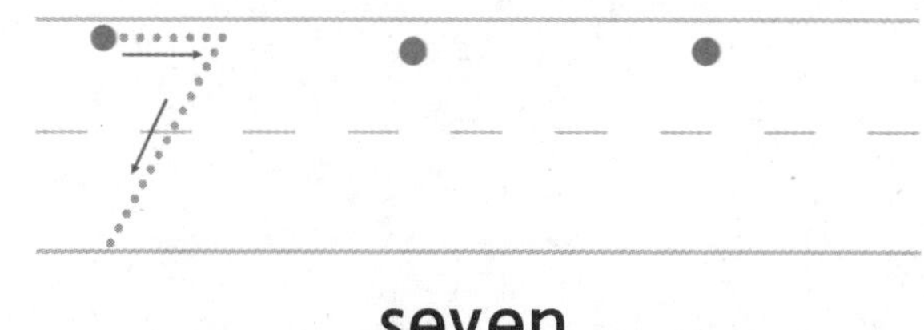

seven

**Directions: 4–6.** Count the objects. Say how many. Use Work Mat 3 and connecting cubes to show the number. Trace the number. Write the number two times.

Processes &Practices

# Brain Builders

7

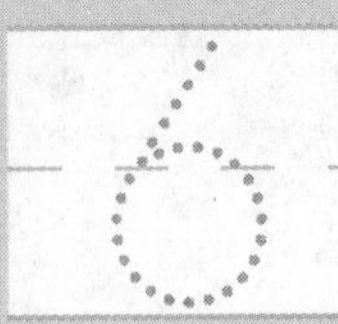
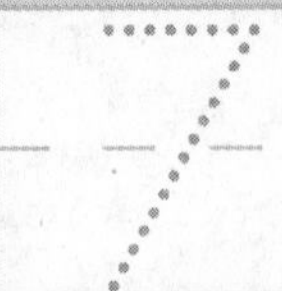
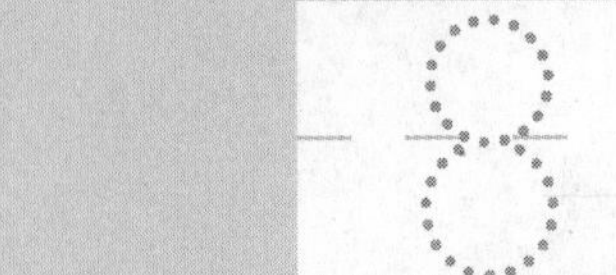

6 7 8

**Directions:** **7**. Trace each number. Draw tomatoes on each plant to show how many. Say how many. Tell a friend how many more tomatoes you would need to draw on the plant with 6 to make 7. Tell a friend how many more tomatoes you would need to draw on the plant with 6 to make 8. Tell a friend how many more tomatoes you would need to draw on the plant with 7 to make 8.

Name

# My Homework

Lesson 3

Read and Write 6, 7, and 8

## Homework Helper

Need help? connectED.mcgraw-hill.com

1

8 8 8

eight

2

six

3

seven

**Directions: 1–3.** Count the objects. Say how many. Trace the number. Write the number two times.

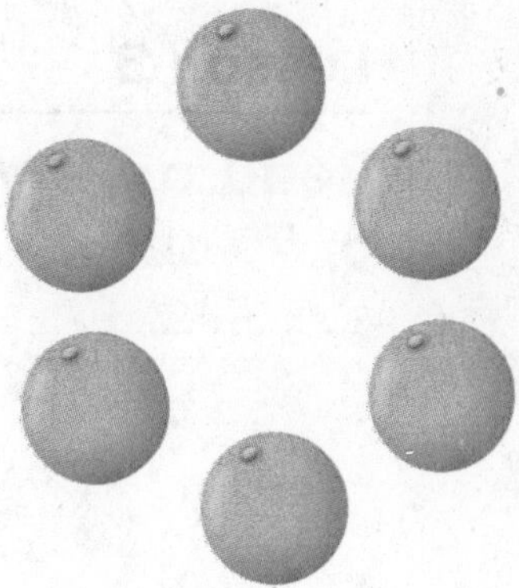

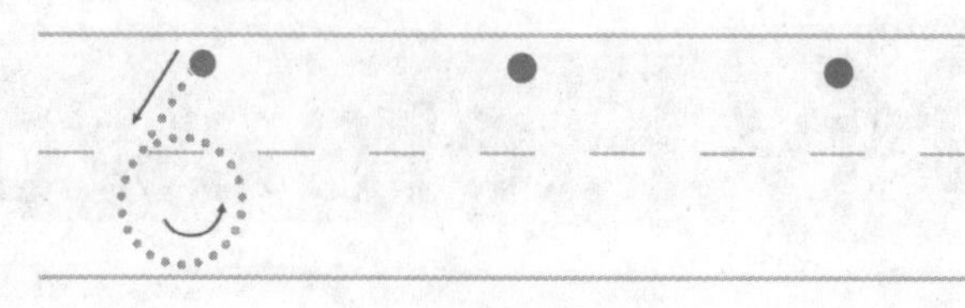

six

## Vocabulary Check

six 6

seven 7

eight 8

**Directions: 4.** Count the objects. Say how many. Trace the number. Write the number two times. **5–7.** Say the word. Draw objects to show how many.

**Math at Home** Using six index cards help your child make two cards with 6 dots, two cards with 7 dots, and two cards with 8 dots. Have your child write the number of dots on each card. Play a matching game.

Name

# Lesson 4
# Number 9

**ESSENTIAL QUESTION**
**What do numbers tell me?**

## Math in My World

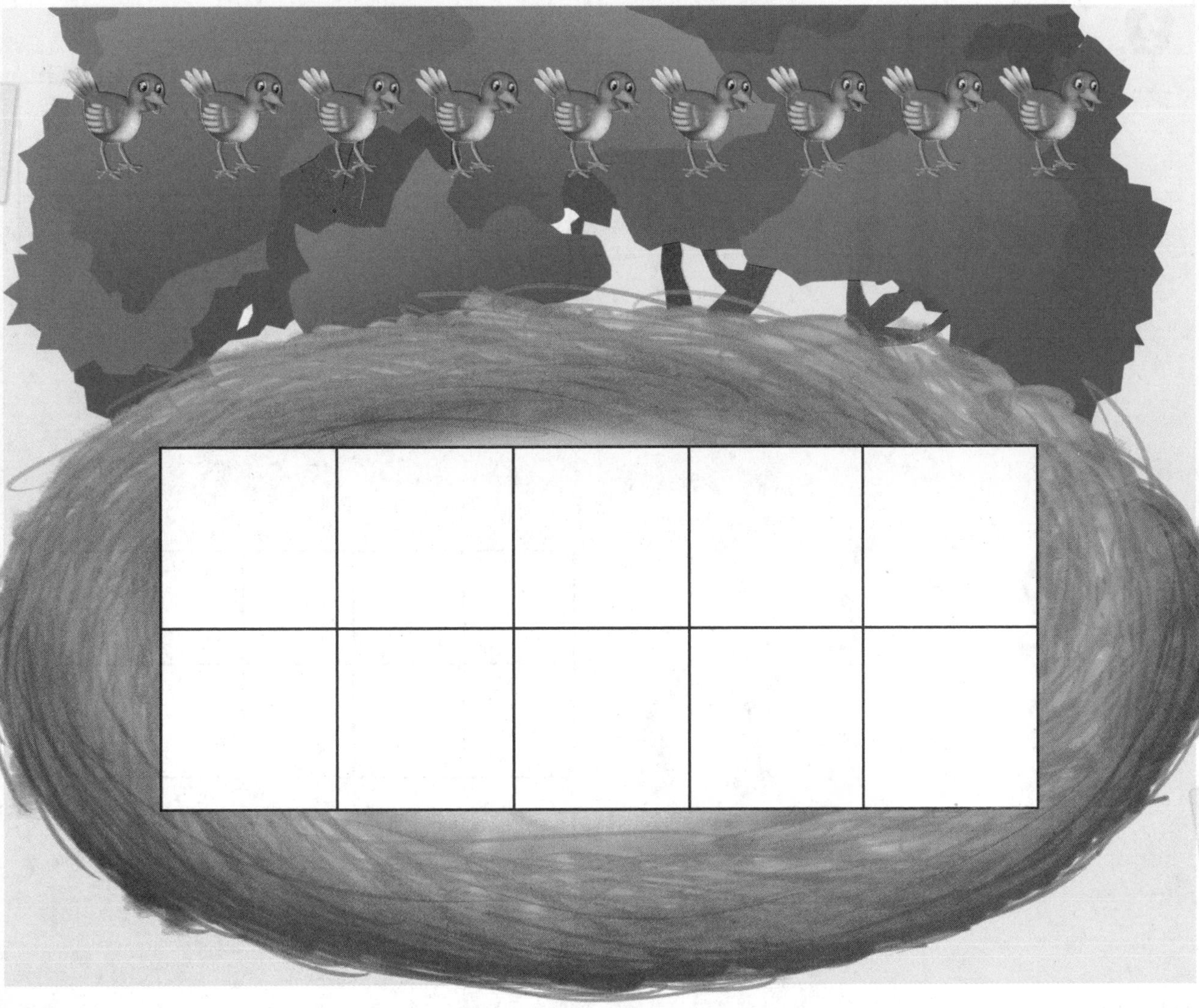

**Teacher Directions:** Count the birds. Say how many. Use to show five birds on the ten-frame. Use to show the other birds. Color the boxes to show how many.

# Guided Practice

Processes & Practices

1

2

3

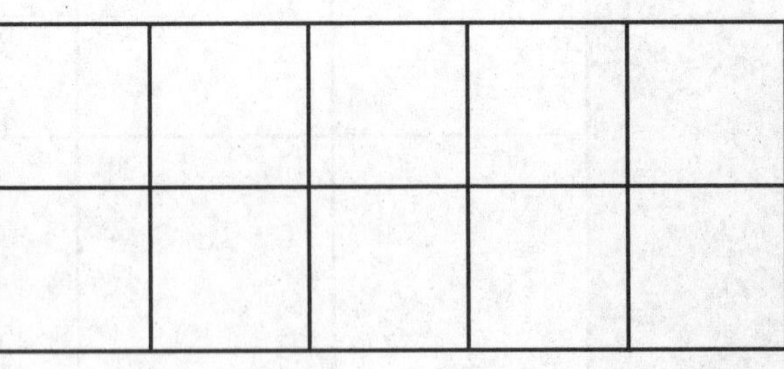

**Directions: 1–3.** Count the objects. Say how many. Use red connecting cubes to show five objects counted. Color one box red for each red cube. Use yellow connecting cubes to show the rest of the objects counted. Color one box yellow for each yellow cube.

Name

## Independent Practice

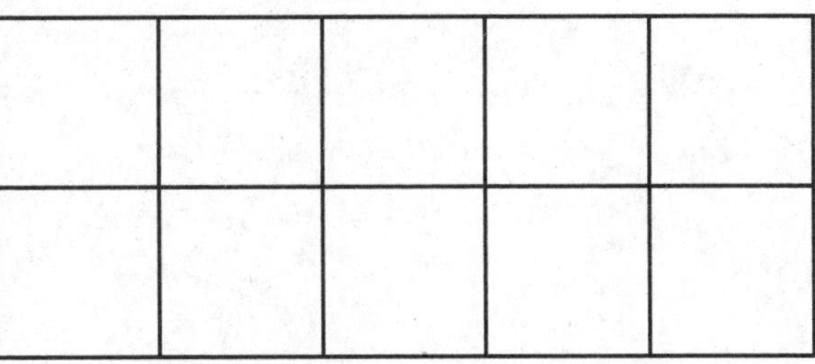

**Directions: 4–6.** Count the objects. Say how many. Use red connecting cubes to show five objects counted. Color one box red for each red cube. Use yellow connecting cubes to show the rest of the objects counted. Color one box yellow for each yellow cube.

# Brain Builders

Processes & Practices

7

9

| ● | ● | ● | ● | ● |
|---|---|---|---|---|
|  |  |  |  |  |

9

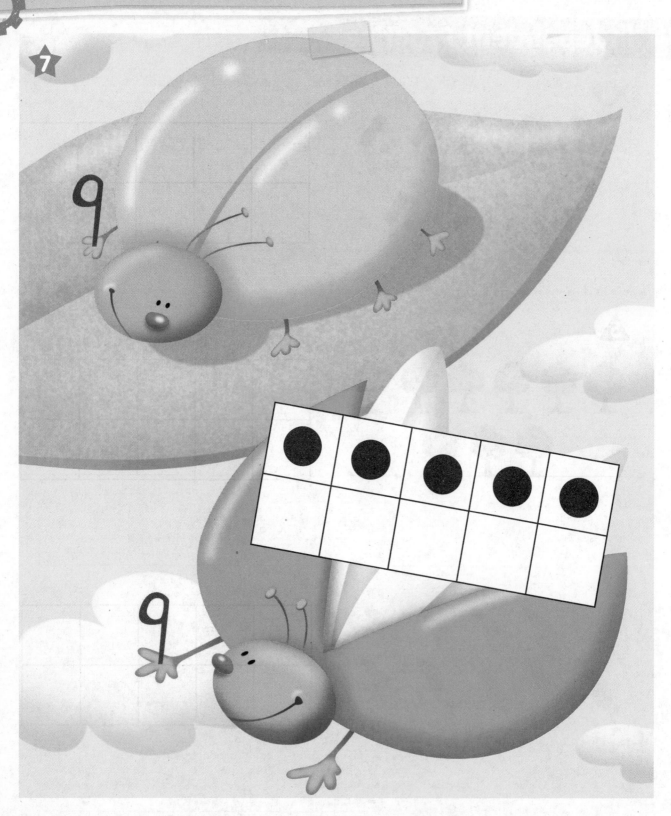

**Directions: 7.** Point to the bug on the leaf. Show nine by drawing spots on the bug's wings. Tell a friend how many spots are on each wing. Point to the bug in the clouds. Show nine by drawing more spots in the boxes. Tell a friend how many spots are in each row.

Name

Lesson 4

Number 9

# My Homework

## Homework Helper

Need help? connectED.mcgraw-hill.com

1

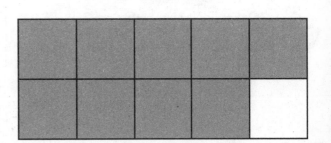

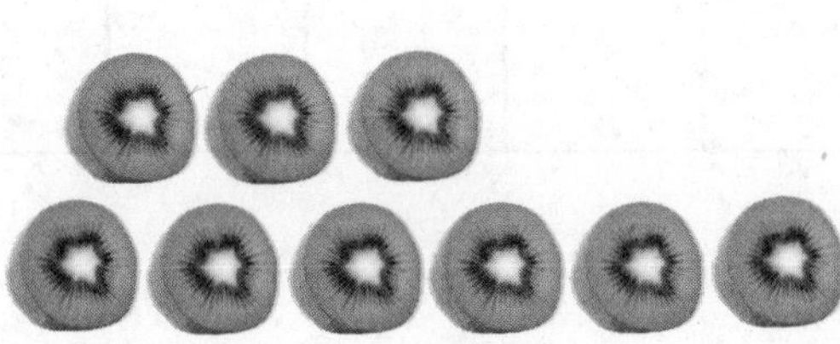

2

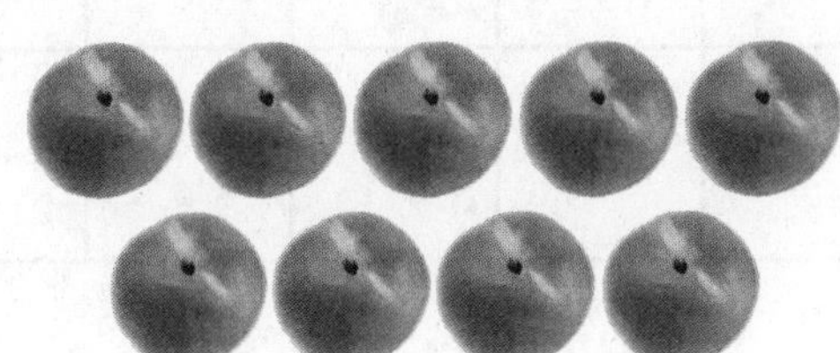

3

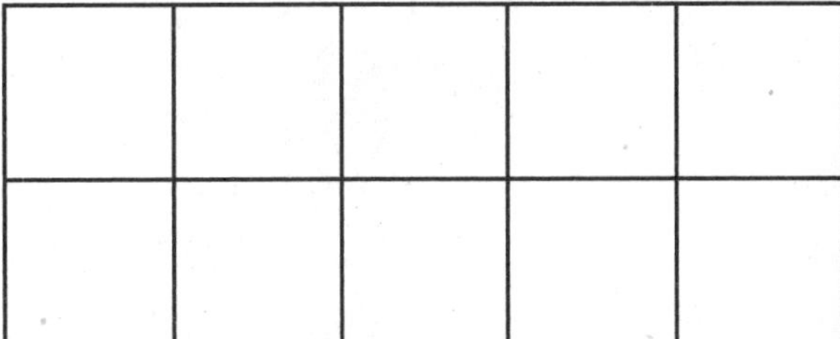

**Directions: 1–3.** Count the objects. Say how many. Use pennies to show how many. Color one box for each object counted.

4

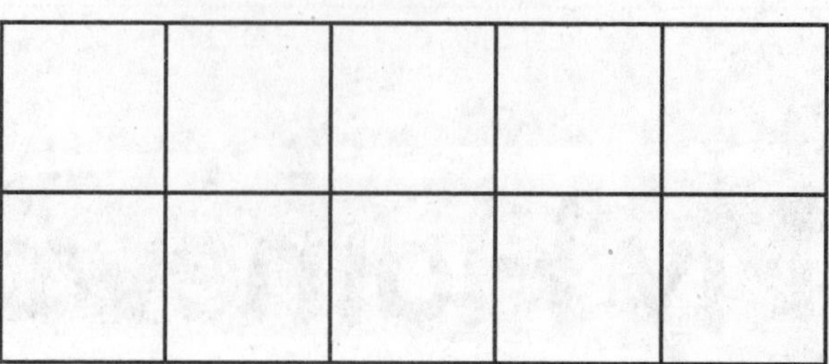

5

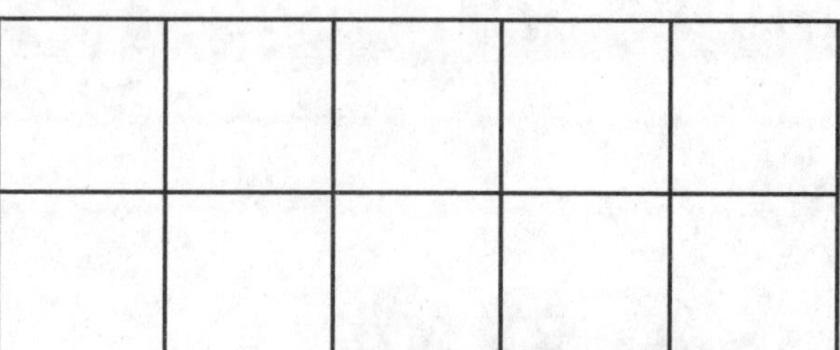

6

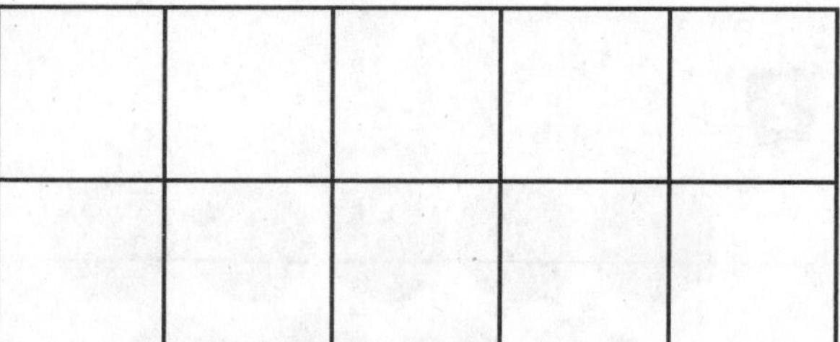

7

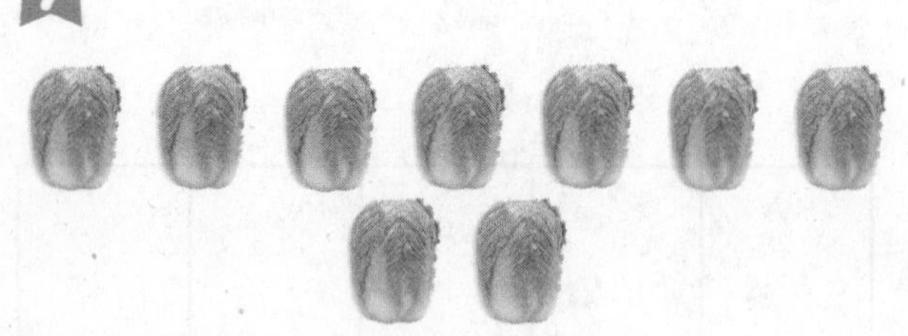

**Directions: 4–7.** Count the objects. Say how many. Use pennies to show how many. Color one box for each object counted.

**Math at Home** Have your child gather nine objects such as eating utensils, socks, or pennies. Count them together.

Name

# Check My Progress

## Vocabulary Check 

1 six 6

2 seven 7

3 eight 8

## Concept Check

4

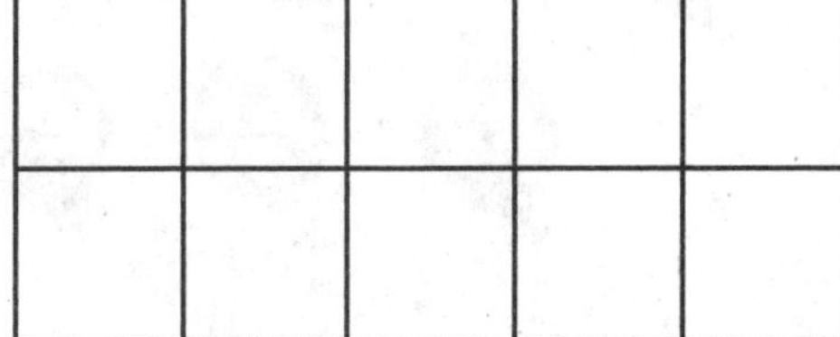

**Directions: 1.** Circle the group with six strawberries. **2–3.** Draw objects to show the number. **4.** Count the objects. Say how many. Color one box for each object counted to show how many.

5

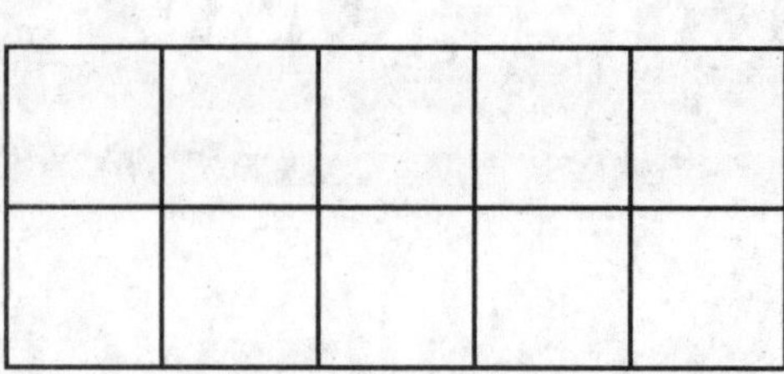

6

7

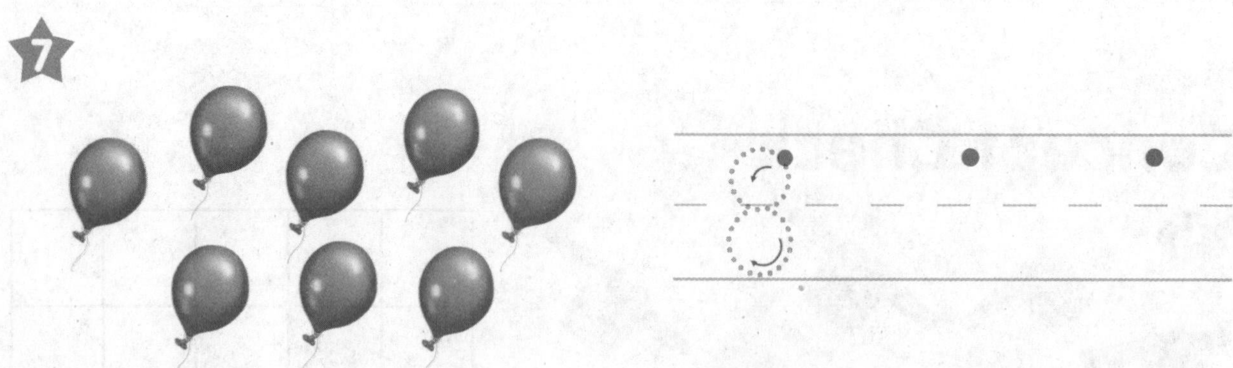

8

**Directions: 5.** Count the objects. Say how many. Color one box for each object counted to show how many. **6–7.** Count the objects. Trace the number. Write the number two times.

Name

# Lesson 5
# Number 10

**ESSENTIAL QUESTION**
What do numbers tell me?

## Math in My World

Tools Watch

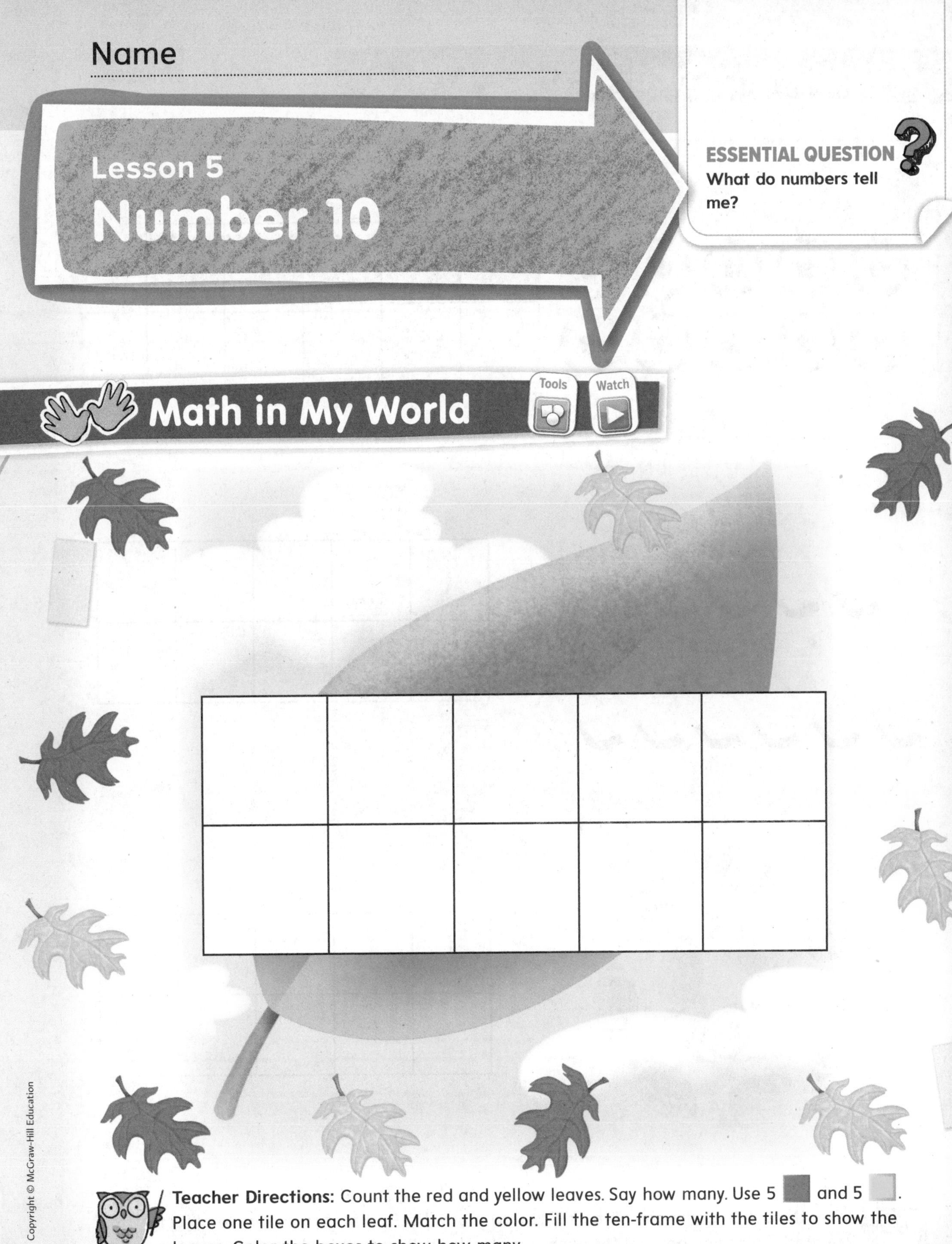

|  |  |  |  |  |
|---|---|---|---|---|
|  |  |  |  |  |
|  |  |  |  |  |

**Teacher Directions:** Count the red and yellow leaves. Say how many. Use 5 [red tiles] and 5 [yellow tiles]. Place one tile on each leaf. Match the color. Fill the ten-frame with the tiles to show the leaves. Color the boxes to show how many.

# Guided Practice

Processes
&Practices

1

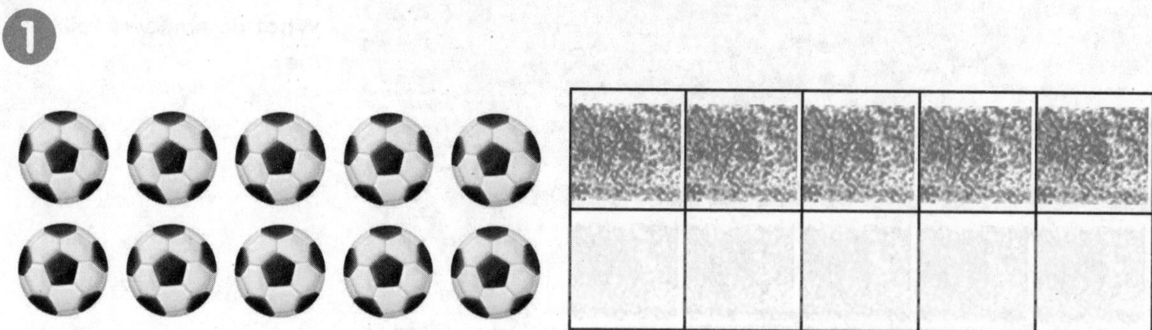

2

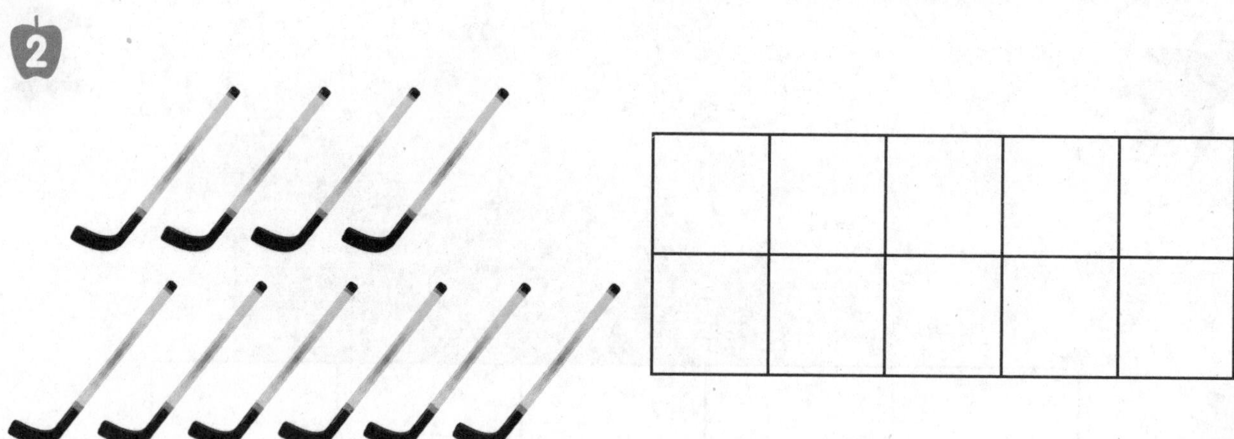

3

**Directions: 1–3.** Count the objects. Say how many. Use red tiles to show five objects counted. Color one box red for each red tile. Use yellow tiles to show the rest of the objects counted. Color one box yellow for each yellow tile.

Name

## Independent Practice

4

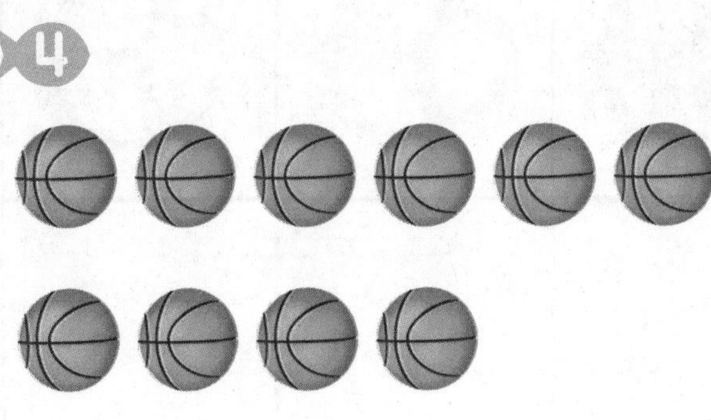

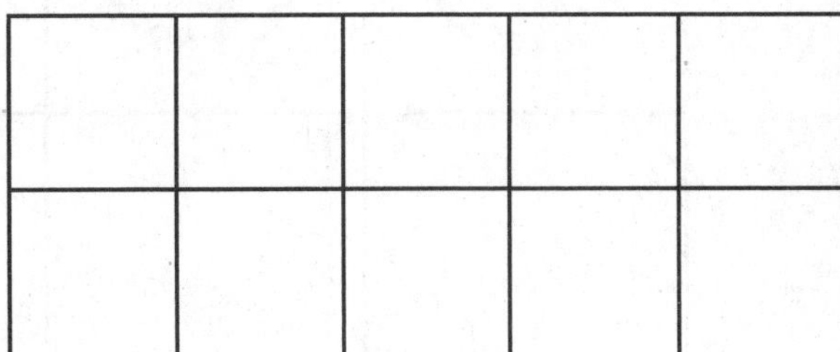

5

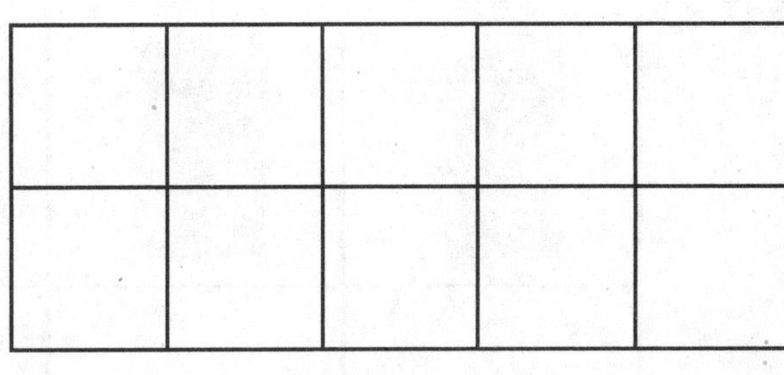

6

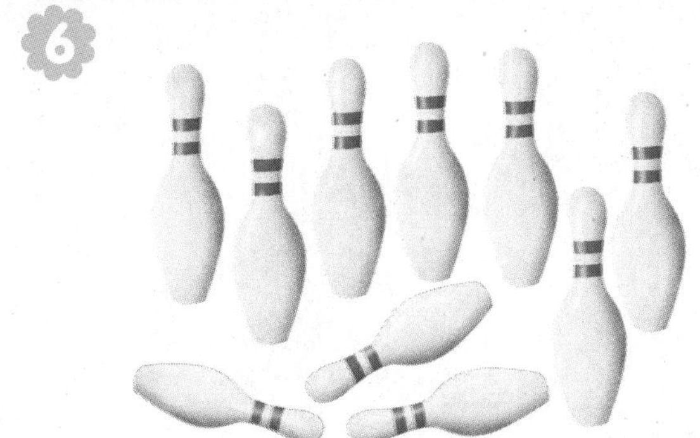

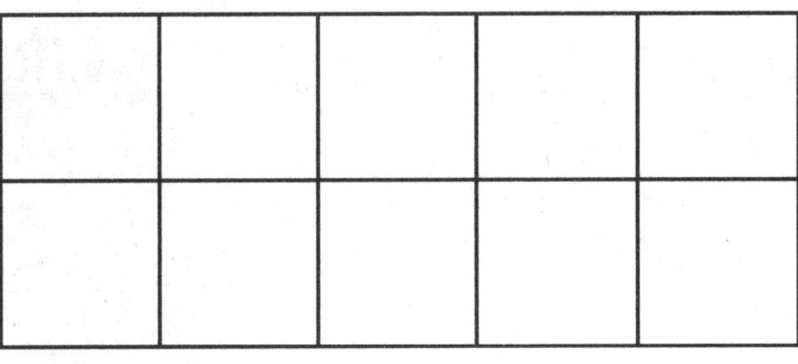

**Directions: 4–6.** Count the objects. Say how many. Use red tiles to show five objects counted. Color one box red for each red tile. Use yellow tiles to show the rest of the objects counted. Color one box yellow for each yellow tile.

# Brain Builders

Processes & Practices

7

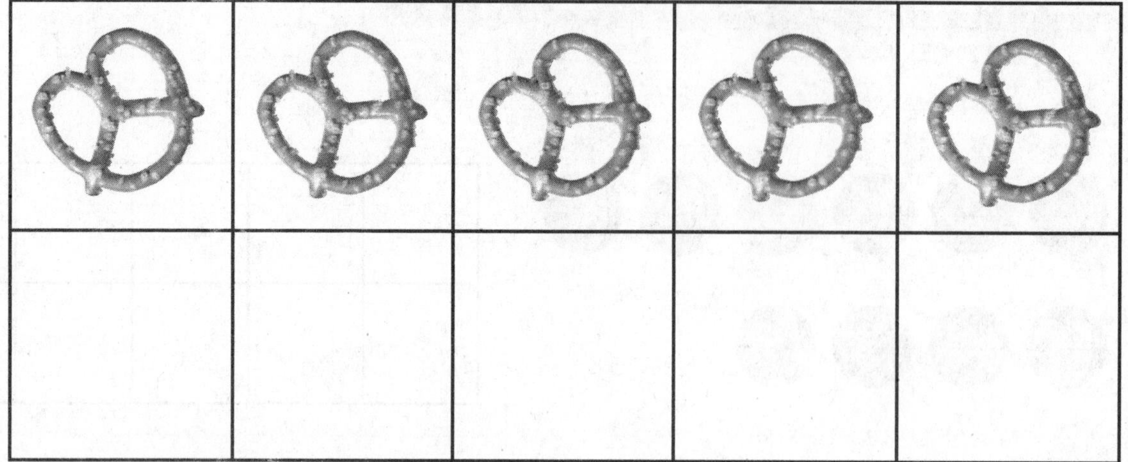

8

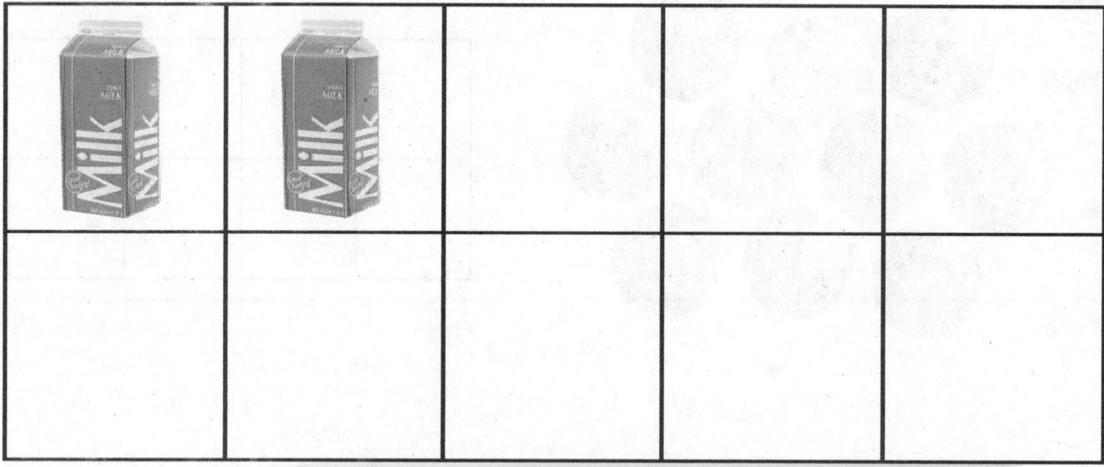

**Directions: 7–8.** Draw more objects to make 10. Explain to a friend the two groups that made each of the 10 objects.

Name

# My Homework

Lesson 5

Number 10

## Homework Helper

Need help? connectED.mcgraw-hill.com

1

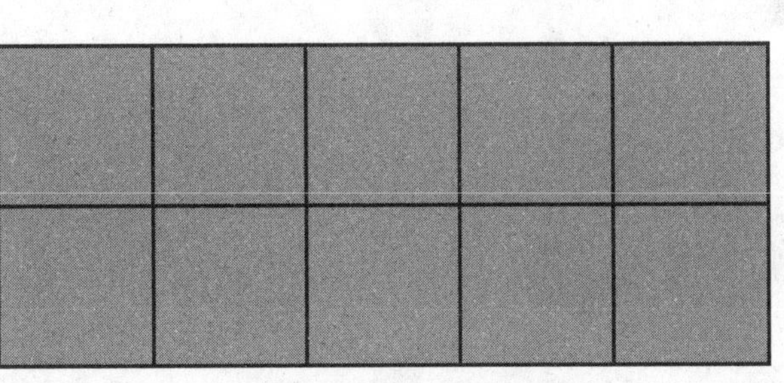

2

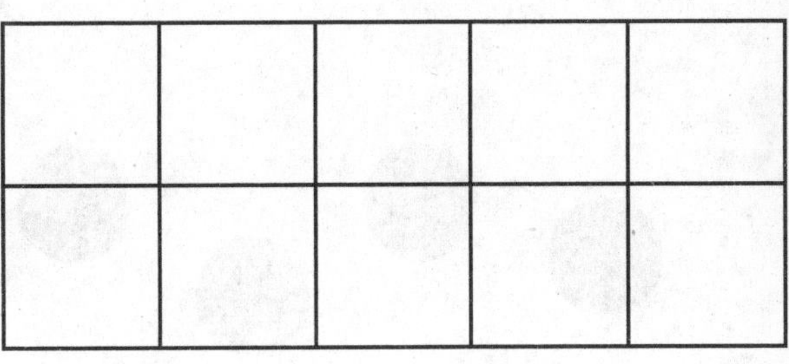

3

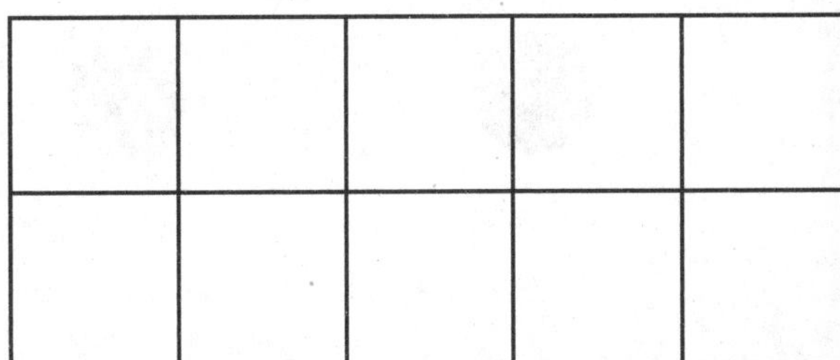

**Directions: 1–3.** Count the objects. Say how many. Use pennies to show how many. Color one box for each object counted.

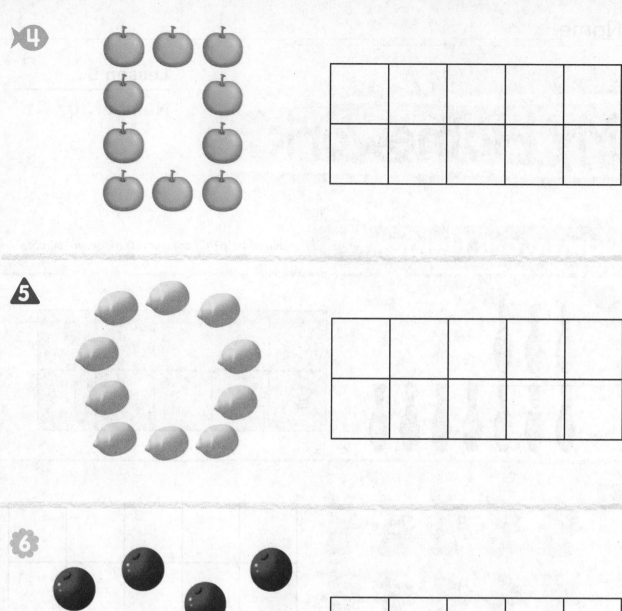

4

5

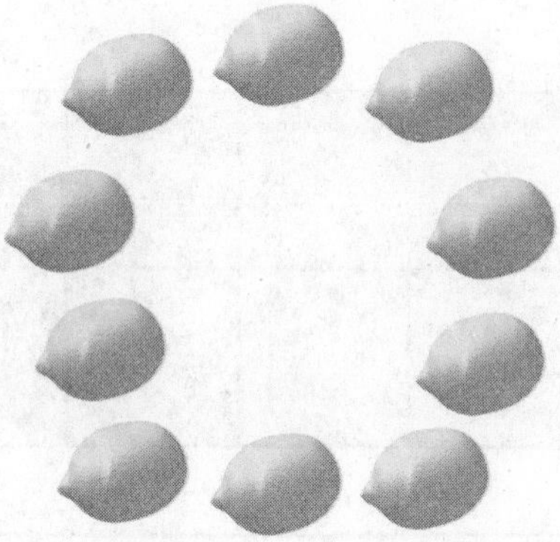

6

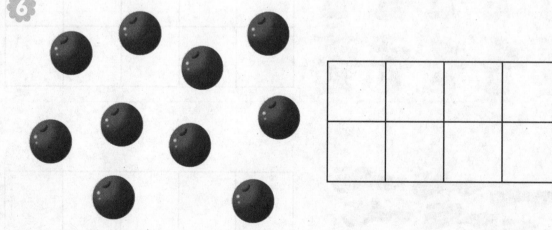

**Directions: 4–6.** Count the objects. Say how many. Use pennies to show how many. Color one box for each object counted.

**Math at Home** Help your child place one piece of tape on each of his or her fingers. Have your child count the number of tape pieces and say how many.

Name

Lesson 6

# Read and Write 9 and 10

**ESSENTIAL QUESTION**

What do numbers tell me?

## Math in My World

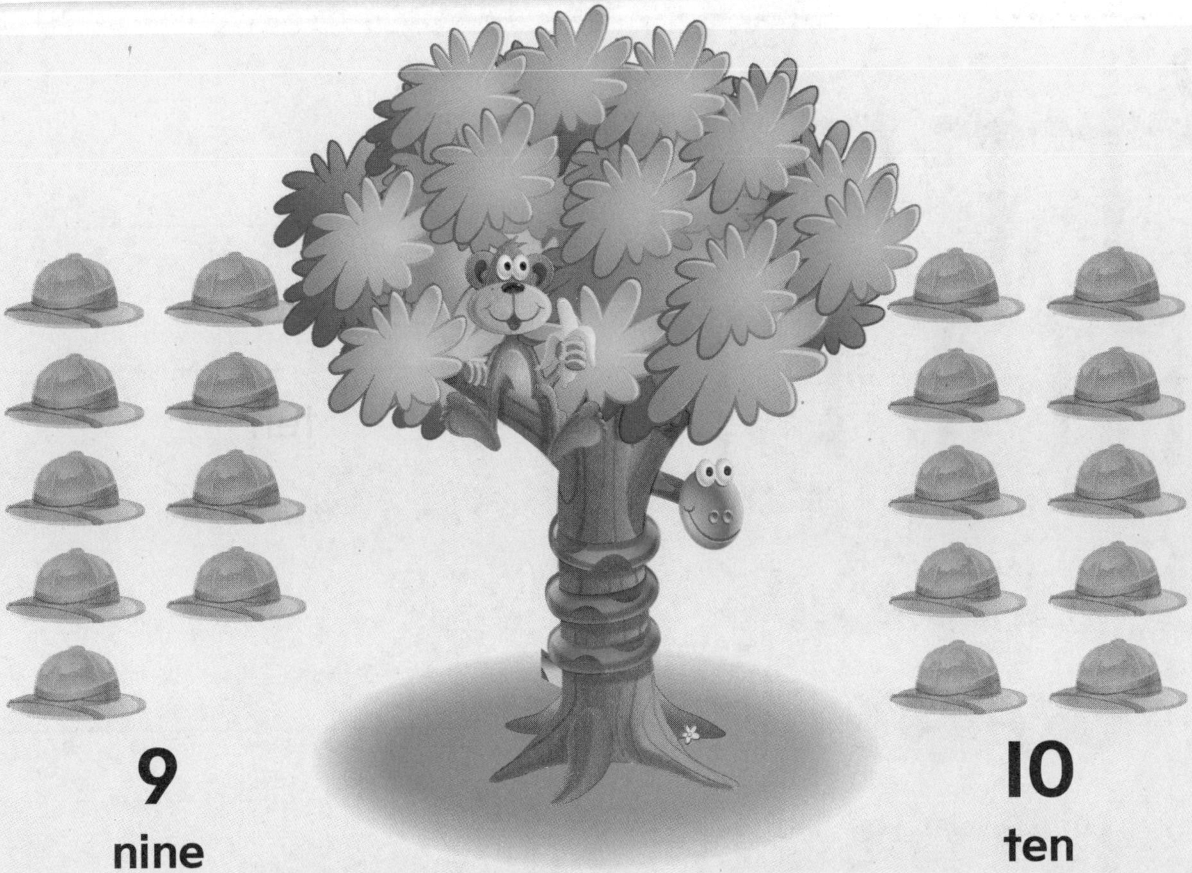

9

nine

10

ten

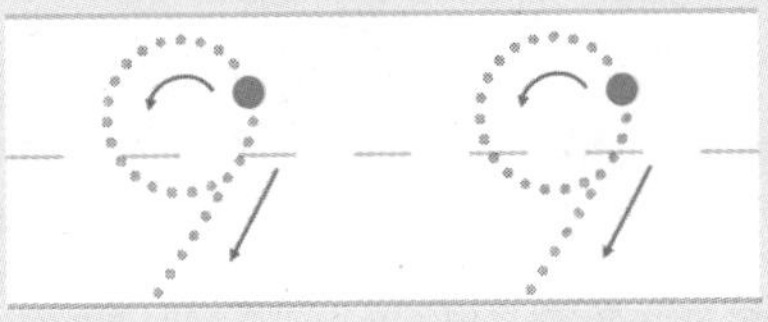

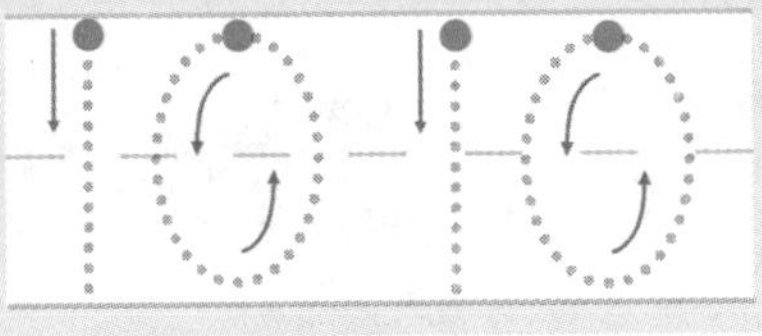

**Teacher Directions:** Count the hats in each group. Say how many. Use ■ to show how many. Trace the numbers.

## Guided Practice

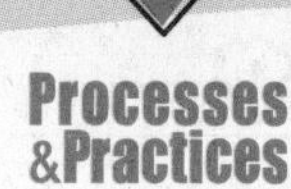


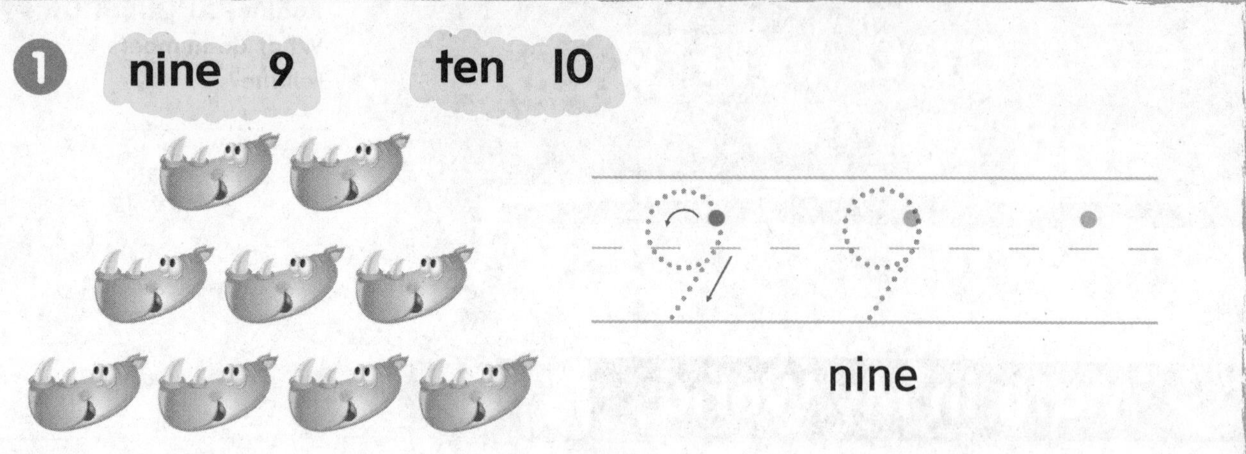


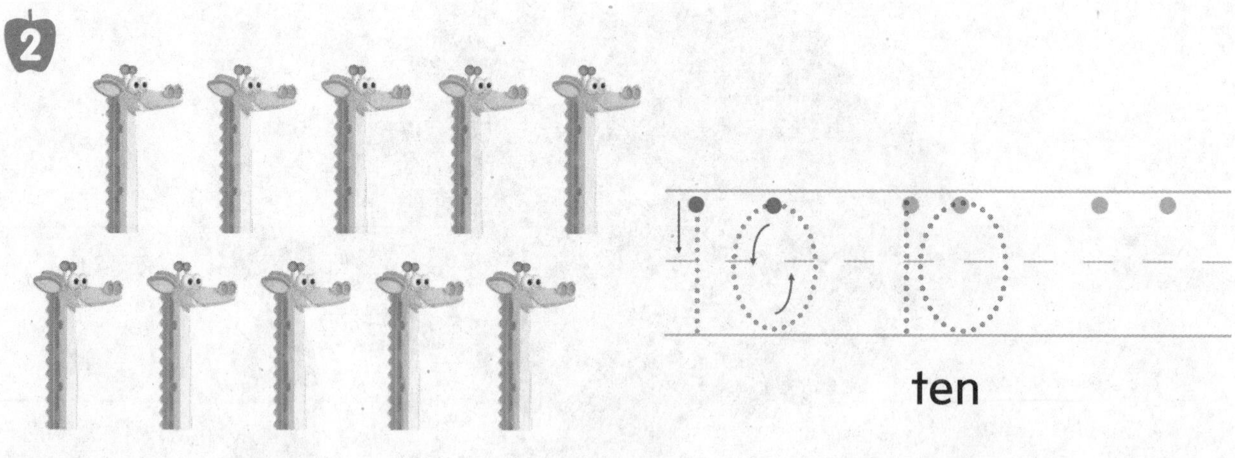


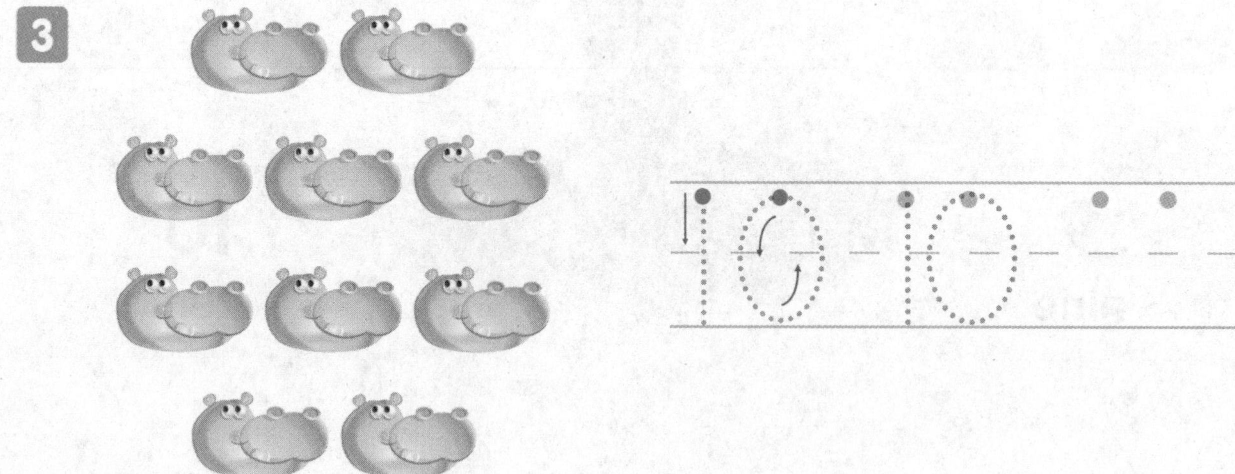

**Directions: 1–3.** Count the animals. Say how many. Use Work Mat 3 and color tiles to show the number. Trace the number two times. Write the number.

Name

## Independent Practice

4

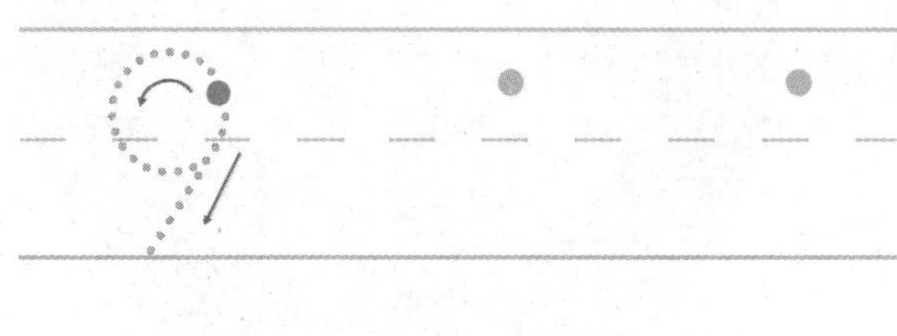

5

6

**Directions: 4–6.** Count the animals. Say how many. Use Work Mat 3 and color tiles to show the number. Trace the number. Write the number two times.

Processes &Practices

# Brain Builders

7

8

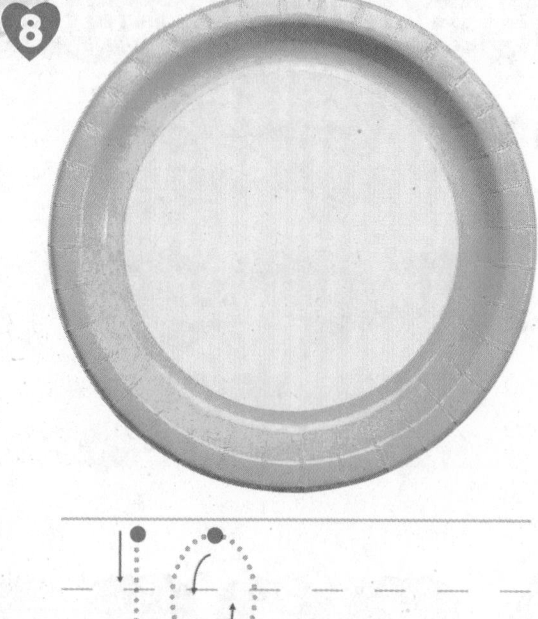

9

10

10

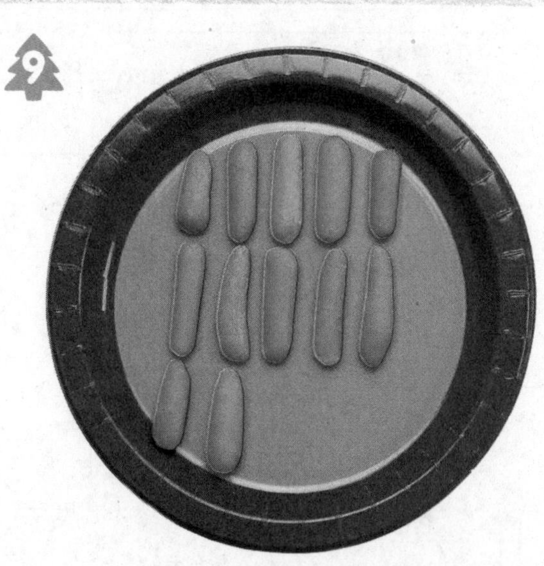

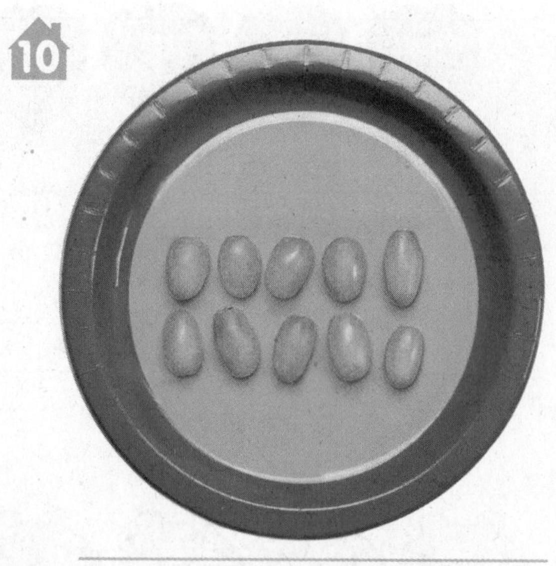

**Directions: 7–8.** Say the number. Trace and write the number. Draw healthful foods to show how many. **9.** Draw a circle around the carrots to show a group of 10. Write the number two times. Explain to a friend if there are more or less than 10 carrots. **10.** Draw a circle around the grapes to show a group of nine. Write the number two times. Tell a friend how many more grapes there are than 9.

Name

# My Homework

Lesson 6

Read and Write 9 and 10

## Homework Helper

Need help? connectED.mcgraw-hill.com

1

10 10 10

2

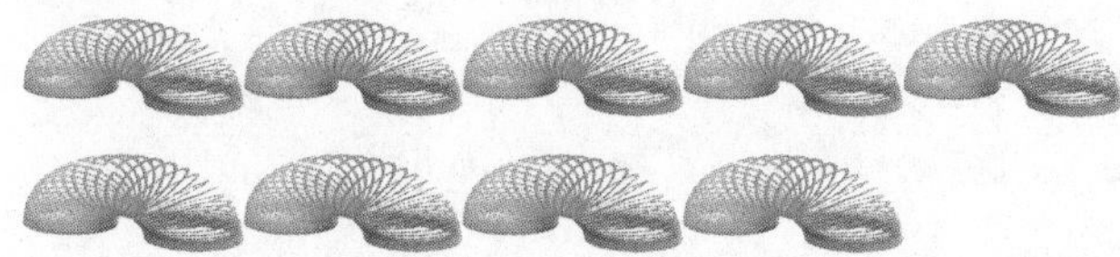

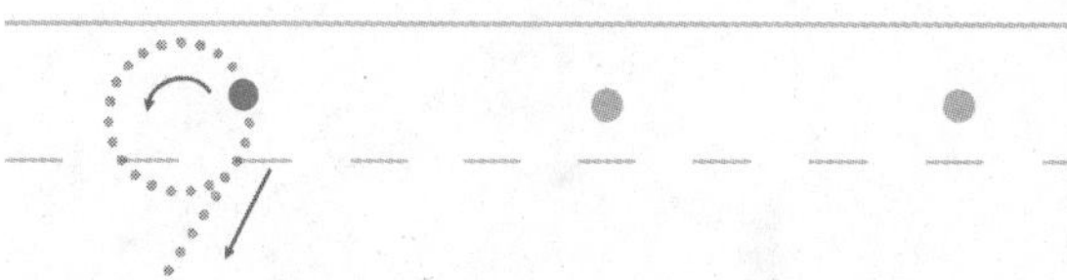

3

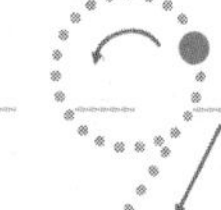

**Directions: 1–3.** Count the objects. Say how many. Trace the number. Write the number two times.

4

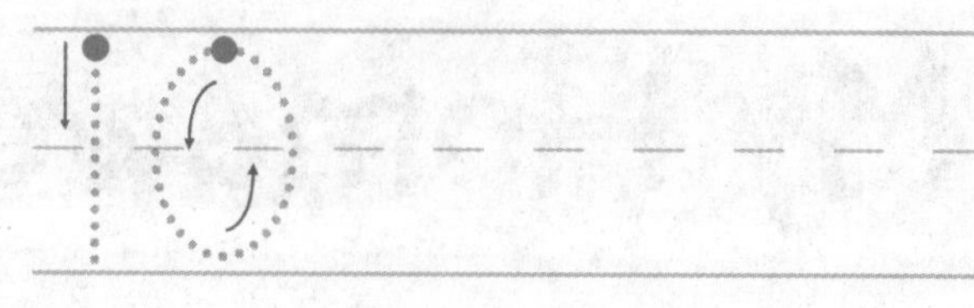

## Vocabulary Check

5 nine 9

6 ten 10

**Directions: 4.** Count the objects. Say how many. Trace the number. Write the number two times. **5–6.** Say the word. Draw objects to show how many.

**Math at Home** Use a book with numbered pages. Ask your child to count the pages to page nine and write nine. Repeat with pages to 10 and write 10.

Name

Lesson 7

# Problem Solving

**STRATEGY: Act It Out**

**ESSENTIAL QUESTION**
What do numbers tell me?

## How many snacks?

## Act It Out

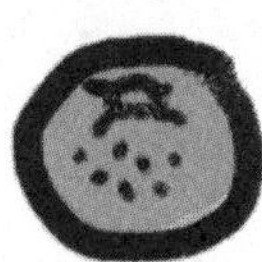

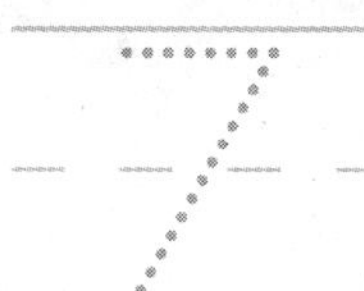

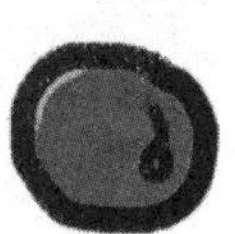

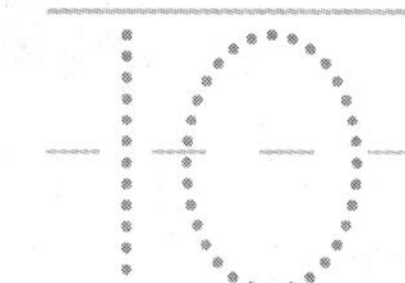

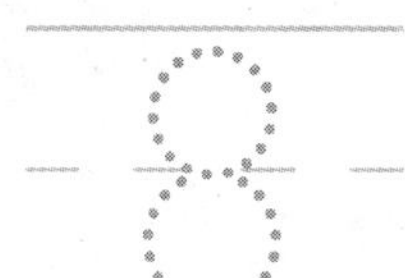

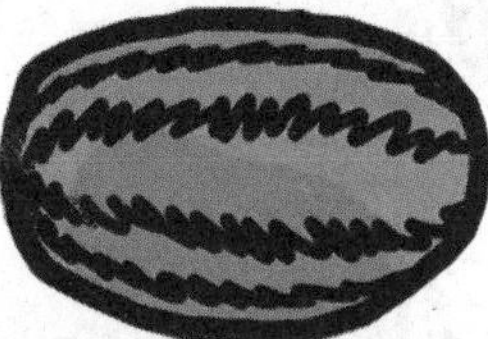

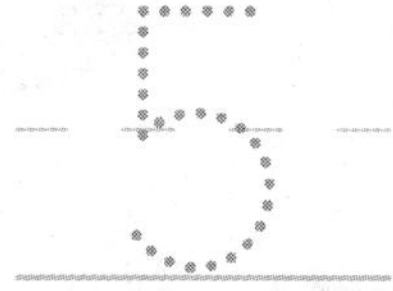

**Teacher Directions: 1–4.** Look at the picture. Use connecting cubes to show how many there are of each fruit. Count the cubes. Say how many. Trace the number.

# How many are at the circus?

## Act It Out

1
2
3
4

**Directions: 1–4.** Look at the picture. Use connecting cubes to show how many there are of each object. Count the cubes. Say how many. Write the number.

Name

## How many are at the tea party?

## Act It Out

1

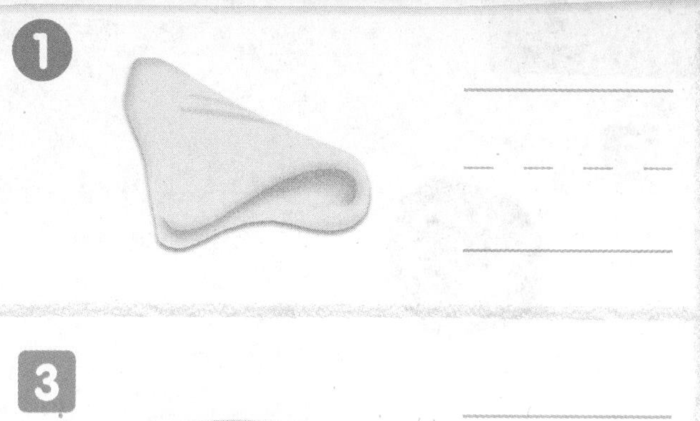

2

3

4

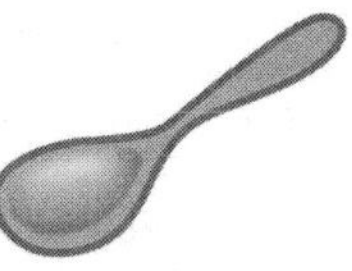

**Directions: 1–4.** Look at the picture. Use connecting cubes to show how many there are of each object. Count the cubes. Say how many. Write the number. Draw another animal at the tea party. Tell a friend if there are enough of each item for all of the animals now.

## How many vegetables?

## Act It Out

2

**Directions: 1–4.** Look at the picture. Use connecting cubes to show how many there are of each vegetable. Count the cubes. Say how many. Write the number.

Name

Lesson 7

Problem Solving: Act It Out

# My Homework

## How many toys?

## Act It Out

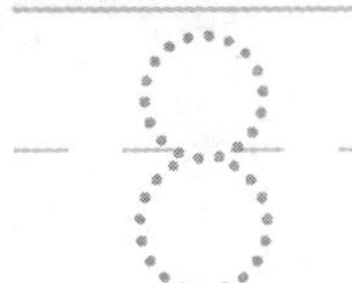

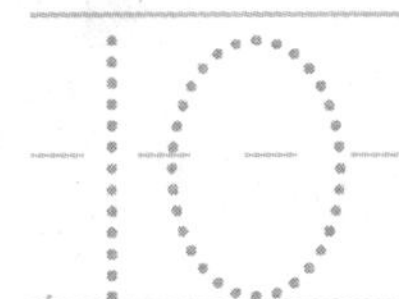

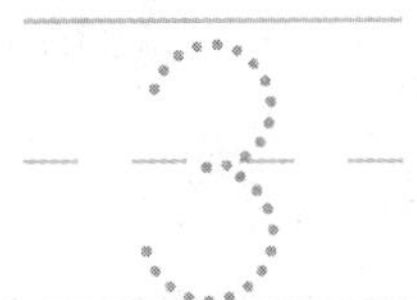

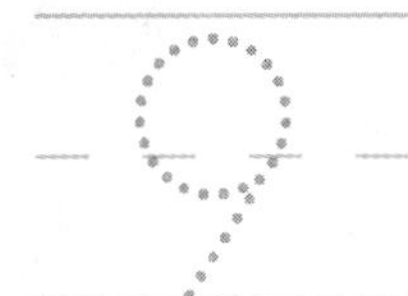

**Directions: 1–4.** Look at the picture. Use pennies to show how many there are of each object. Count the pennies. Trace the number.

## How many bugs?

## Act It Out

2

3

4

**Directions: 1–4.** Look at the picture. Use pennies to show how many there are of each object. Count the pennies. Say how many. Write the number.

**Math at Home** Take advantage of problem-solving opportunities during daily routines such as setting the table. Count the number of plates, forks, spoons, cups, and so on.

Name ..............................

# Lesson 8
# Compare Numbers 0 to 10

**ESSENTIAL QUESTION**
What do numbers tell me?

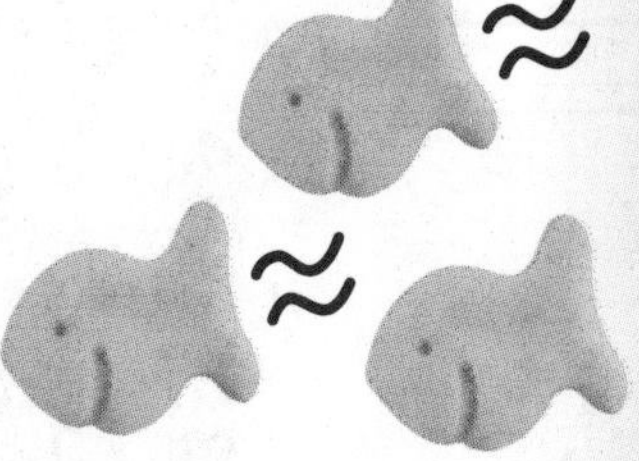

## Math in My World

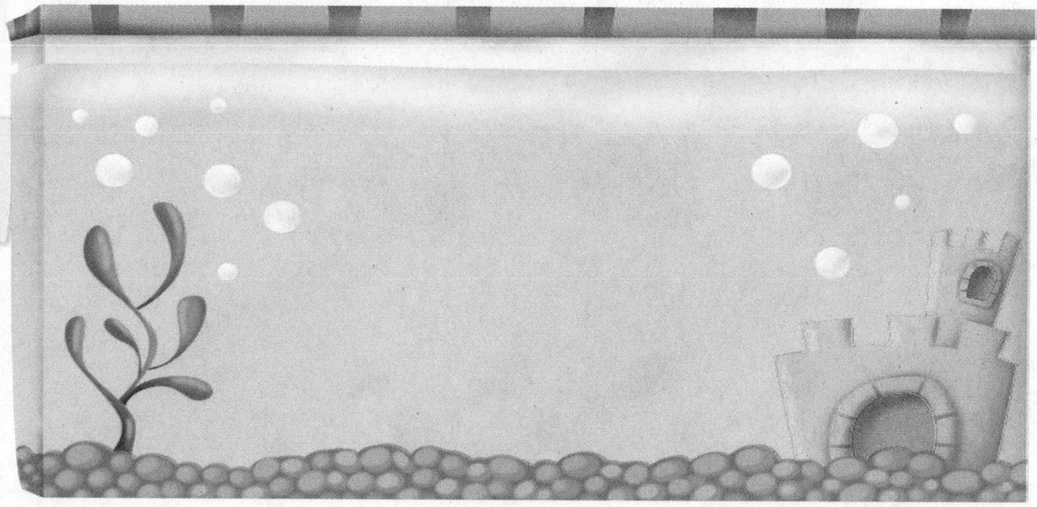

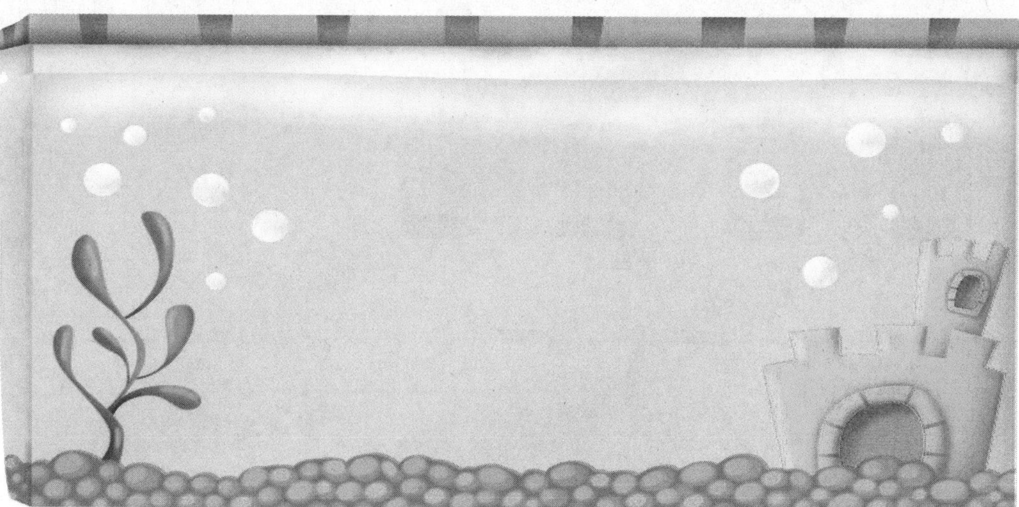

**Teacher Directions:** Use up to 10 [cube] to make a group of fish in each tank. Draw the fish. Write the numbers. Circle the group and number that has more than the other group. Draw an X on the number and group that has less than the other group. Draw a box around the groups if they are equal to each other.

# Guided Practice

Processes &Practices

1

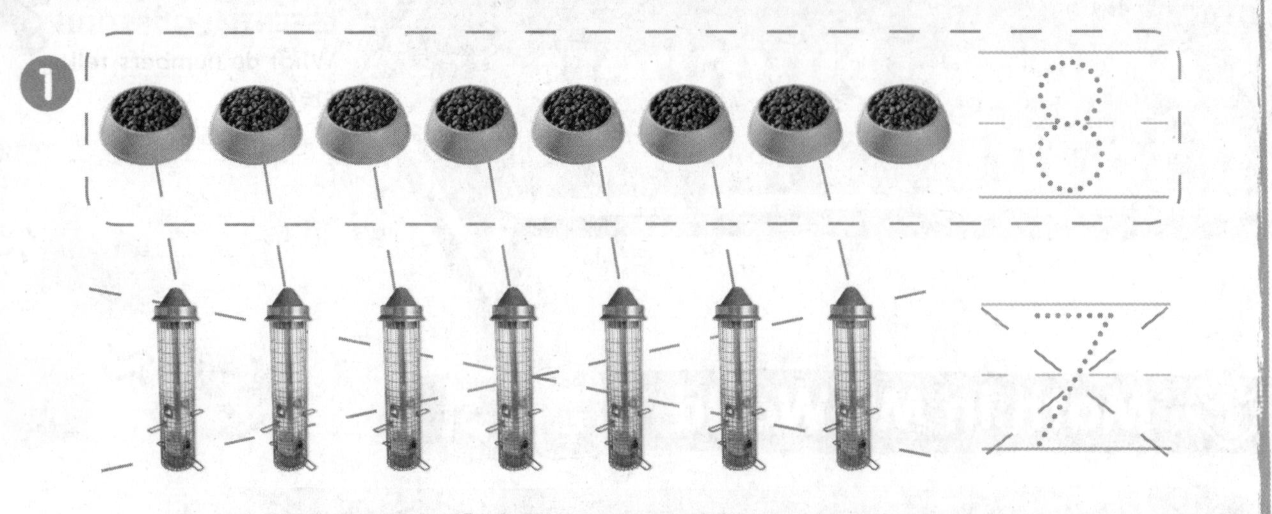

2

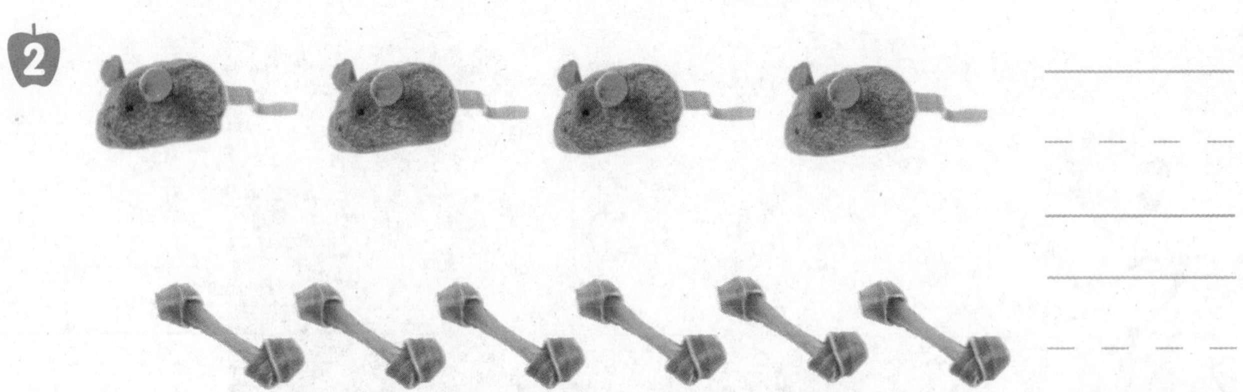

3

**Directions: 1–3.** Count and draw lines to match objects in one group with objects in the other. Write the numbers. Circle the number and group that is greater than the other. Draw an X on the number and group that is less than the other. Draw a box around the numbers and groups that are equal to each other.

Name

## Independent Practice

4

5

6

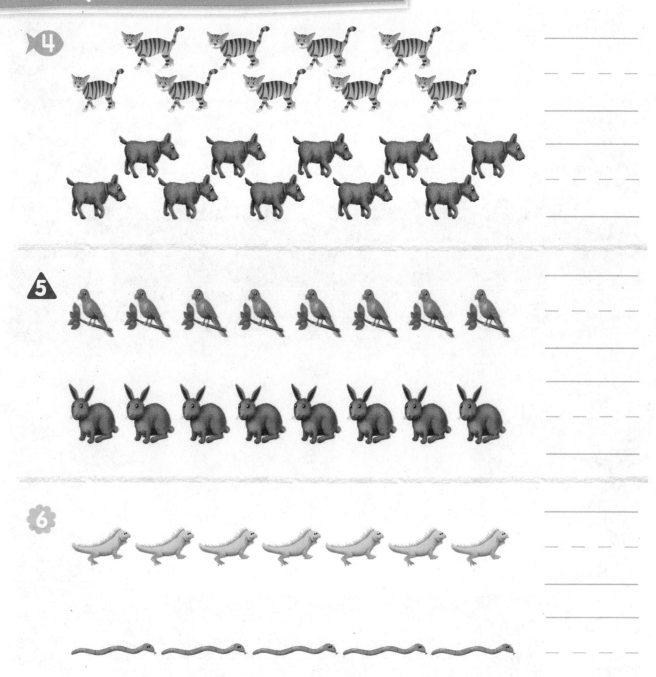

**Directions: 4–6.** Count and draw lines to match objects in one group with objects in the other. Write the numbers. Circle the number and group that is greater than the other. Draw an X on the number and group that is less than the other. Draw a box around the numbers and groups that are equal to each other.

## Brain Builders

7

7

10

**Directions: 7.** Draw seven flowers in the box with the number 7 on it. Draw 10 flowers in the box with the number 10 on it. Circle the box that has a greater number of flowers. Tell a friend which box has less. Then explain how you would find how many less it has.

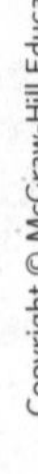

Name

# My Homework

**Lesson 8**

**Compare Numbers 0 to 10**

## Homework Helper

Need help? connectED.mcgraw-hill.com

1

10

9

2

3

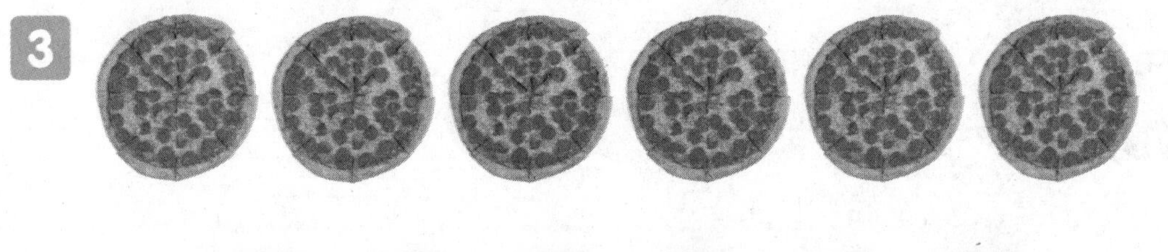

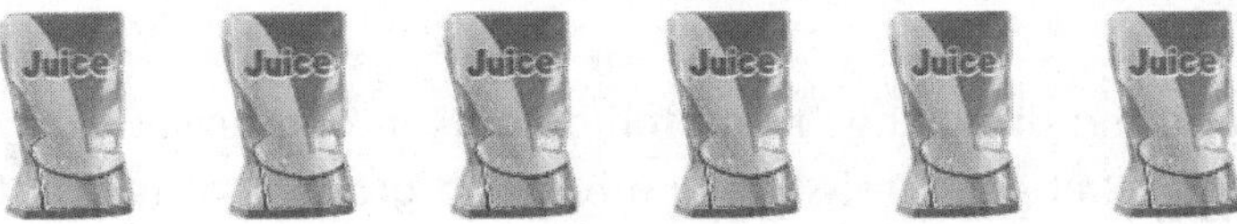

**Directions: 1–3.** Count and draw lines to match objects in one group with objects in the other. Write the numbers. Circle the number and group that is greater than the other. Draw an X on the number and group that is less than the other. Draw a box around the numbers and groups that are equal to each other.

4

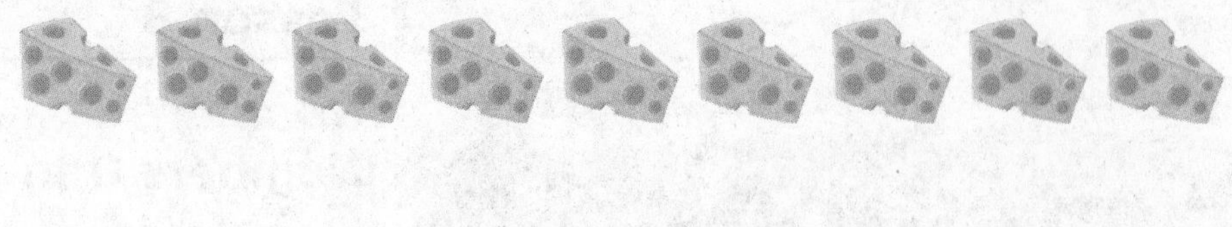

5

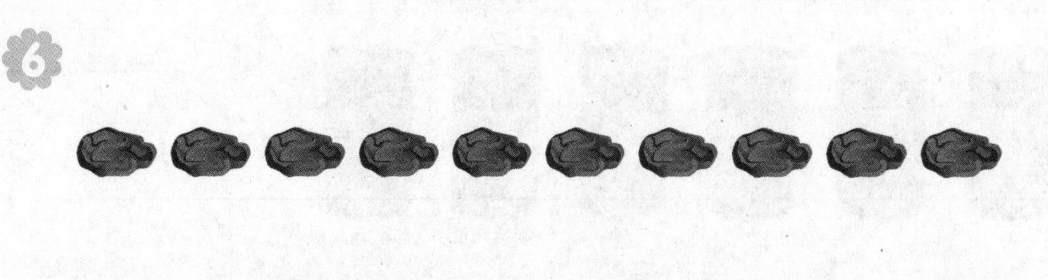

6

**Directions: 4–6.** Count and draw lines to match objects in one group with objects in the other. Write the numbers. Circle the number and group that is greater than the other. Draw an X on the number and group that is less than the other. Draw a box around the numbers and groups that are equal to each other.

**Math at Home** Give your child two groups of objects with up to 10 objects in each group. Have your child count the number of objects and compare the groups by saying which group is greater than, less than, or equal to the other group.

Name

# Check My Progress

## Vocabulary Check

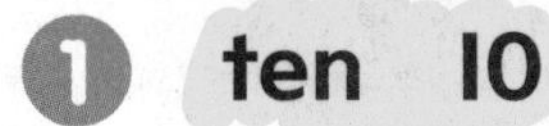

1 ten 10

2 nine 9

## Concept Check

3

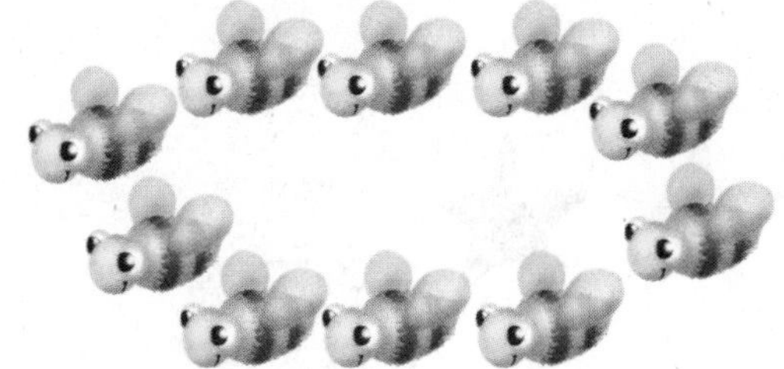

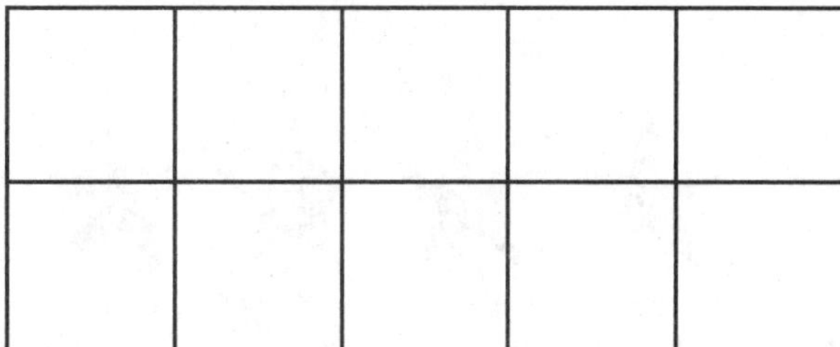

**Directions: 1–2.** Draw a group of objects to show how many. **3.** Count the objects. Say how many. Color one box for each object counted to show how many.

4

5

6

**Directions: 4–5.** Count the objects. Say how many. Write the number. **6.** Count and draw lines to match objects in one group with objects in the other. Write the numbers. Circle the number and group that is greater than the other. Draw an X on the number and group that is less than the other.

Name

# Lesson 9
# One More with Numbers to 10

**ESSENTIAL QUESTION**
What do numbers tell me?

## Math in My World

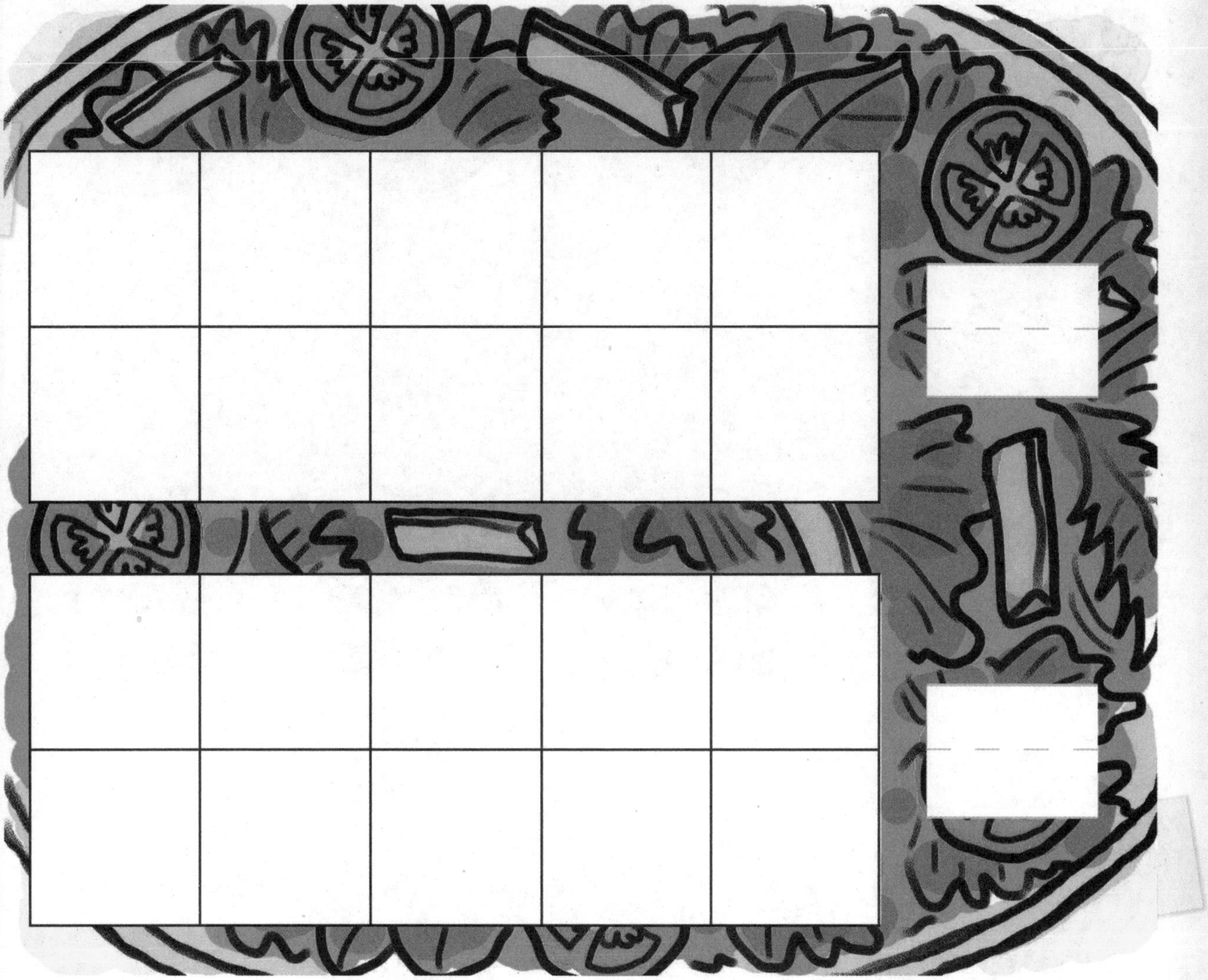

**Teacher Directions:** Use ● to act out the story. A salad has seven peppers. Another salad has seven and one more peppers. Count the counters. Color one box for each counter. Write the numbers. Circle the number that shows one more than seven.

## Guided Practice

Processes &Practices

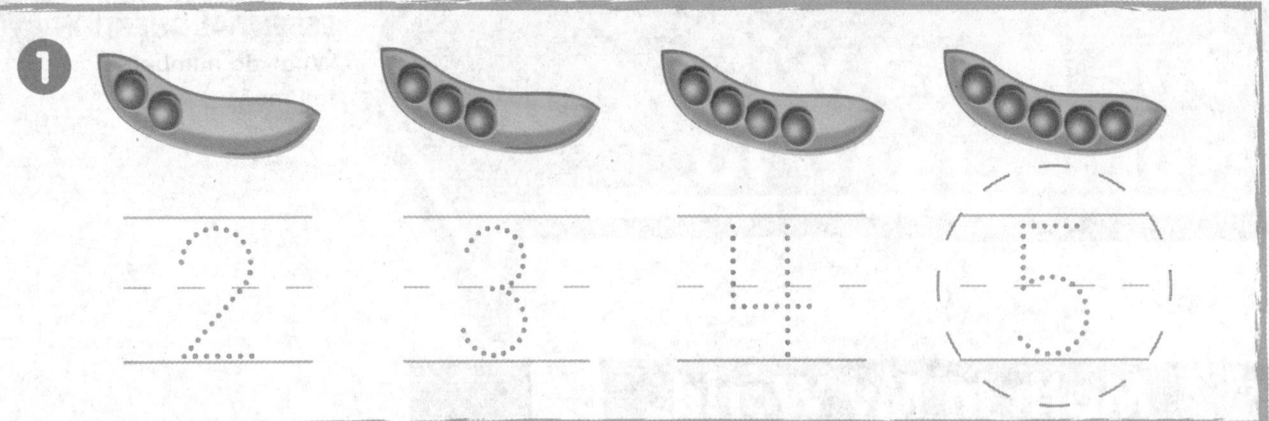

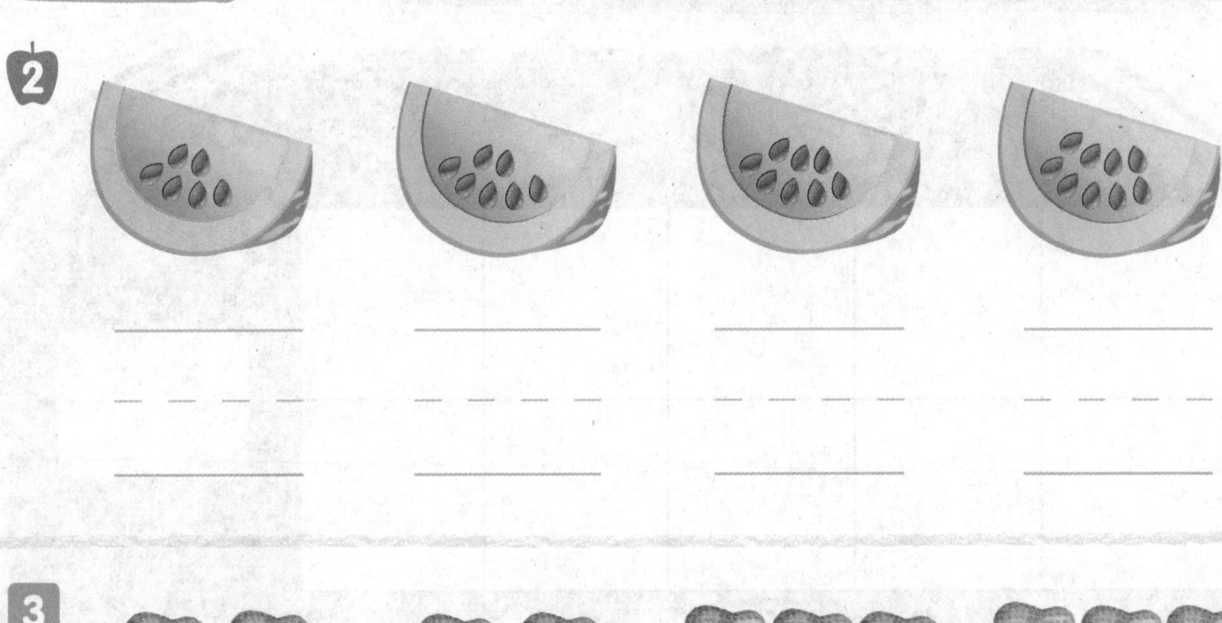

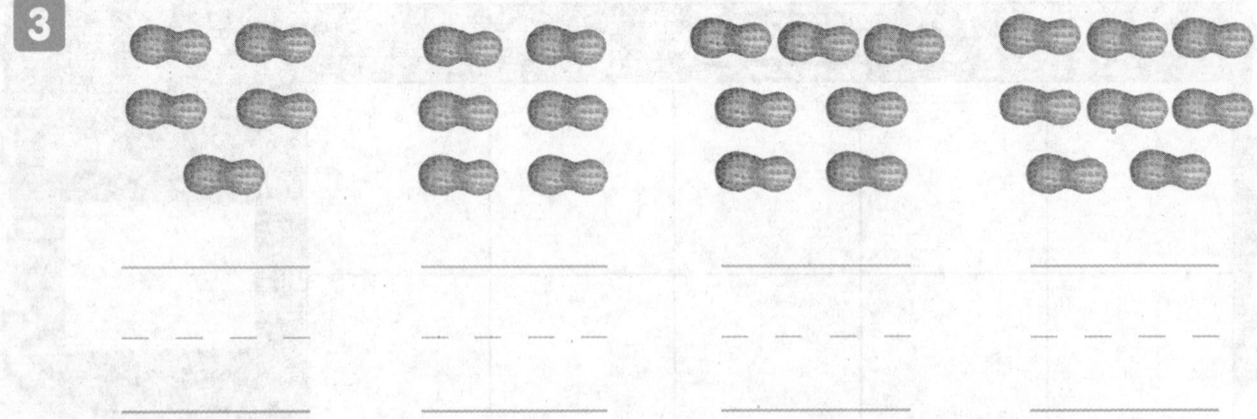

**Directions: 1.** Count the peas in each pod. Trace each number. Trace the circle showing the number that is one more than four. **2.** Count the seeds in each melon. Write how many. Circle the number that is one more than seven. **3.** Count the peanuts. Write how many. Circle the number that is one more than six.

Name

## Independent Practice

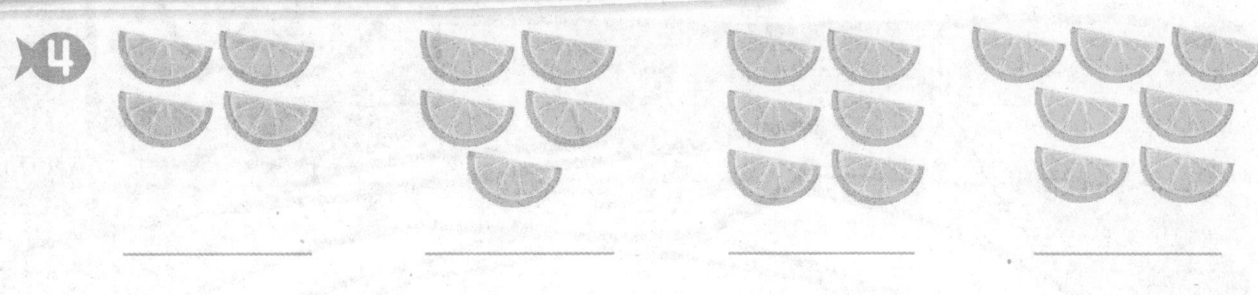

**Directions: 4.** Count the oranges. Write how many. Circle the number that is one more than five. **5.** Count the sandwiches. Write how many. Circle the number that is one more than one. **6.** Count the beans. Write how many. Circle the number that is one more than nine.

## Brain Builders

Processes & Practices

7

**Directions: 7.** Count the apples. Write the number. Draw a group of apples that shows one more. Write the number.

Name

# My Homework

Lesson 9

One More with Numbers to 10

## Homework Helper

eHelp

Need help? connectED.mcgraw-hill.com

1

7 (8) 9 10

2

3

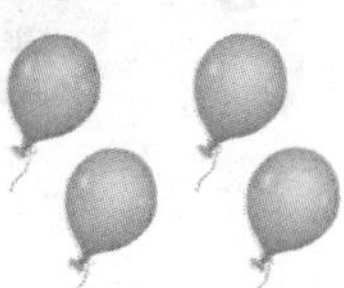

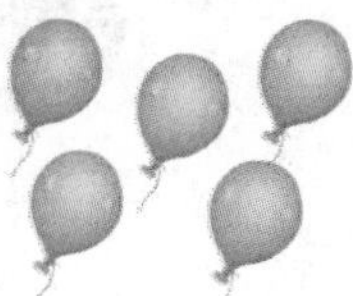

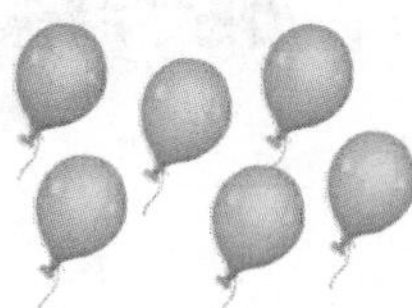

**Directions: 1.** Count the stickers. Write the numbers. Circle the number that is one more than seven. **2.** Count the hats. Write the numbers. Circle the number that is one more than four. **3.** Count the balloons. Write the numbers. Circle the number that is one more than six.

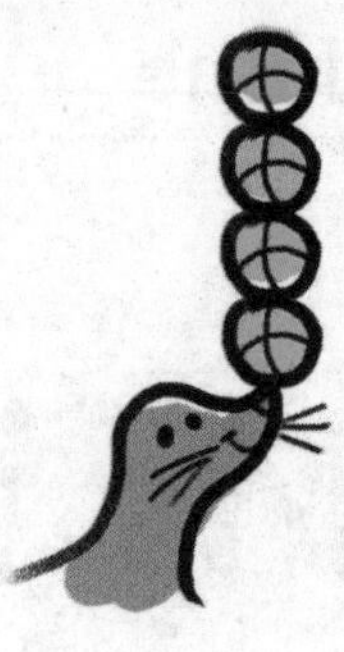

**Directions: 4.** Count the balls. Write the numbers. Circle the number that is one more than two. **5.** Count the balls. Write the numbers. Circle the number that is one more than seven. **6.** Count the flowers. Write the numbers. Circle the number that is one more than five.

**Math at Home** Cut several small squares. Show a group of four. Have your child show a group that is one more. Repeat with other numbers less than 10.

Name

Lesson 10

# Ordinal Numbers to Fifth

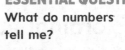

**ESSENTIAL QUESTION**
What do numbers tell me?

## Math in My World

Line up here

Concession

**Teacher Directions:** Use 5 different colored cubes. Place a cube on each box near the slide. Point to the first cube in line for the slide. Color the first box to match that cube. Color the third and fifth boxes to match those cubes. Repeat with the concession stand.

# Guided Practice

Processes & Practices

2

**Directions: 1.** Trace the circle around the joey that is third from the pouch. Trace the line under the joey that is second. Trace the X on the joey that is fourth. **2.** Draw a line under the fifth cat from the yarn. Circle the fourth cat. Draw an X on the second cat.

Name

## Independent Practice

3

4

**Directions: 3.** Draw a box around the third camel from the water. Draw an X on the fifth camel. Draw a line under the fourth camel. **4.** Draw a box around the second bird from the berries. Draw an X on the third bird. Draw a line under the first bird.

# Brain Builders

5

**Directions: 5.** Draw a sandwich on the first plate the ants will find. Draw a circle around the third bird closest to the worm. Draw an X on the second caterpillar from the leaf. Draw a fourth ant in line to reach the plate. Draw five cups starting at the sandwiches. Draw a circle around the fifth cup.

Name

# My Homework

Lesson 10

Ordinal Numbers to Fifth

## Homework Helper

Need help? connectED.mcgraw-hill.com

1

2

**Directions: 1.** Draw a box around the first rabbit near the carrot. Draw an X on the third rabbit. Draw a line under the fifth rabbit. **2.** Draw a box around the second duck from the water. Draw an X on the fourth duck. Draw a line under the first duck.

3

## Vocabulary Check

**ordinal number**

**Directions: 3.** Draw a box around the first monkey from the bananas. Draw an X on the fourth monkey. Draw a line under the second monkey. **4.** Draw a box around the fifth pig from the mud. Draw an X on the third pig. Draw a line under the fourth pig. **5.** Color the first mouse from the cheese blue. Color the fifth mouse orange. Color the third mouse green.

**Math at Home** Show your child a row of five pieces of cereal next to a bowl. Have your child tell you which piece is first, third, and fifth from the bowl.

Name

Lesson 11

# Ordinal Numbers to Tenth

**ESSENTIAL QUESTION**

What do numbers tell me?

## Math in My World

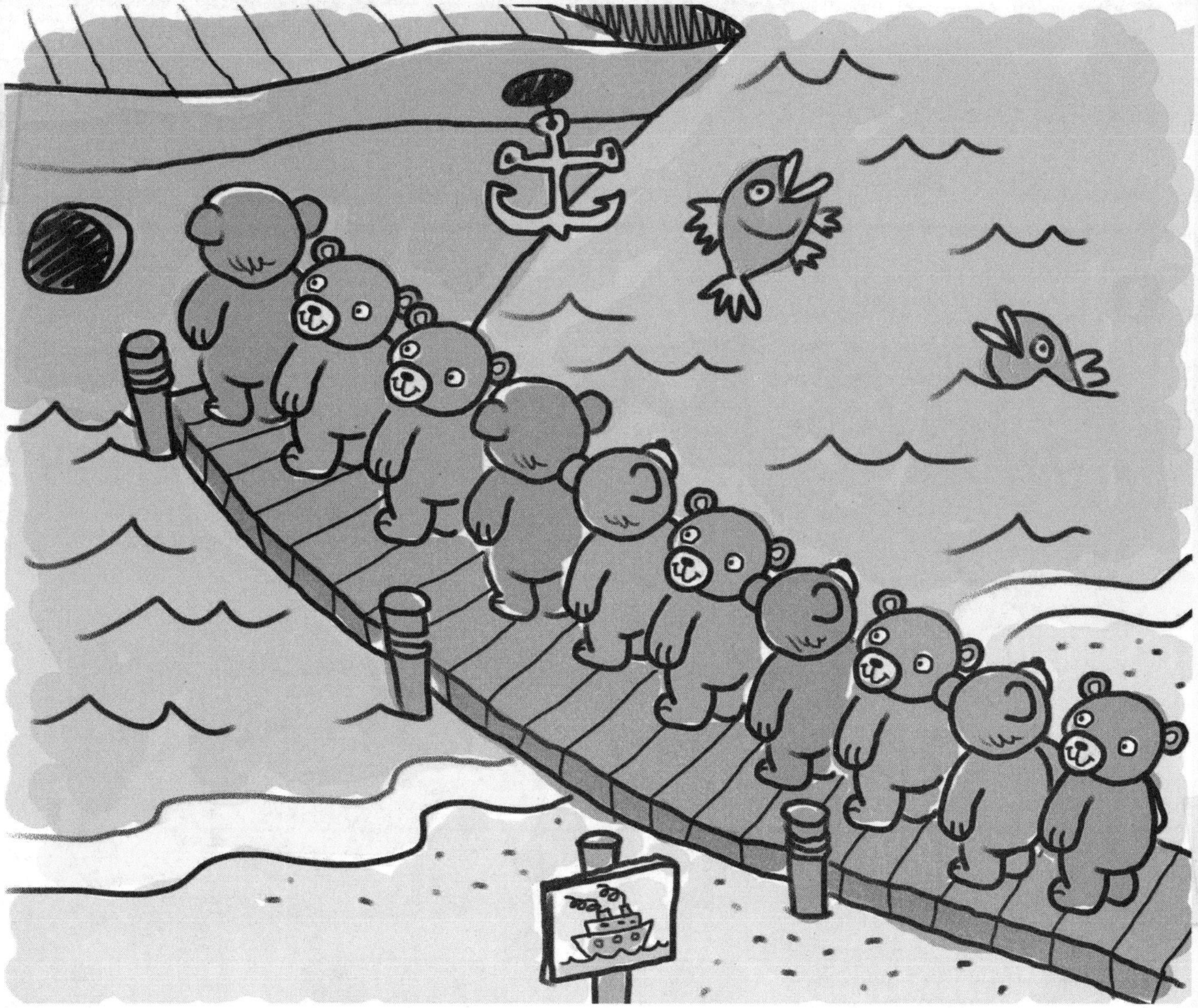

**Teacher Directions:** Point to the first bear to get to the boat. Draw a line above the bear. Draw an X on the second and tenth bears in line. Draw a circle around the fourth and seventh bears in line.

# Guided Practice

Processes &Practices

**Directions: 1.** Draw a line under the first koala near the tip of the branch. Draw an X on the tenth koala. Draw a box around the seventh koala. **2.** Draw a box around the first raccoon on the can. Draw an X on the sixth raccoon. Draw a line under the fourth raccoon.

Name

## Independent Practice

3

4

**Directions: 3.** Draw an X on the seventh horse from the barn. Draw a circle around the third horse. Draw a line under the first horse. **4.** Draw an X on the second mouse from the cheese. Draw a circle around the sixth mouse. Draw a line under the tenth mouse.

# Brain Builders

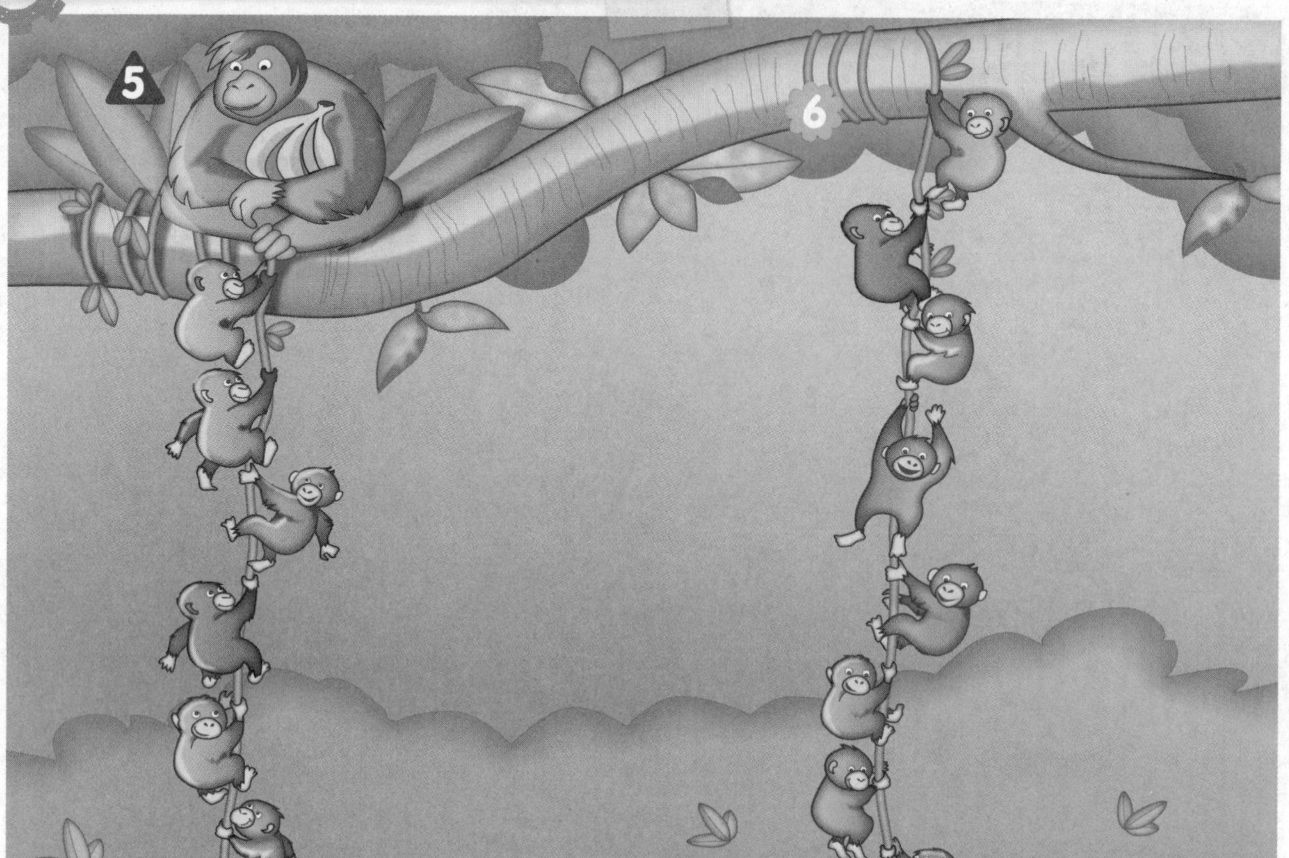

5

6

**Directions: 5.** Draw an X on the sixth baby orangutan climbing up to the mother. Draw a circle around the tenth baby orangutan climbing up. Draw a box around the second baby orangutan climbing up. **6.** Draw an X on the tenth baby orangutan climbing down to the father. Draw a box around the eighth baby orangutan climbing down. Draw a circle around the third baby orangutan climbing down. Explain to a friend which is the first and which is the tenth baby orangutan in each line.

Name

# My Homework

Lesson 11

Ordinal Numbers to Tenth

## Homework Helper

Need help? connectED.mcgraw-hill.com

1

2

3

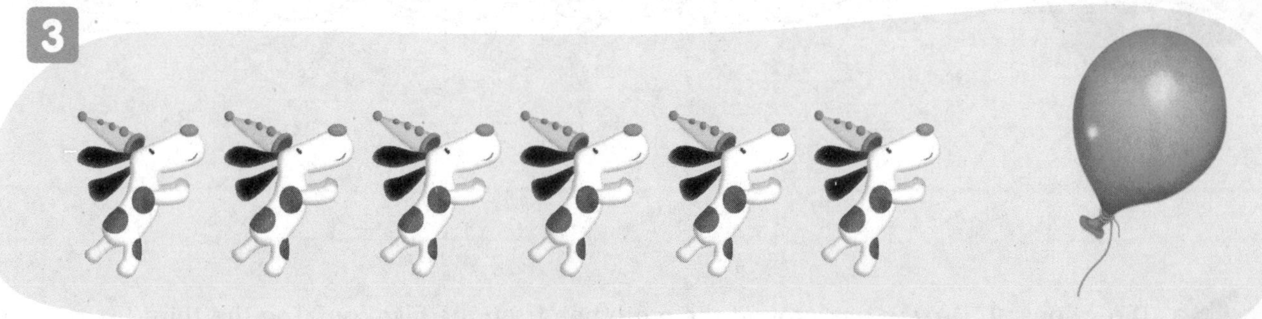

**Directions: 1.** Draw a box around the seventh butterfly from the flower, an X on the fourth, and a line under the first. **2.** Underline the sixth frog from the log, draw a box around the second, and circle the fifth. **3.** Underline the third dog from the balloon, draw a box around the fourth, and circle the first.

4

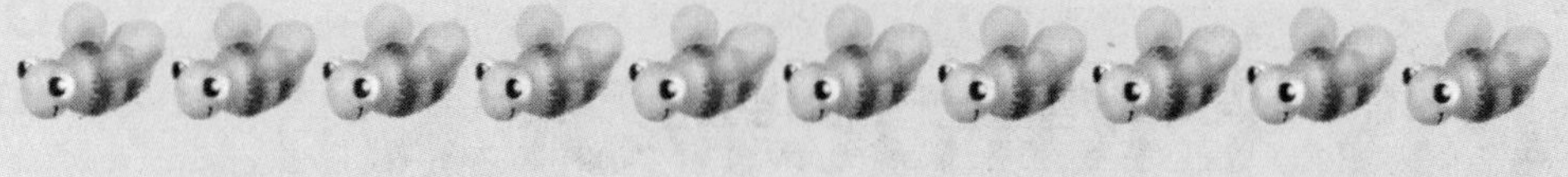

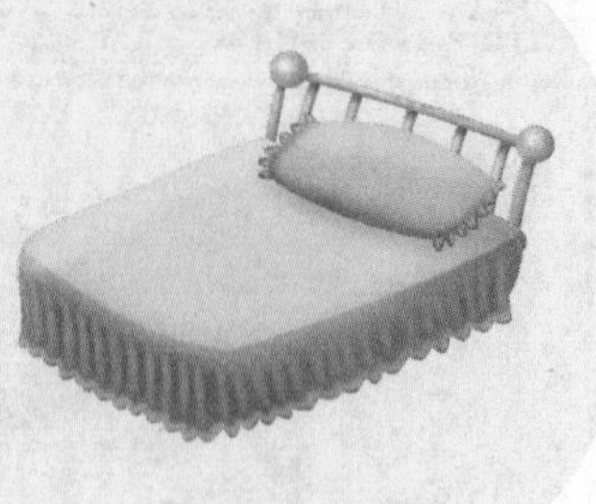

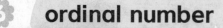

## Vocabulary Check

6 
**ordinal number**

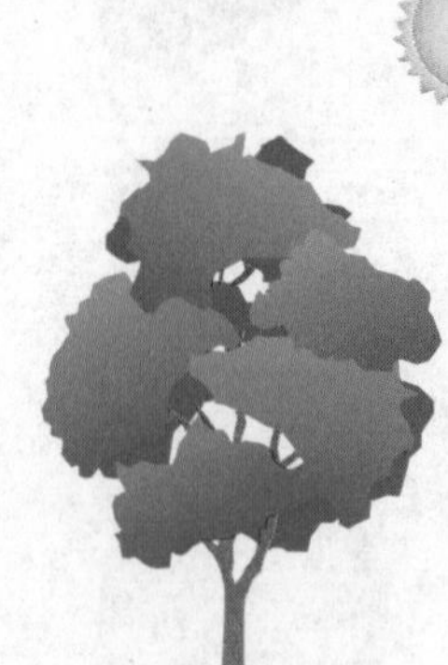

**Directions: 4.** Draw a box around the ninth bee from the hive, an X on the third bee, and a line under the fifth bee. **5.** Draw a box around the second sheep from the bed, an X on the ninth sheep, and a line under the fourth sheep. **6.** Point to the first bird flying to the tree. Color it green. Point to the third bird. Color it blue.

**Math at Home** Line up 10 crayons beside a sheet of paper. Have your child choose the second, fifth, ninth, and tenth crayons from the paper to draw a picture.

Name

# Write Numbers 0 to 5

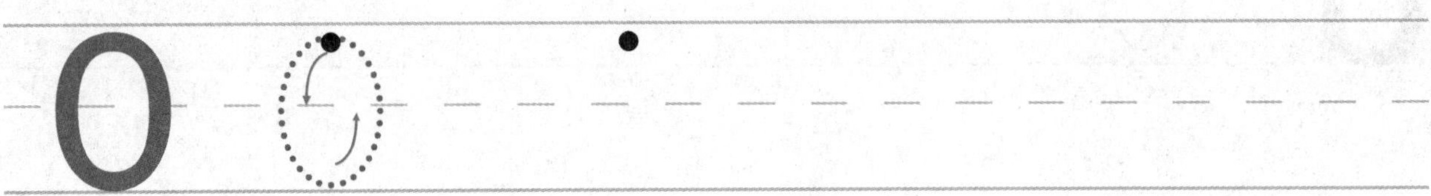


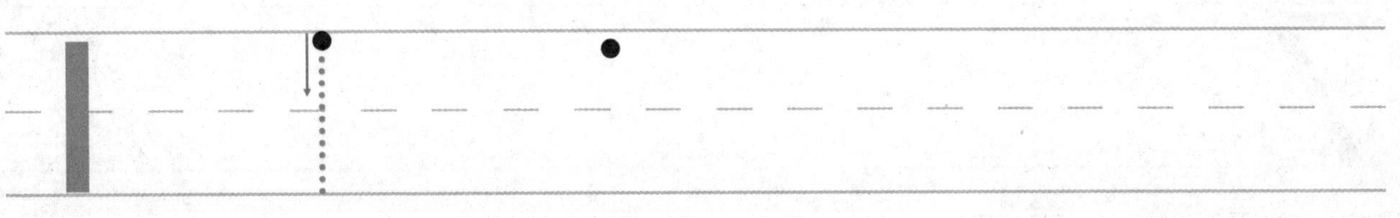


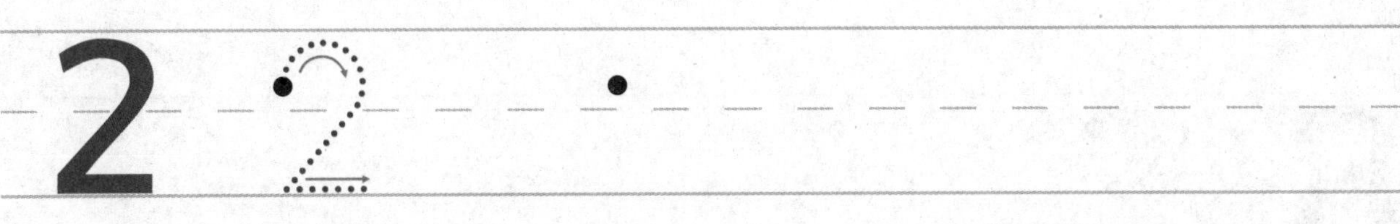


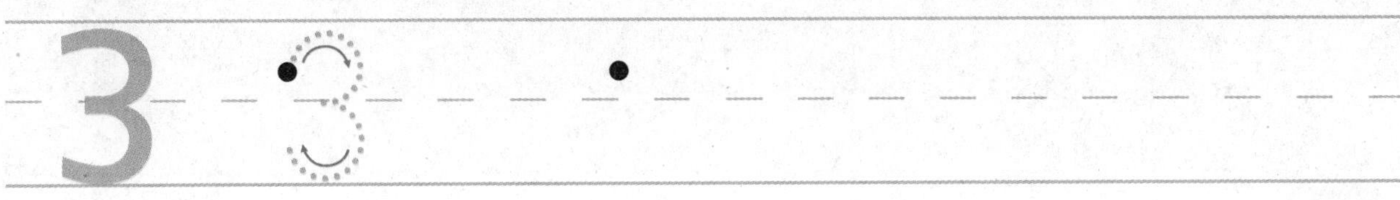


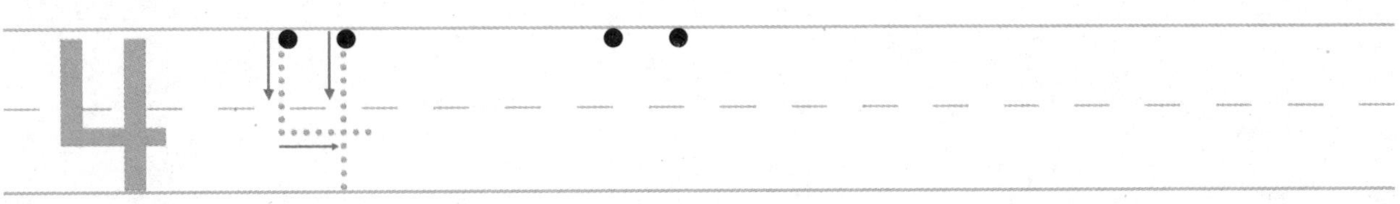


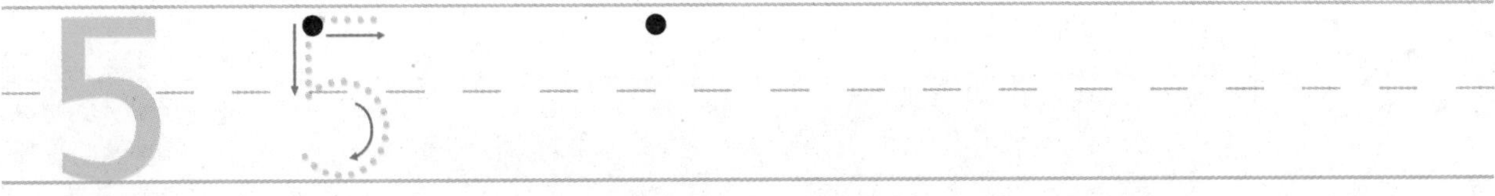

# Write Numbers 6 to 10

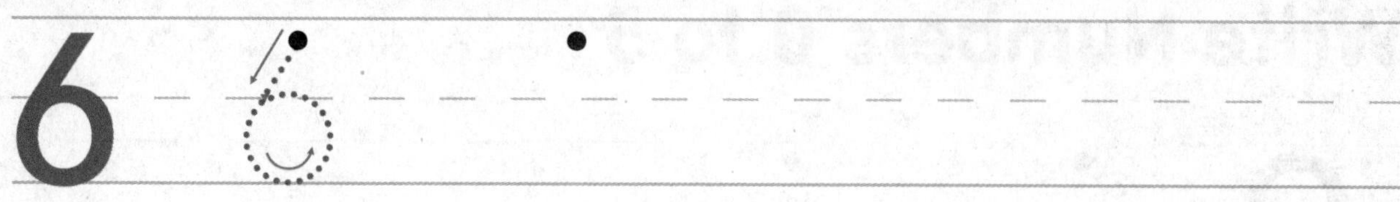

6

7

8

9

10

Name



# My Review

## Vocabulary Check 

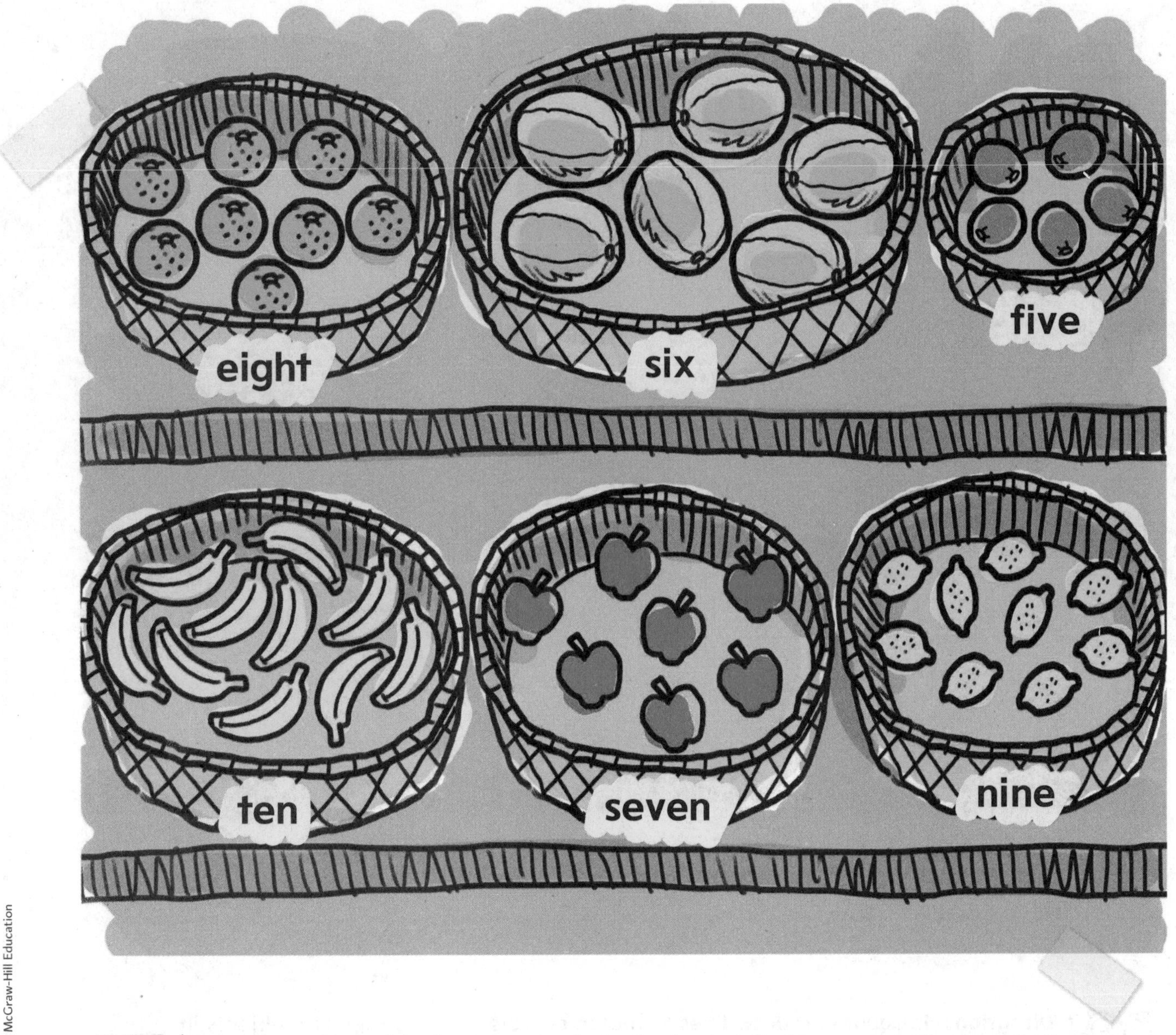

eight

six

five

ten

seven

nine

**Directions:** Use a black crayon to circle the basket that has seven pieces of fruit. Use a purple crayon to circle the basket that has eight pieces of fruit. Use a green crayon to circle the basket that has nine pieces of fruit. Use a red crayon to circle the basket that has ten pieces of fruit.

# Concept Check

1

2

**Directions: 1.** Count and draw lines to match objects in one group with objects in the other. Write the numbers. Circle the number and group that is greater than the other. Draw an X on the number and group that is less than the other. **2.** Draw a box around the fourth chicken from the food. Draw an X on the fifth chicken. Draw a line under the first chicken.

Name

## Brain Builders

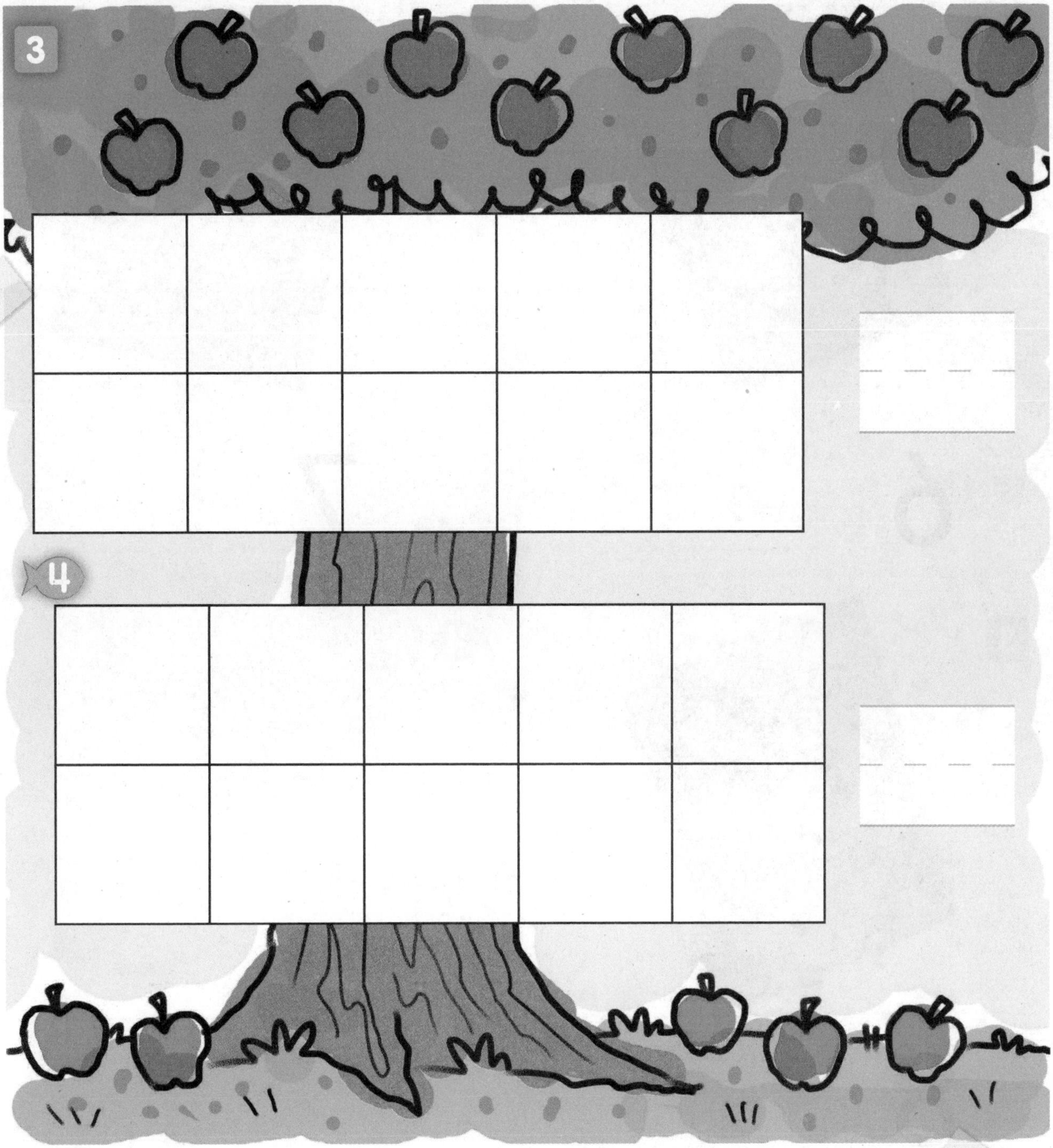

3

4

**Directions: 3.** Count the apples in the tree. Color one box for each apple counted. Write the number. **4.** Count the apples on the ground. Color one box for each apple counted. Write the number. Circle the number and group that is greater than. Draw more apples to make the groups equal.

# Reflect

**Chapter 2**

**ESSENTIAL QUESTION**

**What do numbers tell me?**

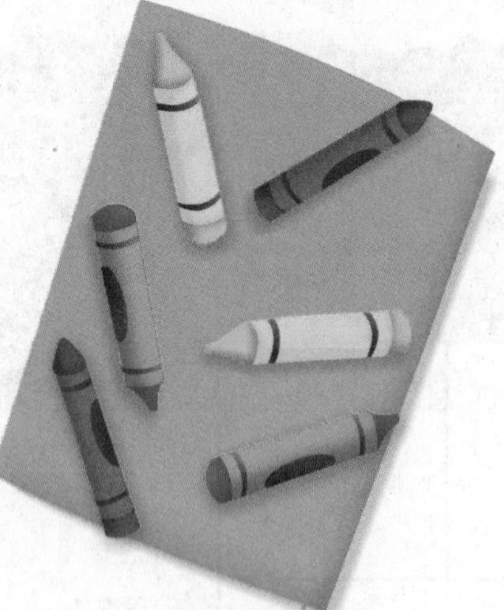

6

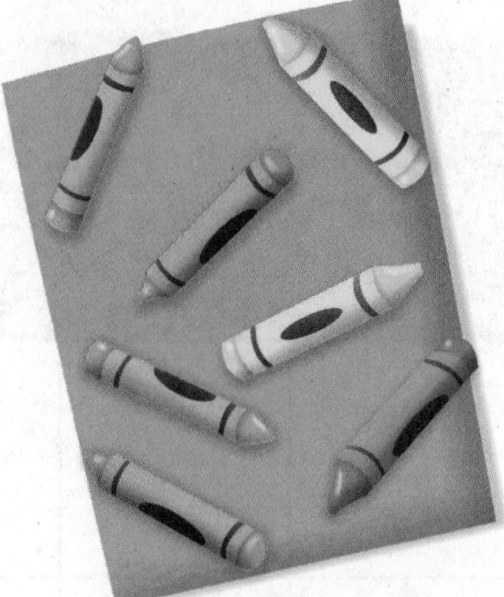

7

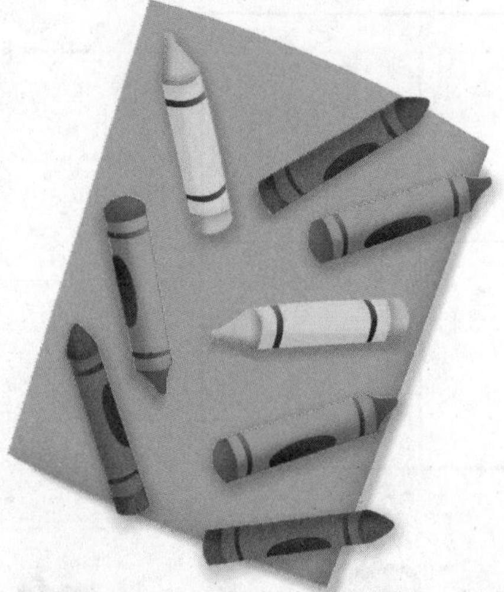

8

**Directions: 1–3.** Count the crayons. Say how many. Draw one more crayon. Count how many. Write the number. **4.** Draw dots to show one more than nine. Write the number.

Name Date

Score

# Performance Task

## Grocery Shopping

You are shopping for snacks at the grocery store.

**Show all your work to receive full credit.**

### Part A

You buy apples for a snack. Count the apples. Write the number.

### Part B

You have some bags for the apples. Count the bags. Write the number.

## Part C

Draw lines to match the apples to the bags.

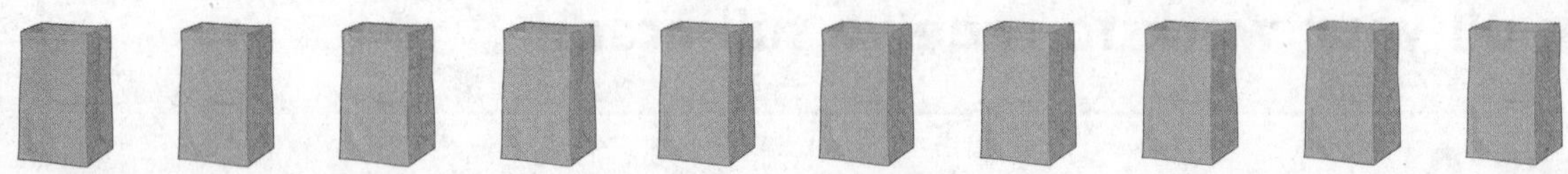

## Part D

Circle the group that is less than the other group in Part C.

## Part E

Draw one orange below for each apple in Part C.

Chapter 3

# Numbers Beyond 10

**ESSENTIAL QUESTION**

How can I show numbers beyond 10?

We're Off to the Playground!

Watch a video!

# Chapter 3 Project

## My Counting Cards

1. Write numbers 1–10 in separate groups below. Draw groups of dots to show each of the numbers.
2. Write the numbers 11–20 on separate index cards. Draw a matching number of dots on each of the cards.
3. Circle each group of ten objects on the cards to help you count.
4. Make sure your group checks each other's work.

Name

# Am I Ready?

1

2

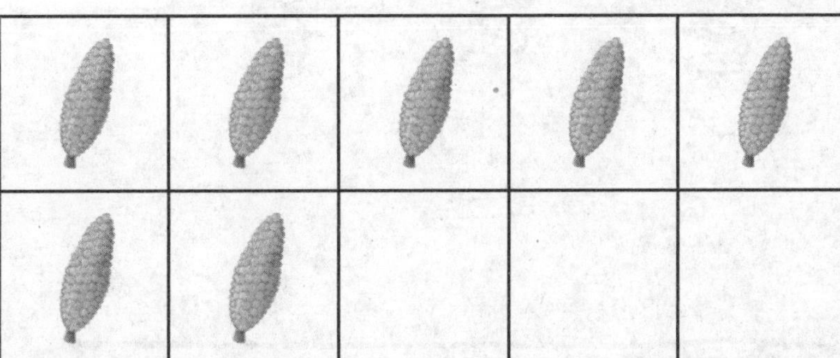

3

4

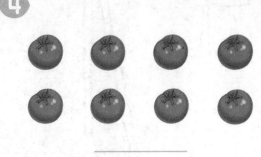

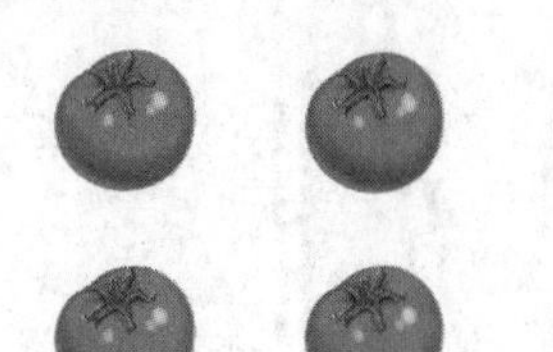

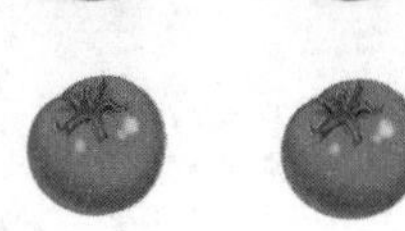

**Directions:** **1.** Count the carrots. Write the number. **2.** Count the ears of corn. Write the number. **3–4.** Count the objects. Write the number.

Name

# My Math Words

## Review Vocabulary

1

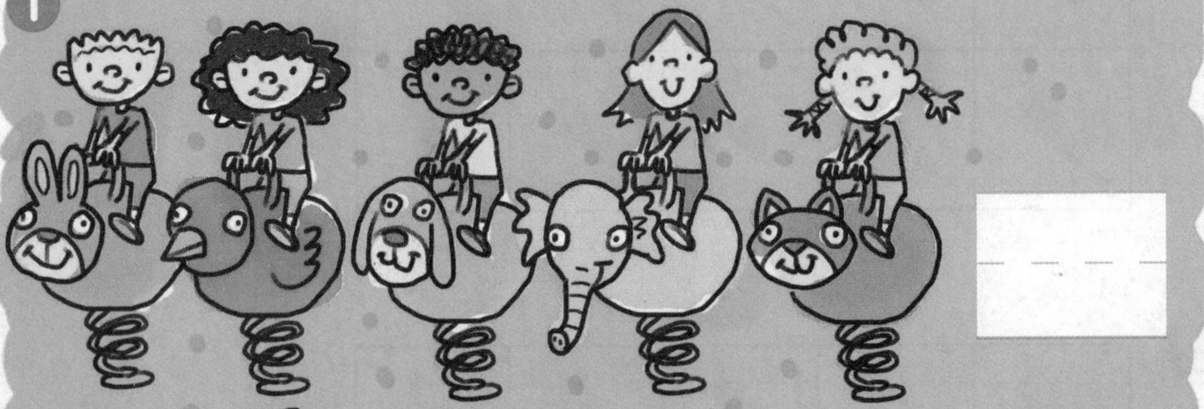

2

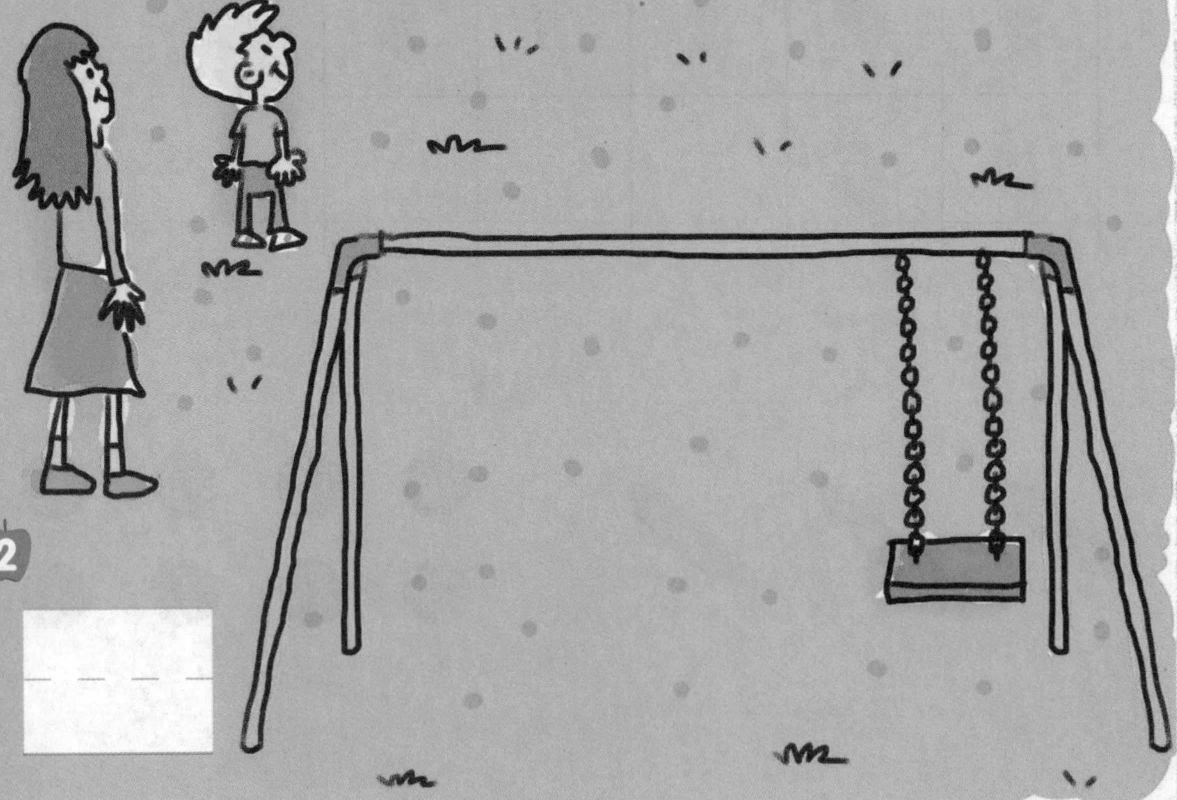

**Directions: 1.** Count the animals. Write the number. **2.** Draw more swings to show four. Write the number.

# My Vocabulary Cards

eighteen 18

eleven 11

fifteen 15

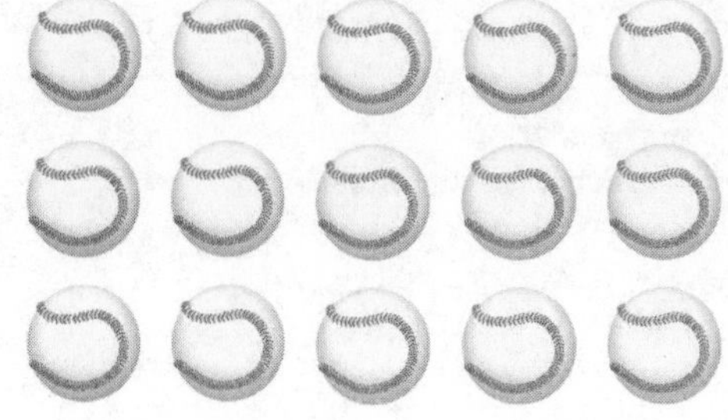

fourteen 14

nineteen 19

seventeen 17

**Teacher Directions:**
**Ideas for Use**

- Write an X on each card every time you read or write the word in this chapter. Try to write at least 10 Xs for each card.
- Create riddles for each word. Have a classmate guess the word for each riddle.

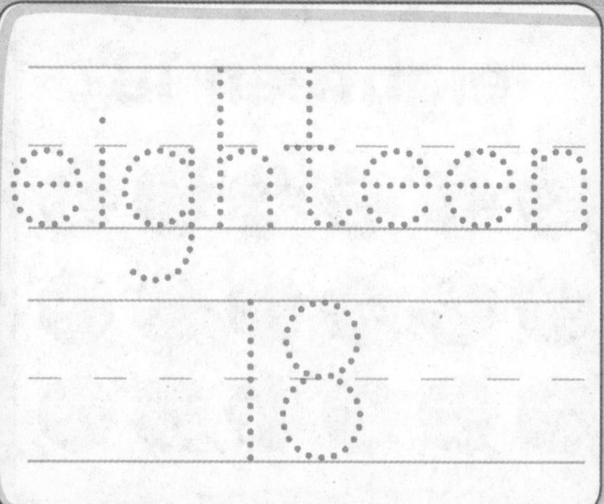


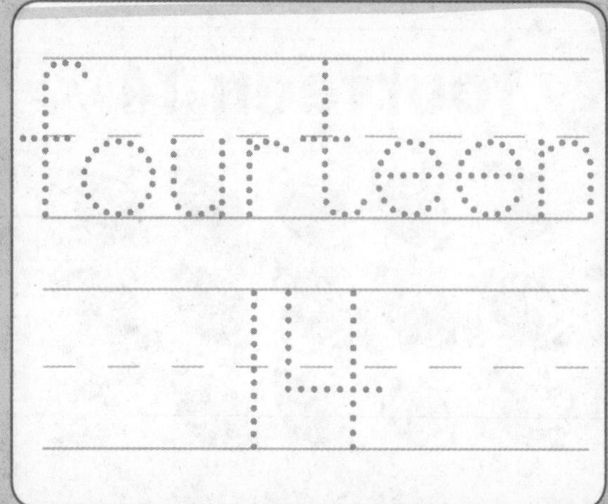


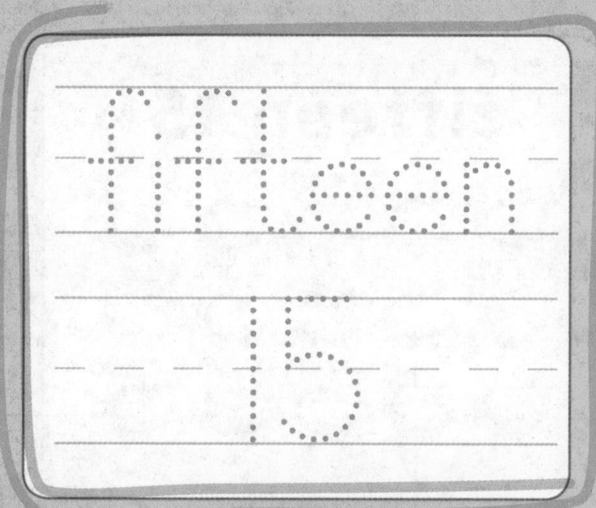


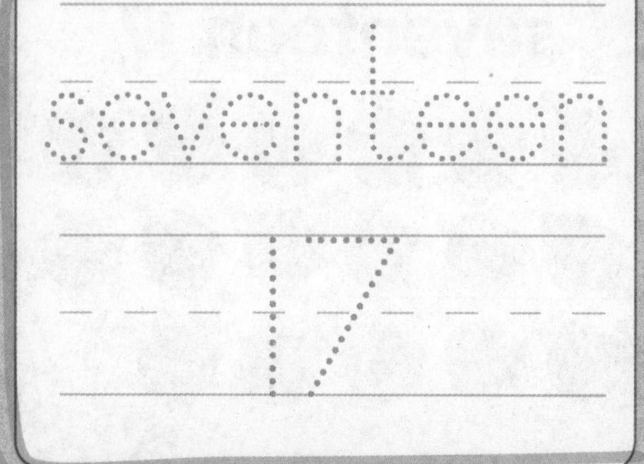


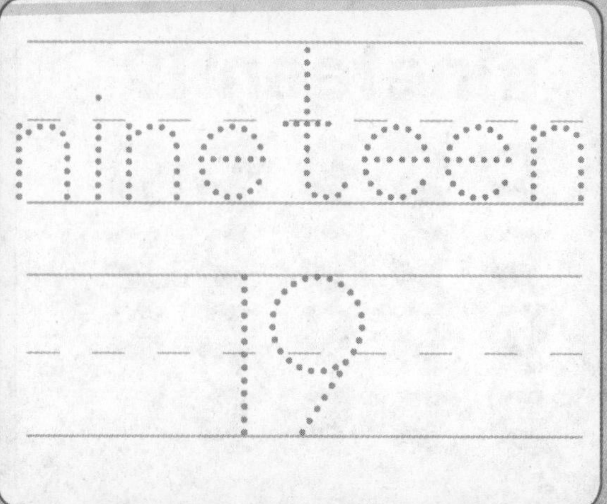

# My Vocabulary Cards

Processes &Practices

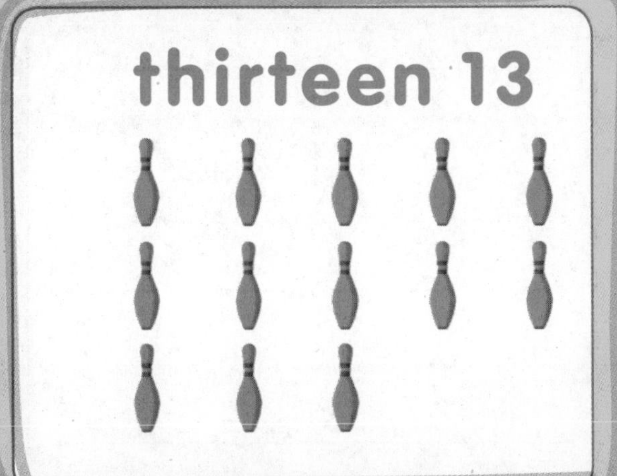


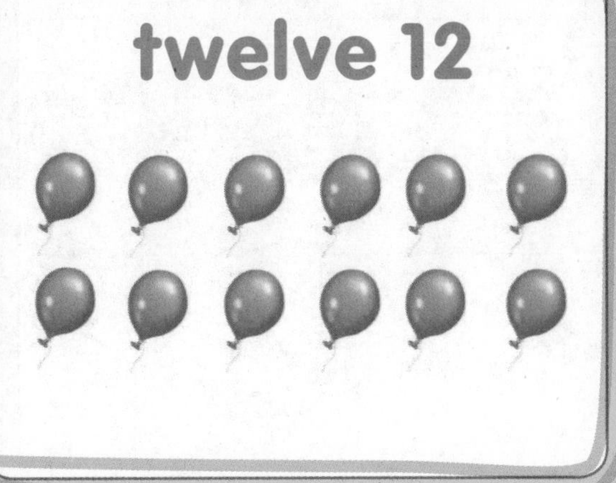

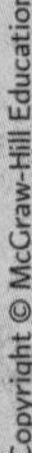

**Teacher Directions:**
**More ideas for Use**

1. Draw an X on each card every time you read or write the word in this chapter.

2. Draw a picture that shows objects in the amount of each number. Have a classmate count the objects in the picture.

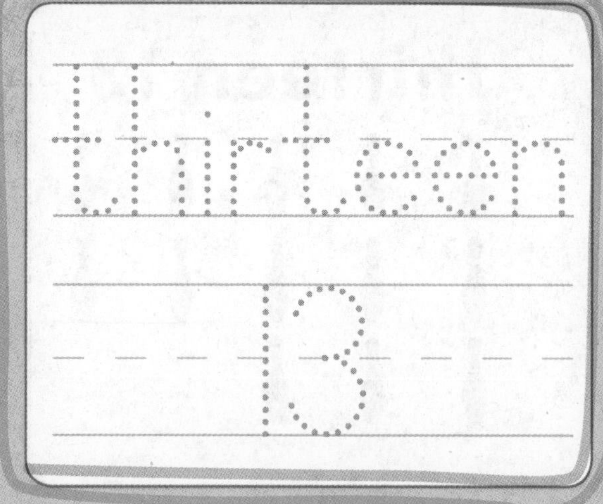


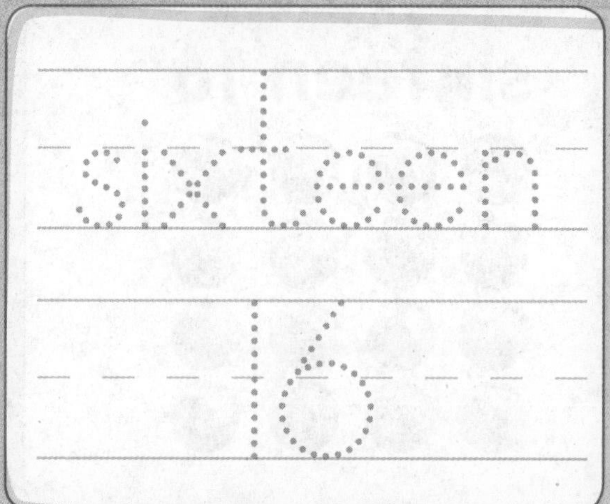

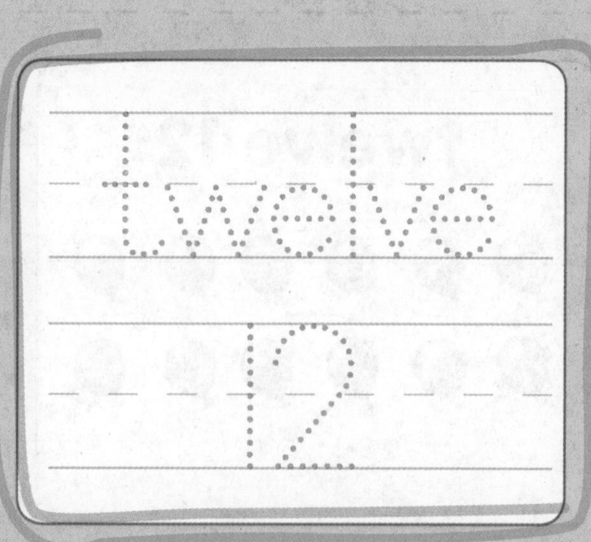

# My Foldable

FOLDABLES® Follow the steps on the back to make your Foldable.

FOLDABLES®
Study Organizer

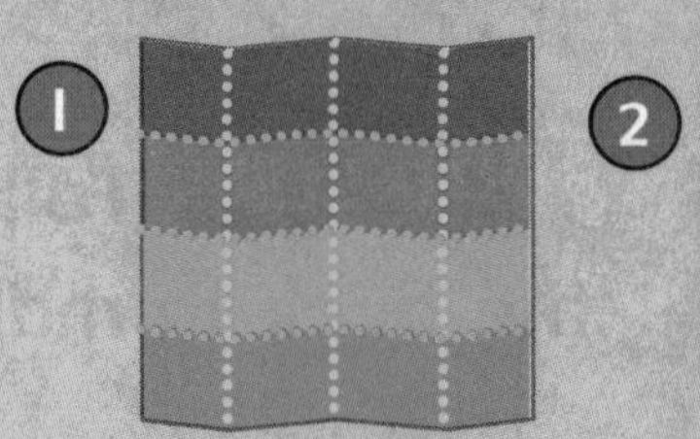
1

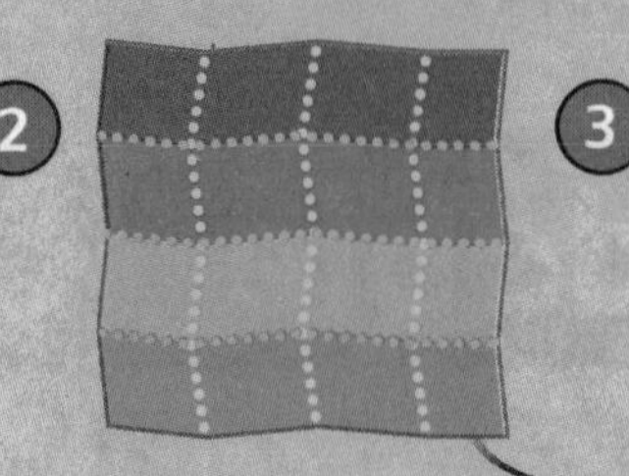
2

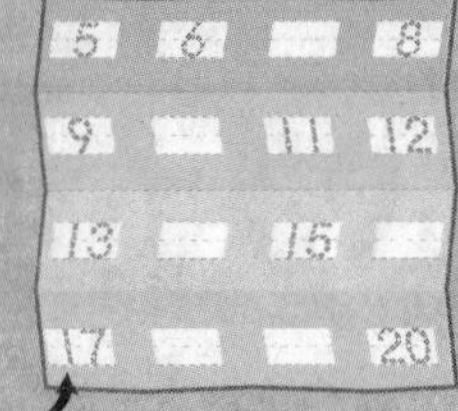
3
5
6
8
9
11
12
13
15
17
20

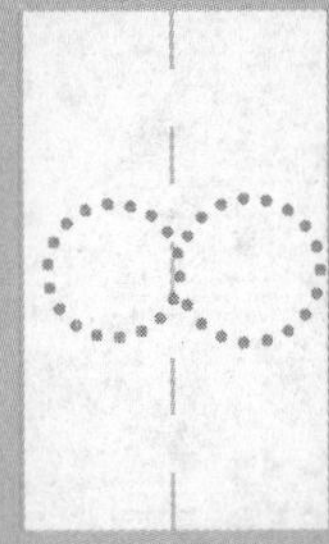

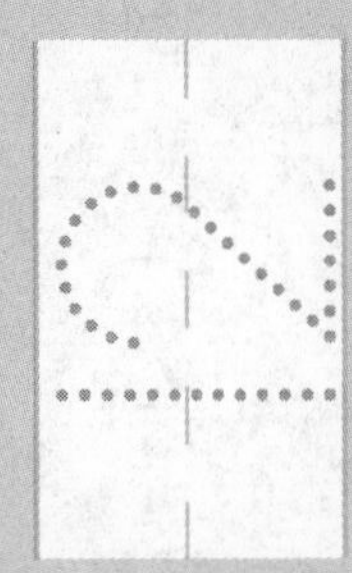

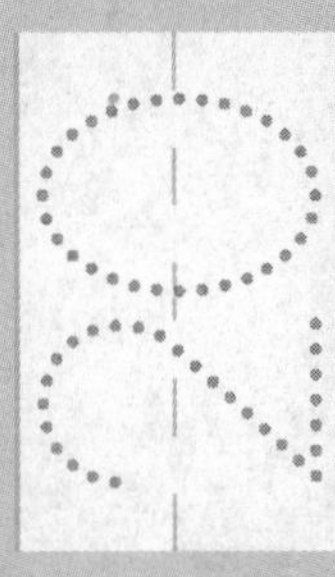

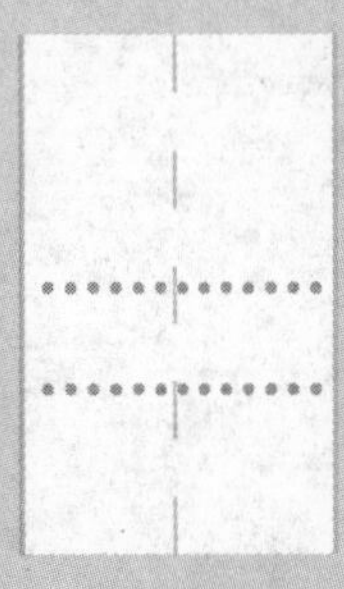

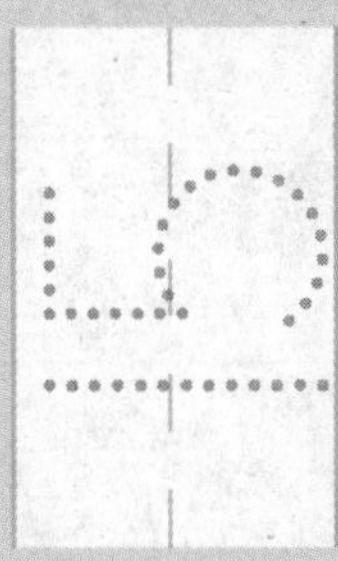

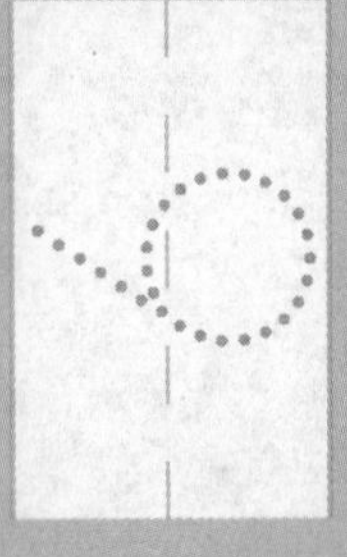

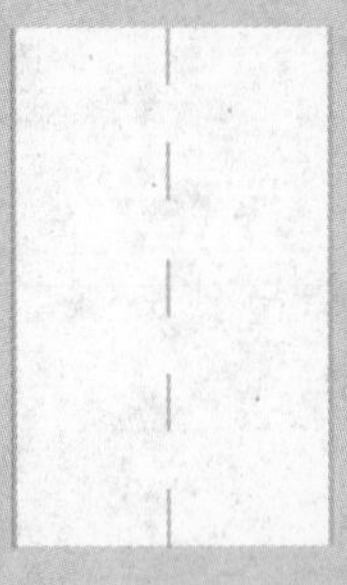

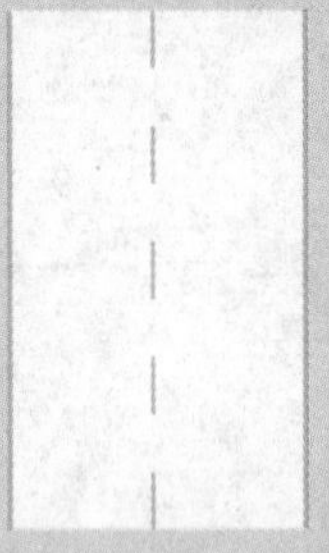

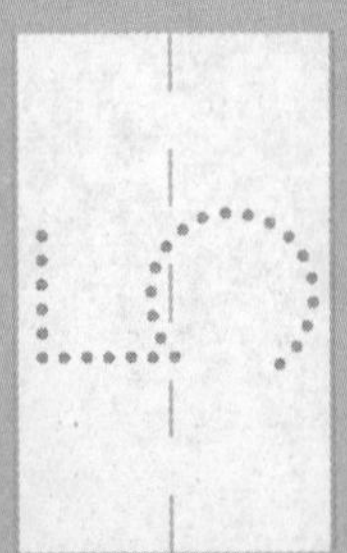

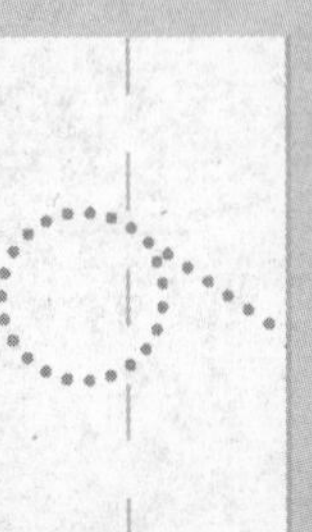

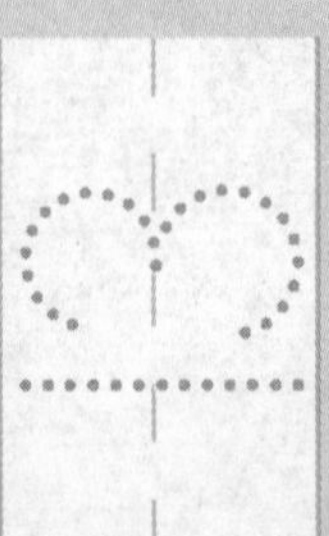

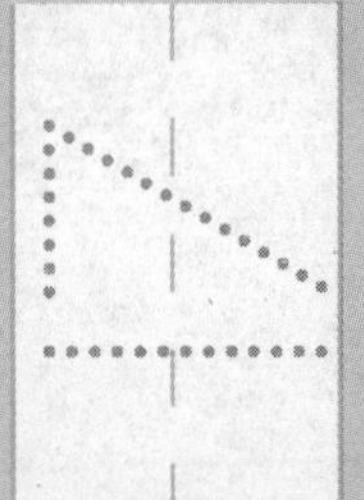

Name

Lesson 1

# Numbers 11 and 12

**ESSENTIAL QUESTION**
How can I show numbers beyond 10?

##  Math in My World  

| | | | | |
|---|---|---|---|---|
| | | | | |
| | | | | |

12

| | | | | |
|---|---|---|---|---|
| | | | | |
| | | | | |

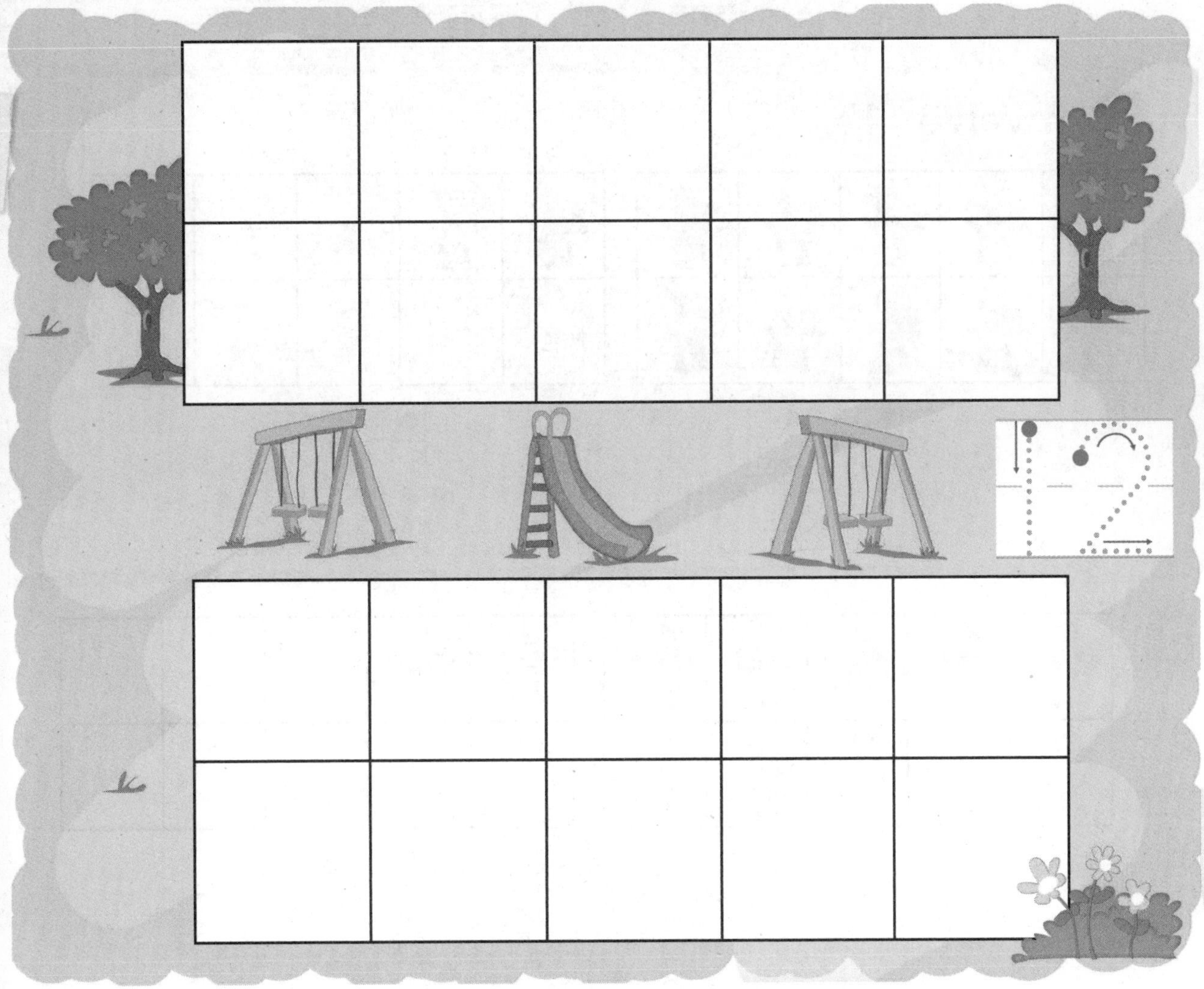

**Teacher Directions:** Starting with the ten-frame above the playground, fill the ten-frames with ● to show number 11 and 12. Count and say the number of objects. Trace a counter in each ten-frame box to show the number 12. Trace the number.

## Guided Practice

Processes &Practices

1 eleven 11

11 11 11

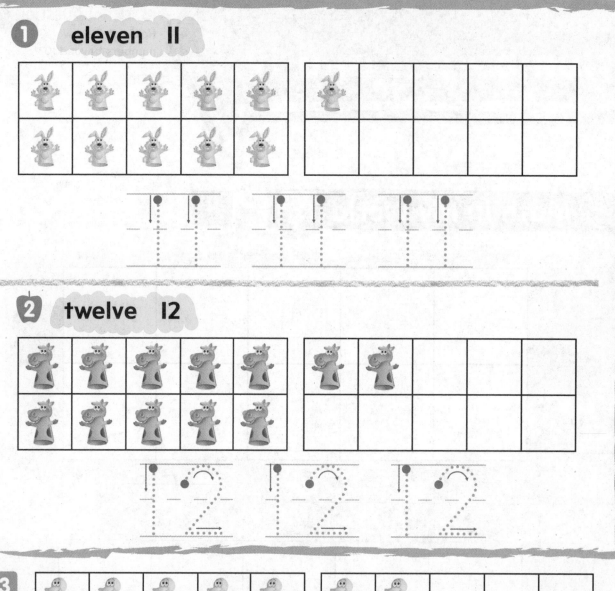

2 twelve 12

12 12 12

3

**Directions: 1–2.** Count the objects. Say how many. Use Work Mat 4 and counters to show the number. Trace the number three times. **3.** Count the objects. Say how many. Use Work Mat 4 and counters to show the number. Write the number three times.

Name

## Independent Practice

4

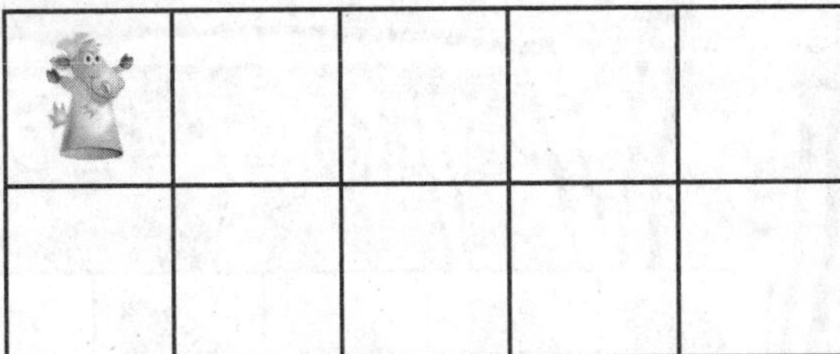

5

11 11 11

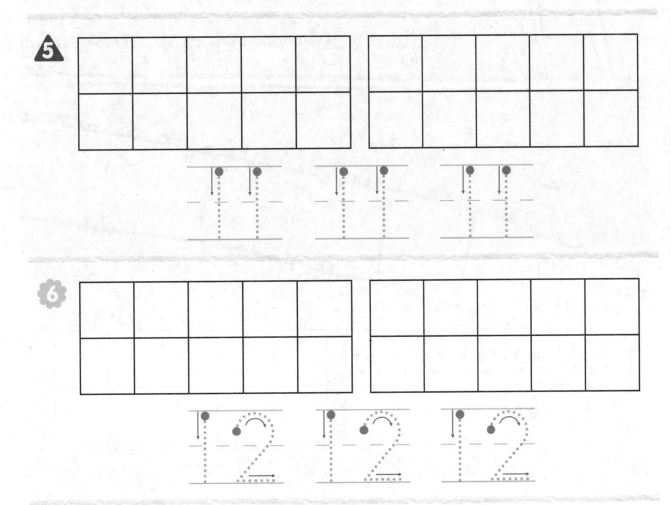

6

12 12 12

**Directions: 4.** Count the objects. Say how many. Use Work Mat 4 and counters to show the number. Write the number three times. **5–6.** Trace the number three times. Say the number. Use Work Mat 4 and counters to show the number. Draw circles for counters to show the number.

Processes &Practices

## Brain Builders

7

10 20 30 40 50

10 20 30 40 50 60

8

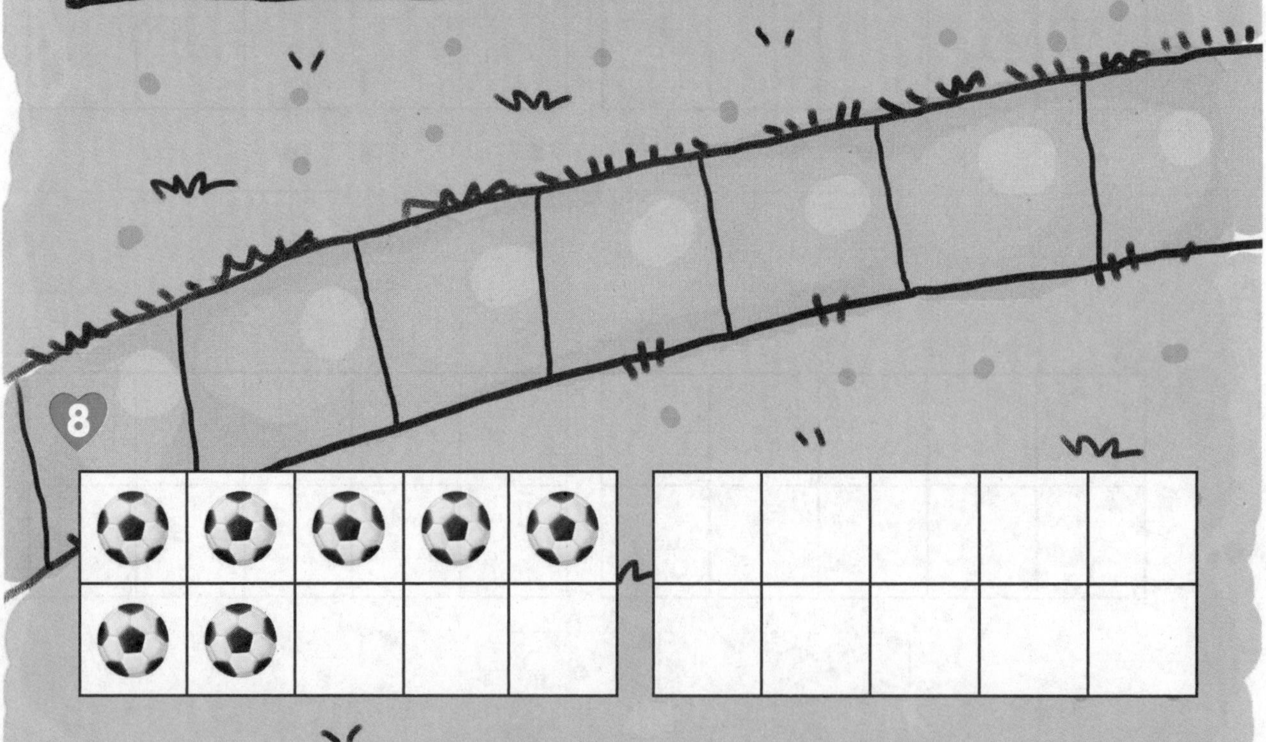

**Directions: 7.** Count the objects. Say how many. Draw more to make eleven. Then draw one more ball. Tell a friend how many balls there are now. **8.** Count the balls. Say how many. Draw more balls to make twelve. Put an X on balls until there are only eleven.

Name

# My Homework

Lesson 1

Numbers 11 and 12

## Homework Helper

Need help? connectED.mcgraw-hill.com

1

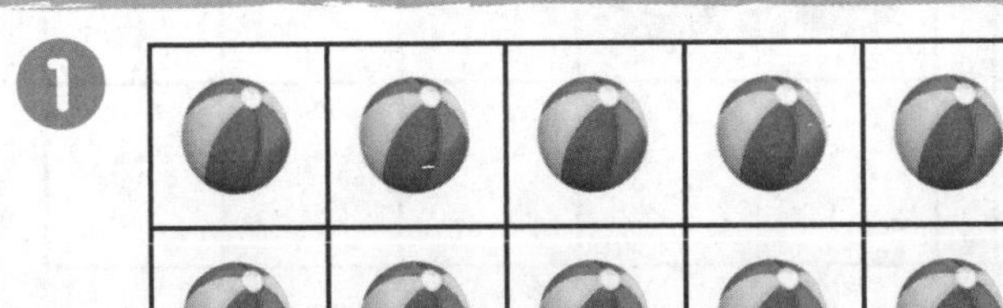

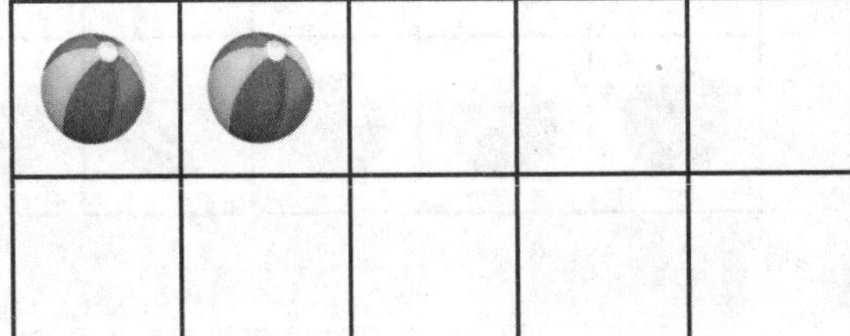

12 12 12

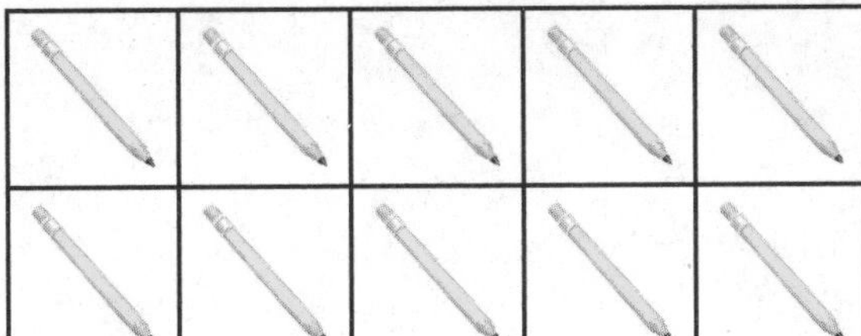

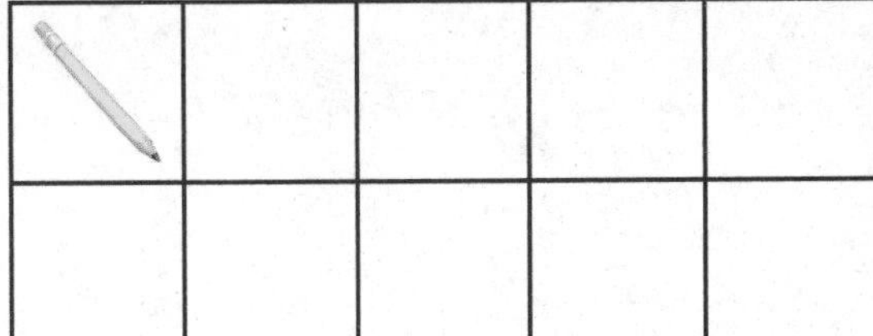

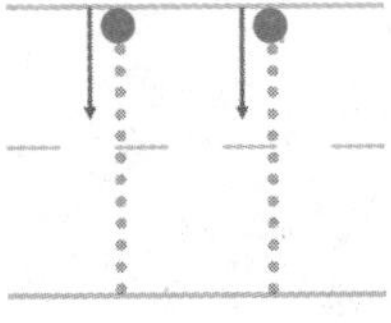

**Directions: 1–2.** Count the objects. Say how many. Trace the number. Write the number two times.

3

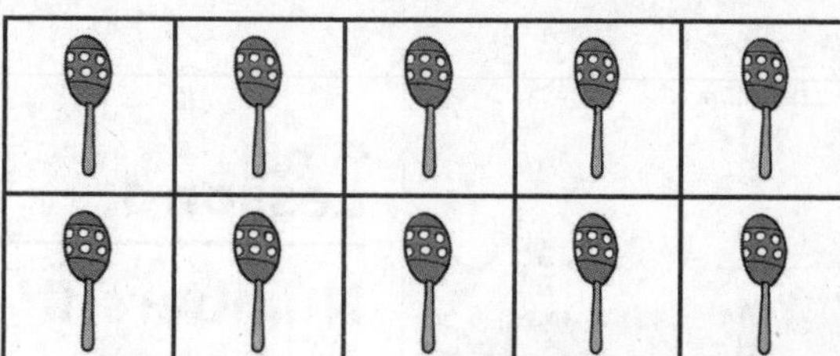

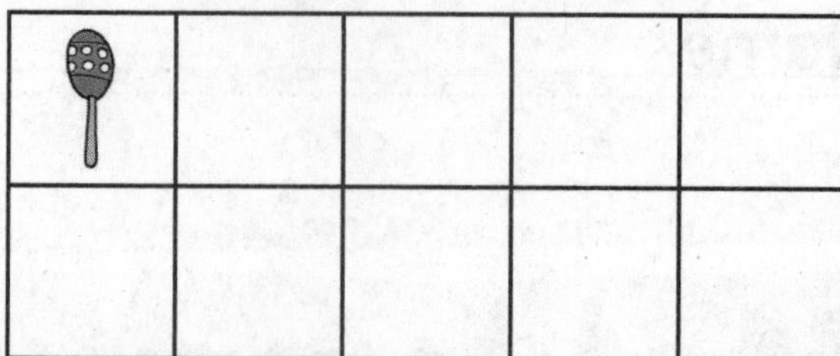

4

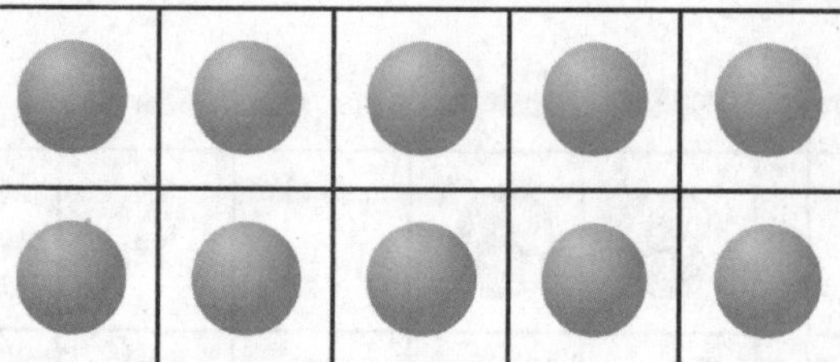

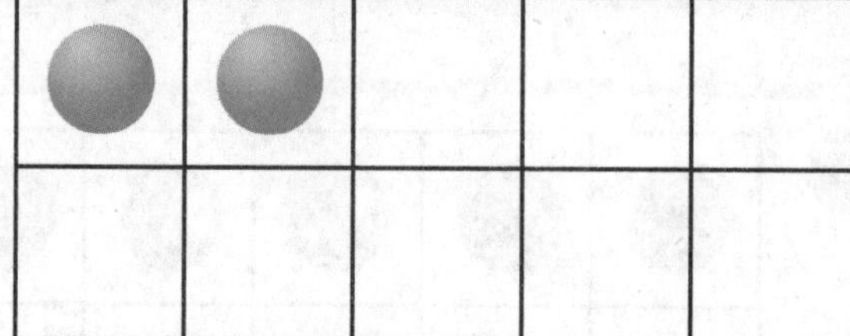

## Vocabulary Check

5 eleven 11

6 twelve 12

**Directions: 3–4.** Count the objects. Say how many. Write the number three times.
**5–6.** Draw circles to show the number.

**Math at Home** Use an empty egg carton. Ask your child to fill each egg holder with one item. Use the items to create a group of 11 and 12. Write each number.

Name

Lesson 2

# Numbers 13 and 14

**ESSENTIAL QUESTION**
How can I show numbers beyond 10?

## Math in My World

|  |  |  |  |  |
|---|---|---|---|---|
|  |  |  |  |  |

|  |  |  |  |  |
|---|---|---|---|---|
|  |  |  |  |  |

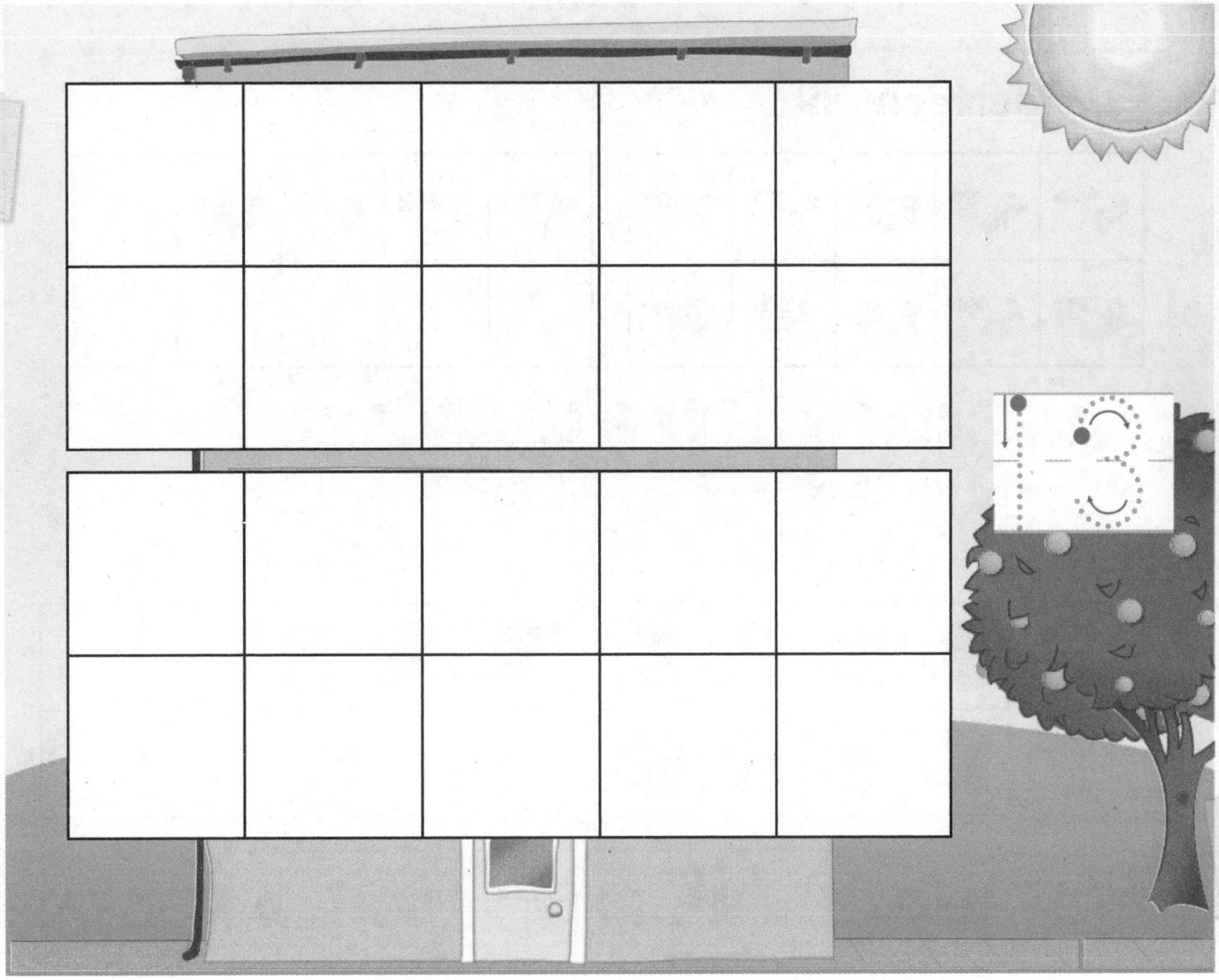

**Teacher Directions:** Trace the number. Use ● to show the number in the ten-frames. Count and say the number of objects. Trace the counters to show the number.

# Guided Practice

Processes & Practices

**1** thirteen 13

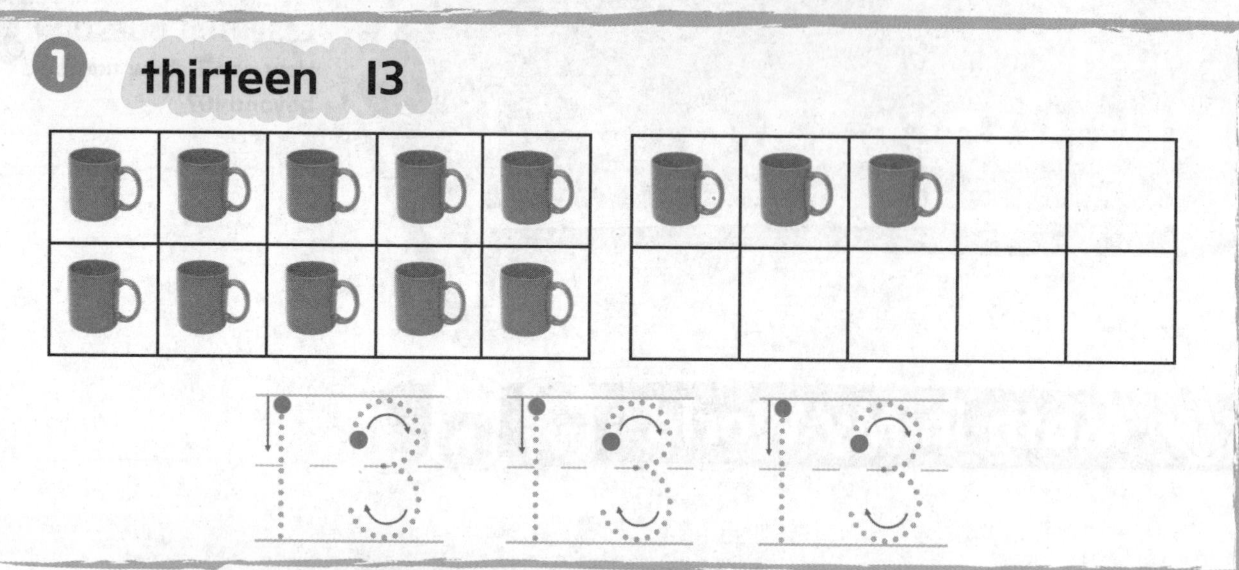

**2** fourteen 14

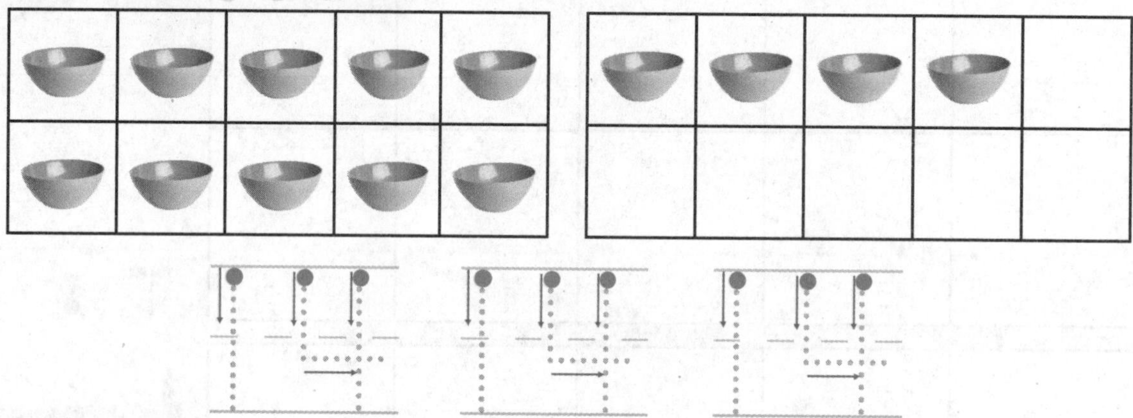

**3**

**Directions: 1–2.** Count the objects. Say how many. Use Work Mat 4 and counters to show the number. Trace the number three times. **3.** Count the objects. Say how many. Use Work Mat 4 and counters to show the number. Write the number three times.

Name

## Independent Practice

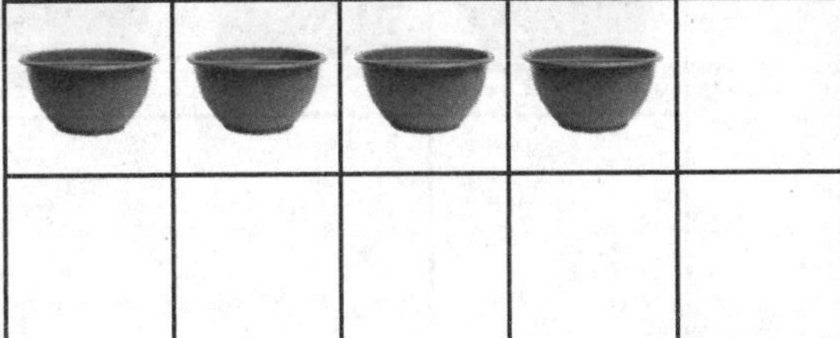

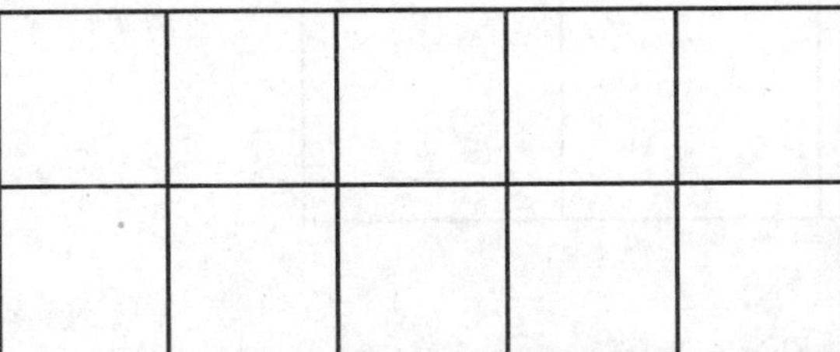

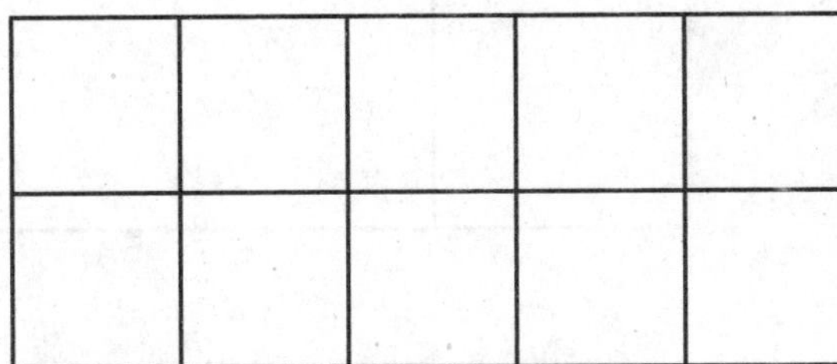

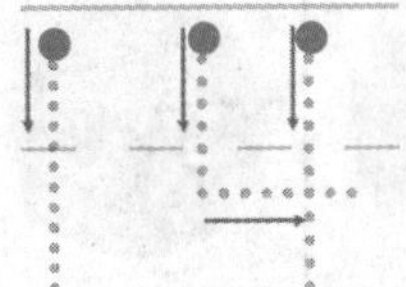

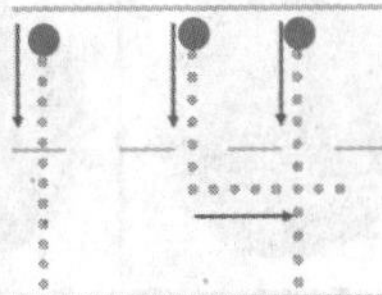

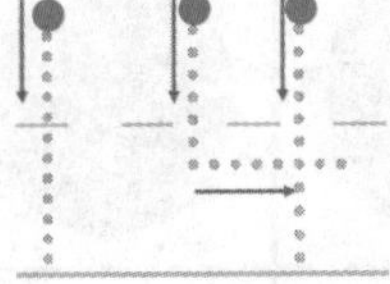

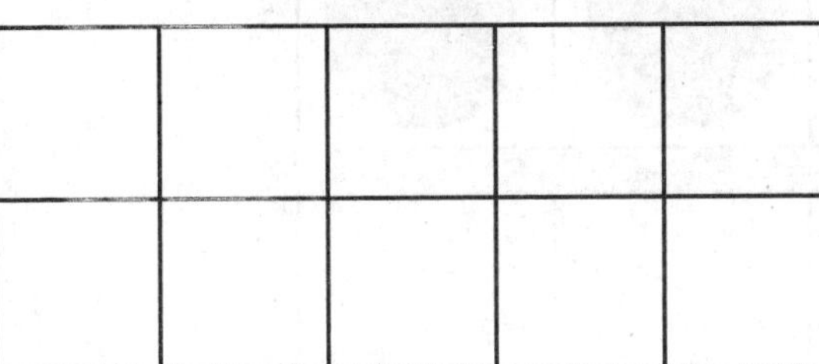

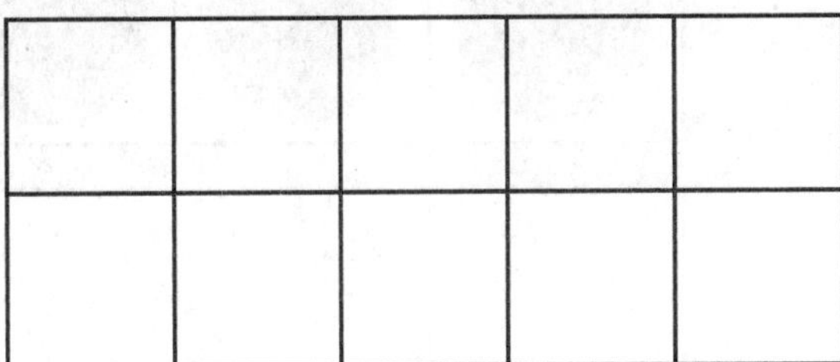

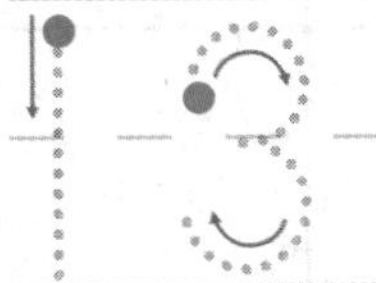

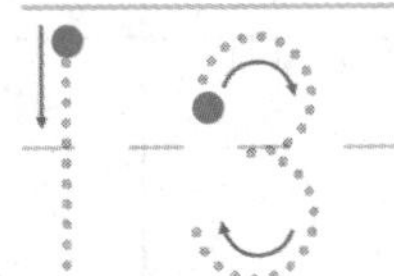

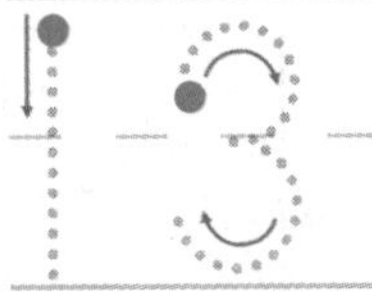

**Directions: 4.** Count the objects. Say how many. Use Work Mat 4 and counters to show the number. Write the number three times. **5–6.** Trace the number three times. Say how many. Use Work Mat 4 and counters to show the number. Draw circles for counters to show the number.

## Brain Builders

7

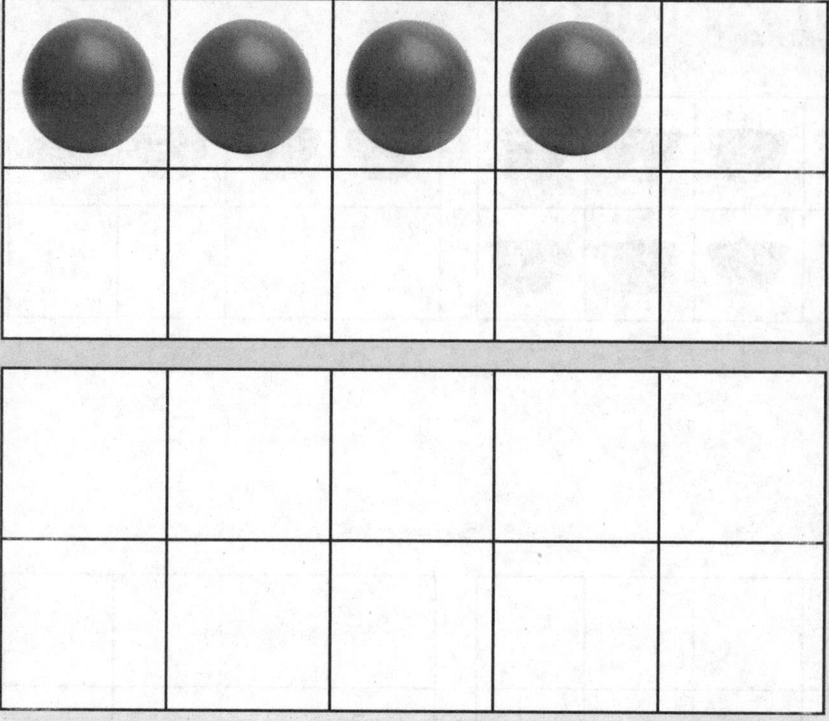

8

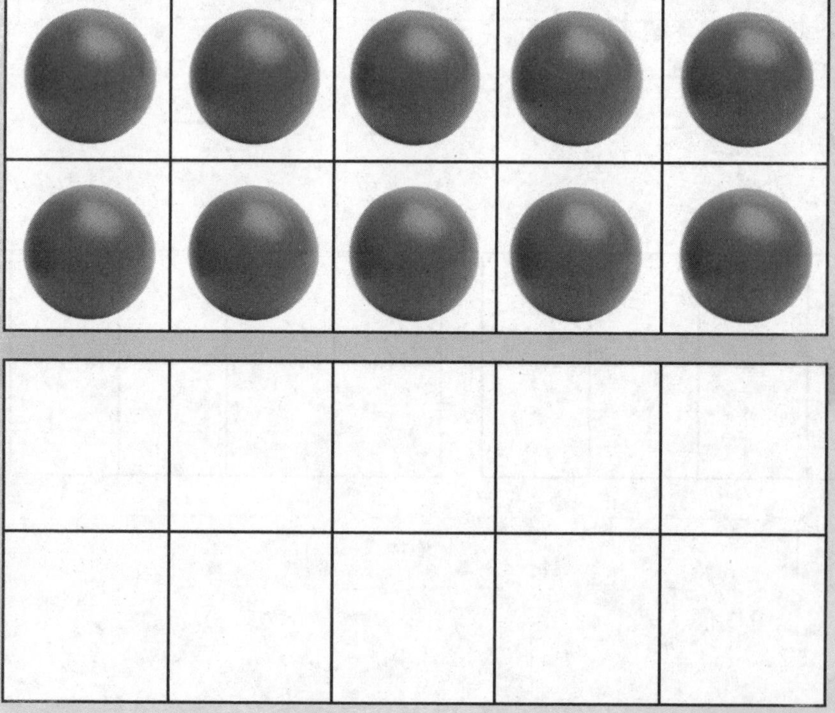

**Directions: 7.** Count the objects. Say how many. Draw more to make thirteen. Tell a friend how many there are in each ten-frame.
**8.** Count the objects. Say how many. Draw more to make fourteen. Tell a friend how many there are in the bottom ten-frame.

Name

# My Homework

Lesson 2

Numbers 13 and 14

## Homework Helper

Need help? connectED.mcgraw-hill.com

1

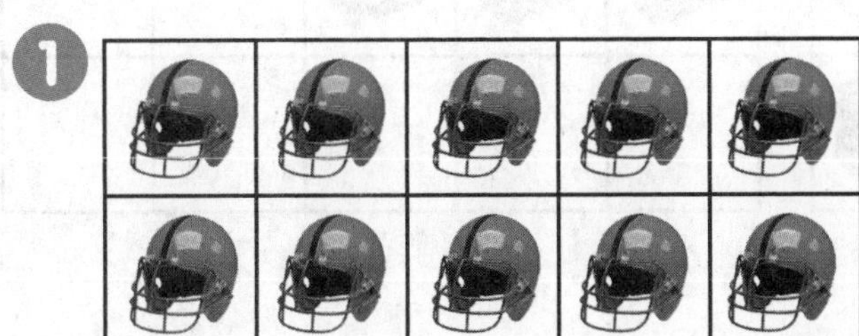

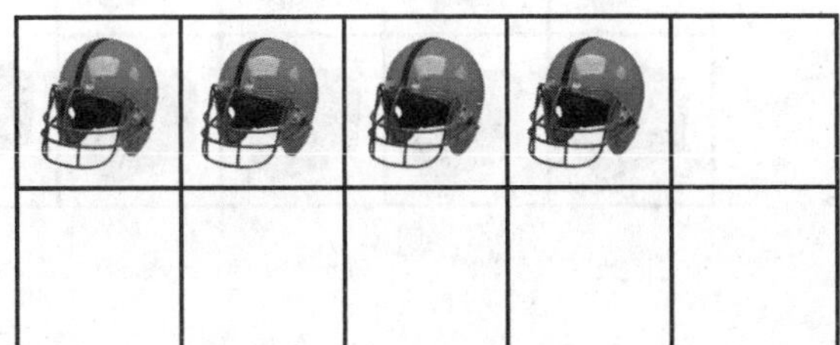

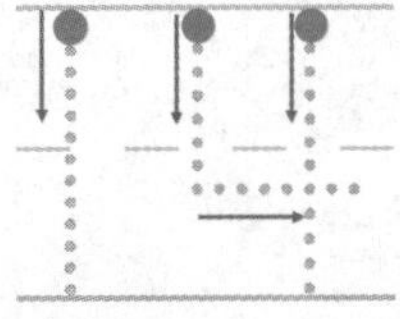

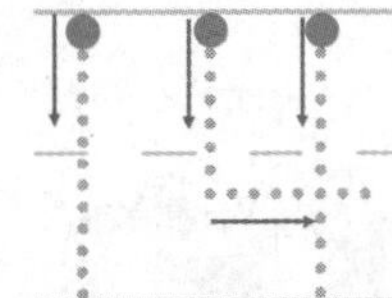

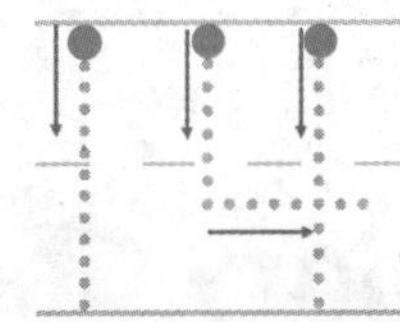

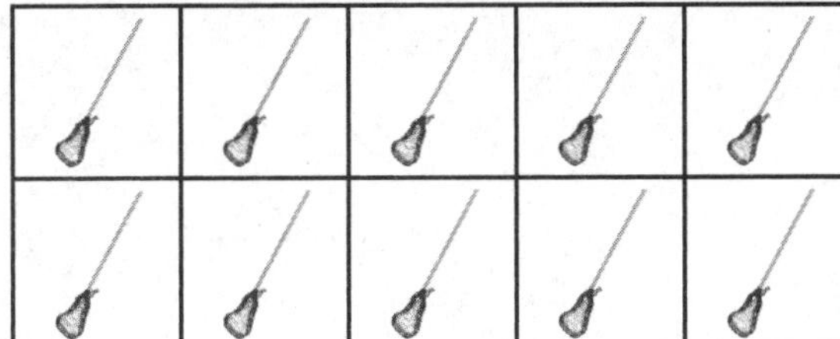

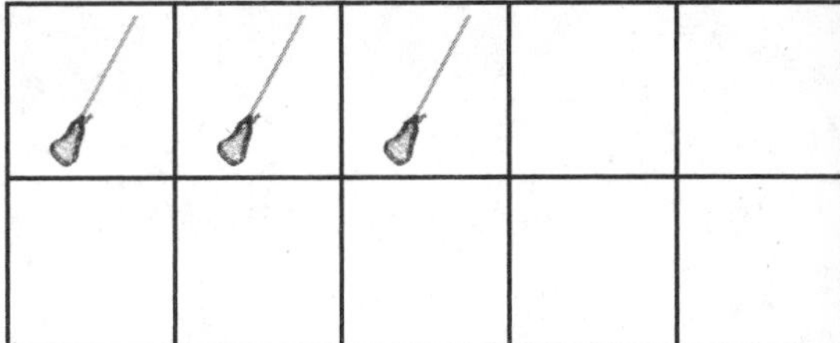

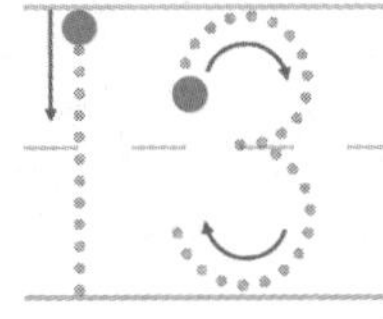

**Directions: 1–2.** Count the objects. Say how many. Trace the number. Write the number two times.

3

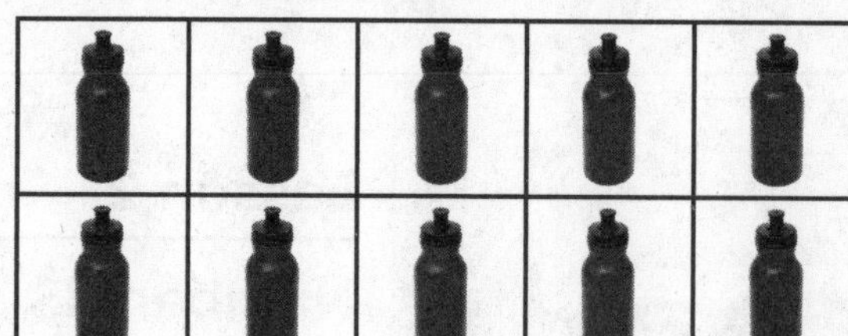

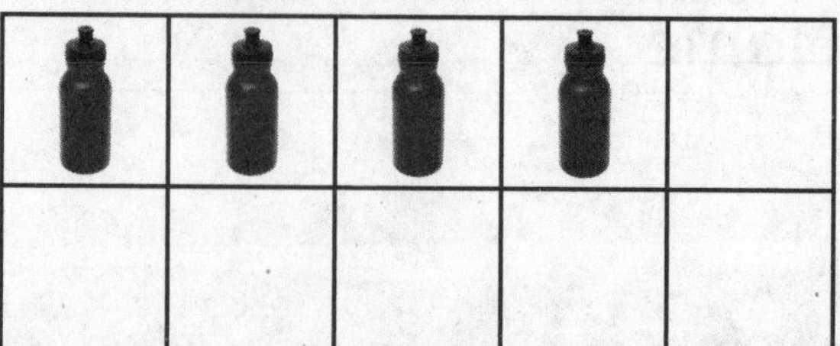

4

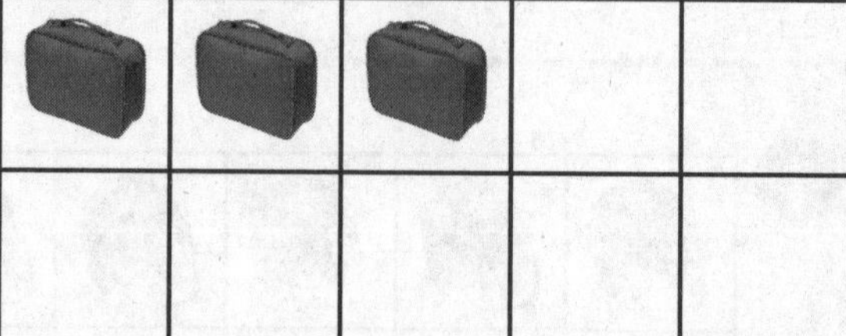

## Vocabulary Check

5 thirteen 13

6 fourteen 14

**Directions: 3–4.** Count the objects. Say how many. Write the number three times.
**5–6.** Draw circles to show the number.

**Math at Home** Have your child draw dots to create groups of 13 and 14. Then ask your child to write the corresponding numbers.

Name

Lesson 3

# Number 15

**ESSENTIAL QUESTION**

How can I show numbers beyond 10?

## Math in My World

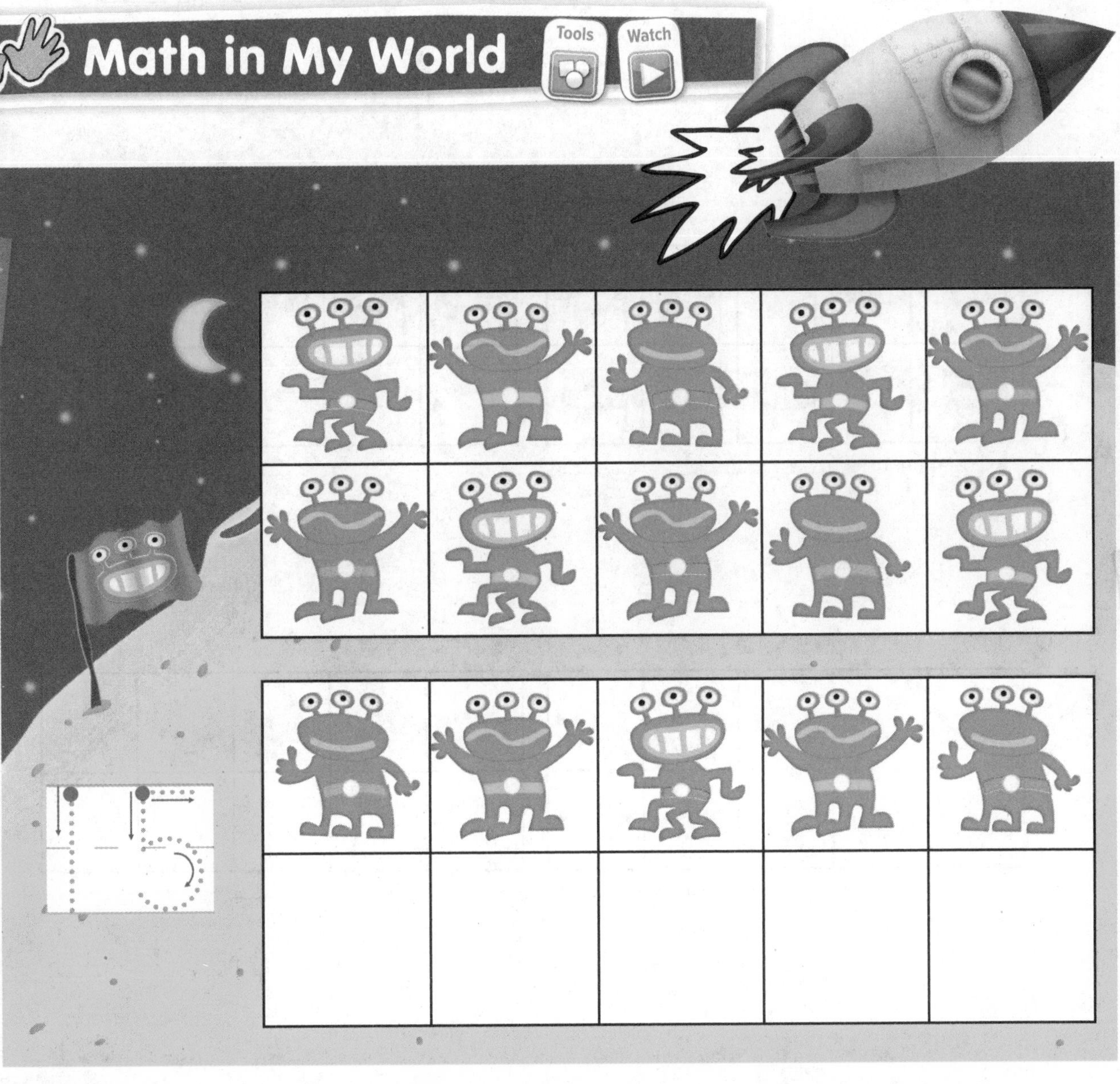

**Teacher Directions:** Place [counter] on each object. Count the objects. Say the number. Trace the number.

# Guided Practice

1 fifteen 15

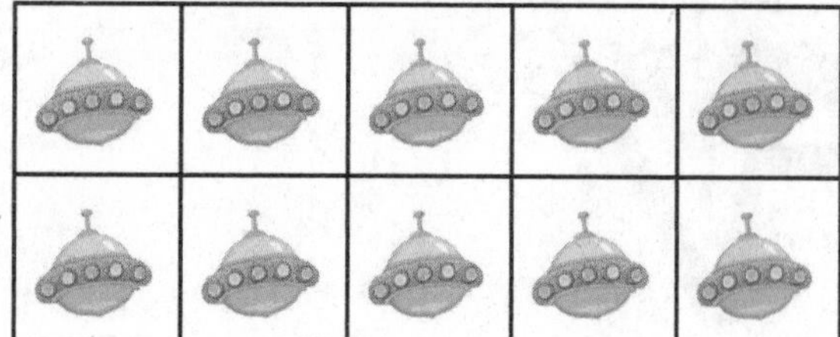

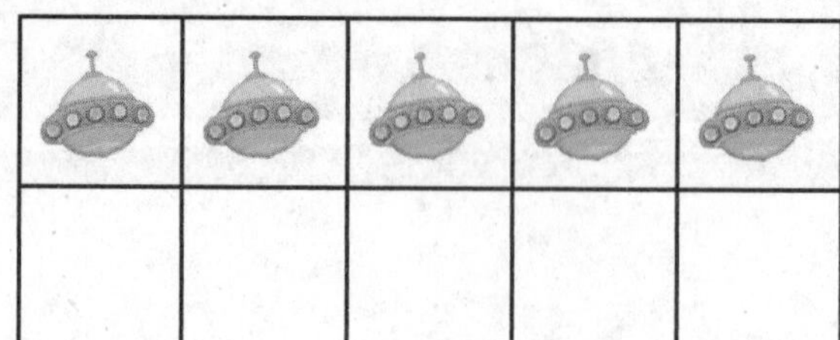

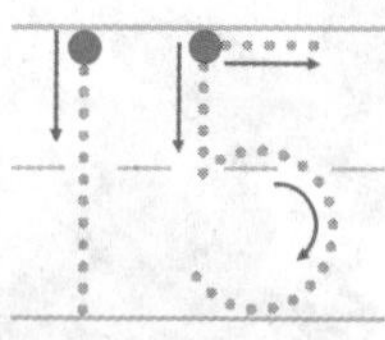

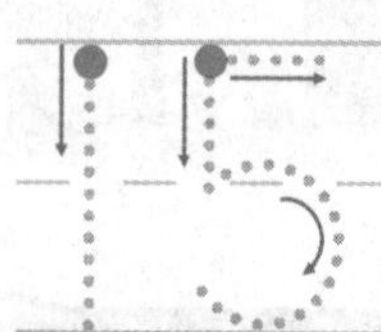

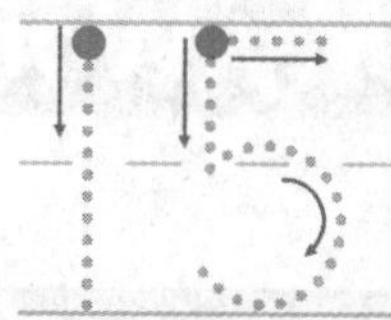

2

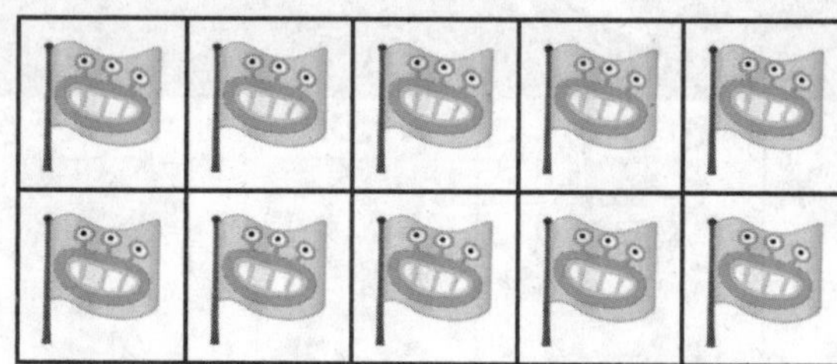

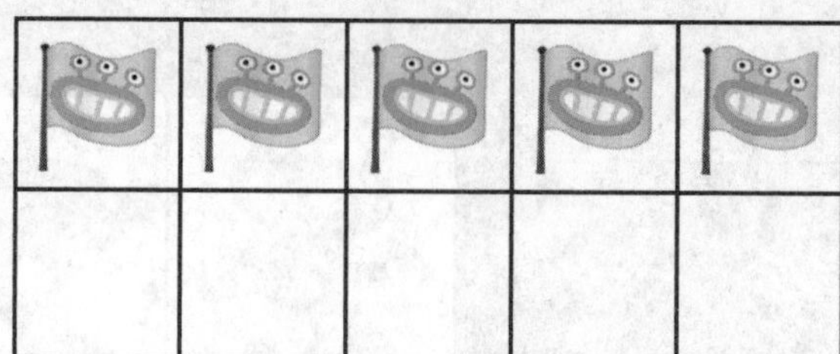

3

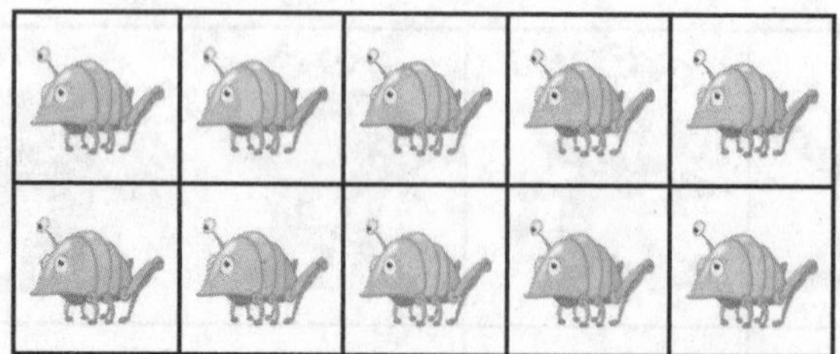

**Directions:** **I.** Count the objects. Say how many. Use Work Mat 4 and counters to show the number. Trace the number three times. **2–3.** Count the objects. Say how many. Use Work Mat 4 and counters to show the number. Write the number three times.

Name

## Independent Practice

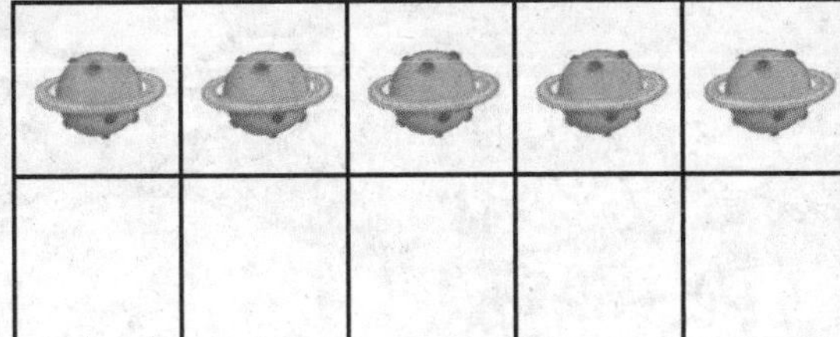

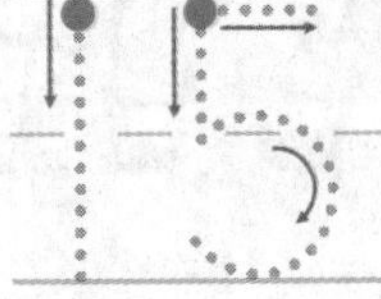

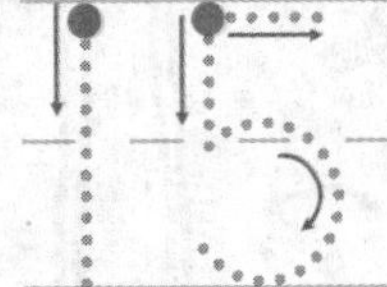

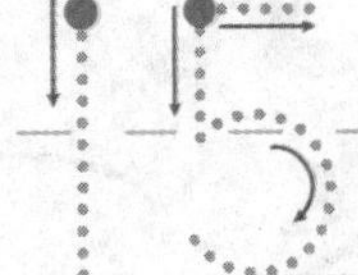

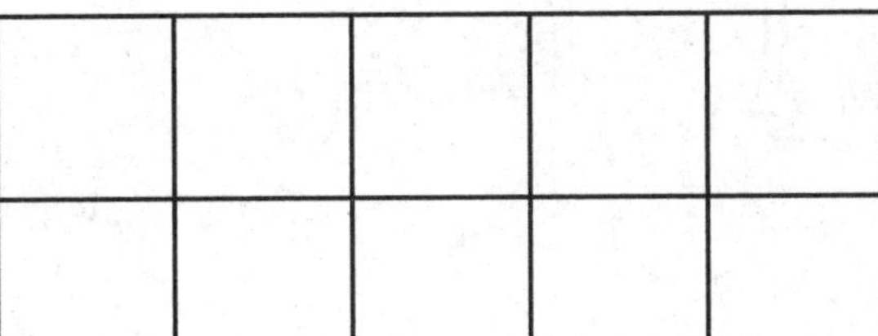

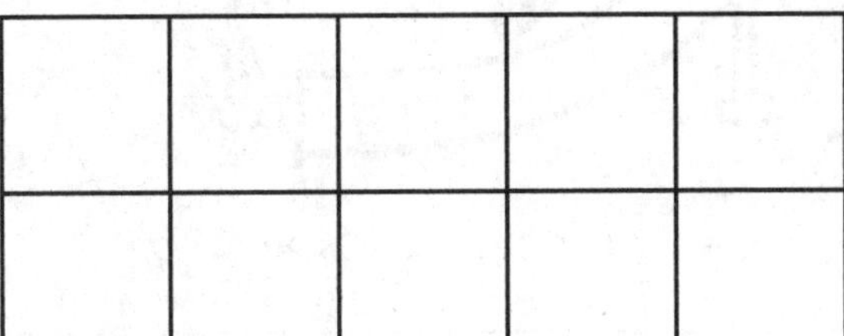

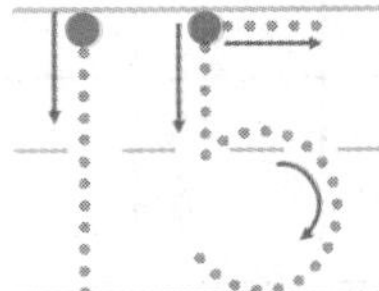

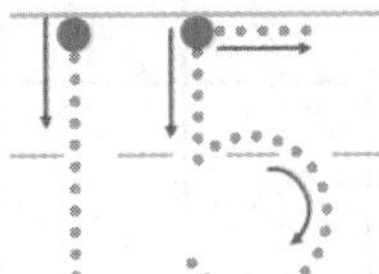

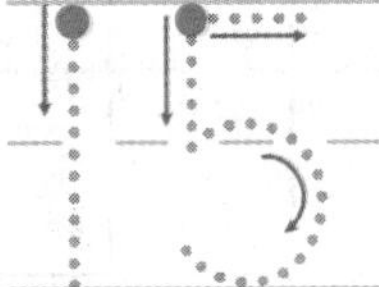

**Directions: 4.** Count the objects. Say how many. Use Work Mat 4 and counters to show the number. Write the number three times. **5–6.** Trace the number three times. Say how many. Use Work Mat 4 and counters to show the number. Draw circles for counters to show the number.

# Brain Builders

7

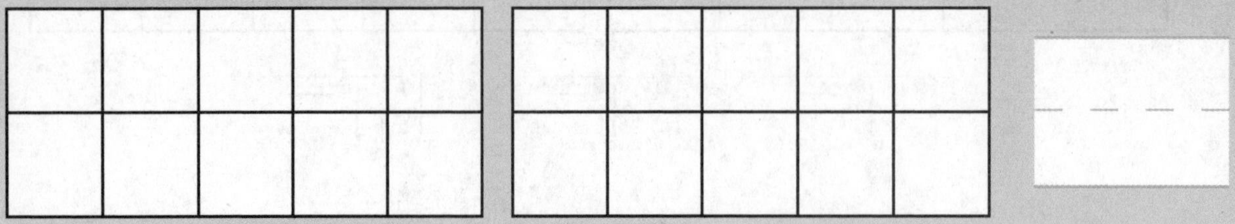

**Directions: 7.** Use Work Mat 4 and counters. Count the aliens on the page. Say how many. Draw circles in the ten-frames for every alien counted. Write the number. Tell a friend how you know there are that many.

Name

# My Homework

Lesson 3

Number 15

## Homework Helper

Need help? connectED.mcgraw-hill.com

1

15 15 15

2

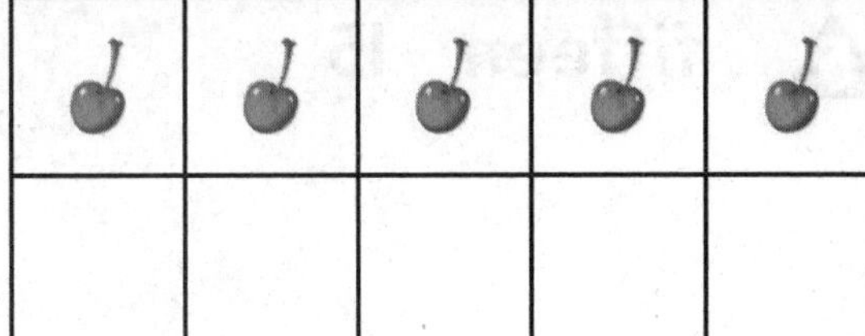

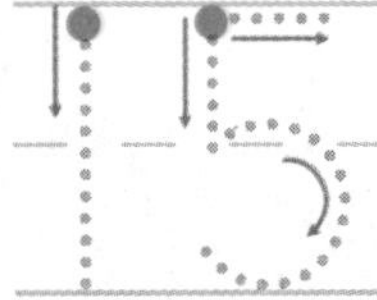

**Directions: 1–2.** Count the objects. Say how many. Trace the number. Write the number two times.

3

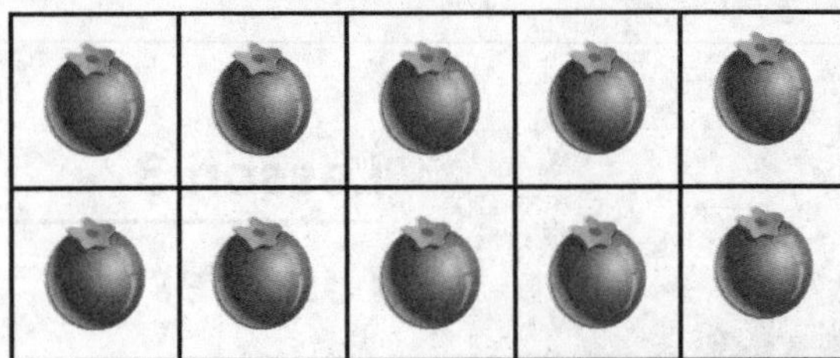
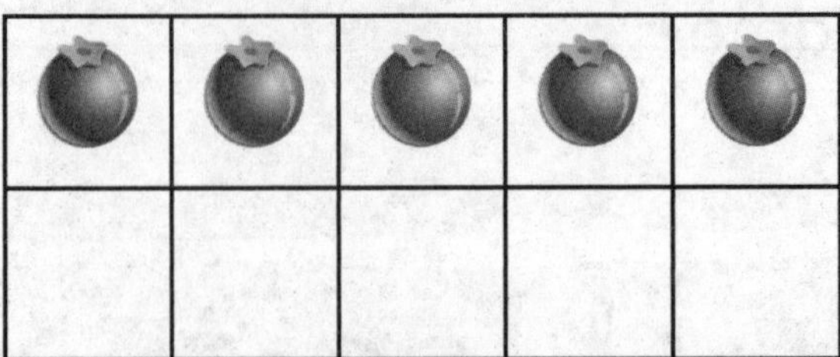

4

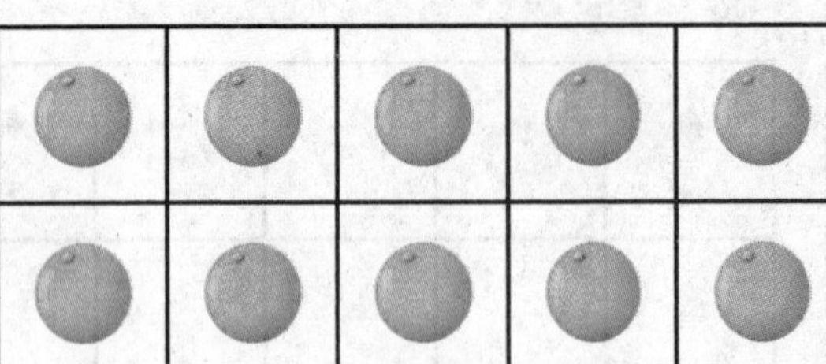
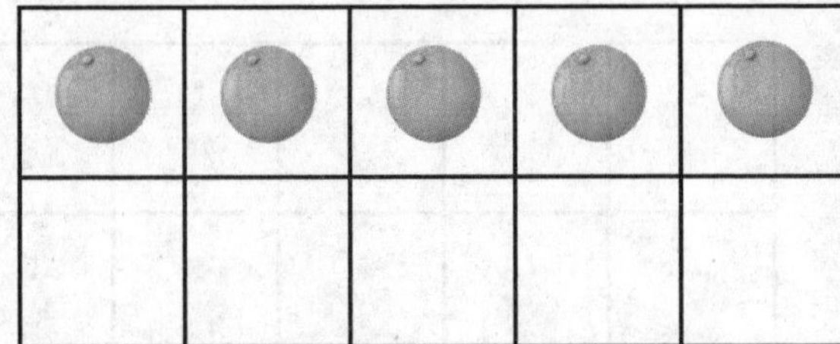

## Vocabulary Check

 fifteen 15

**Directions: 3–4.** Count the objects. Say how many. Write the number three times.
**5.** Draw boxes to show the number.

**Math at Home** Show your child objects such as pennies or dry cereal. Have your child count the objects to create a group of 15.

Name

Lesson 4

# Numbers 16 and 17

**ESSENTIAL QUESTION**
How can I show numbers beyond 10?

## Math in My World

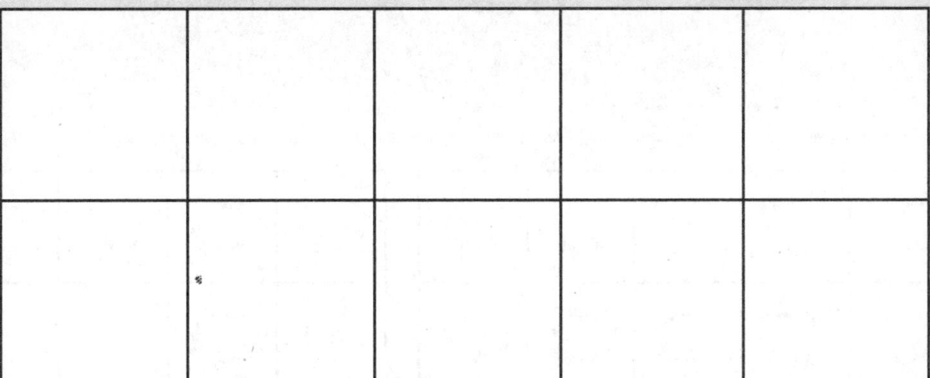

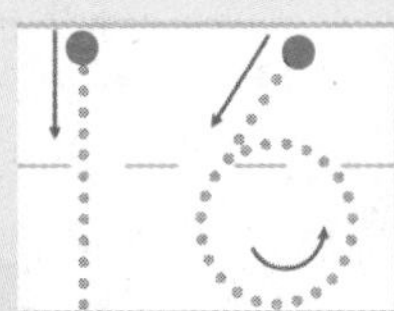

**Teacher Directions:** Place counters in the ten-frames to show 16 and 17. Count and say the number of objects. Draw circles in the ten-frames to show 16. Trace the number.

## Guided Practice

Processes & Practices

1 sixteen 16

2 seventeen 17

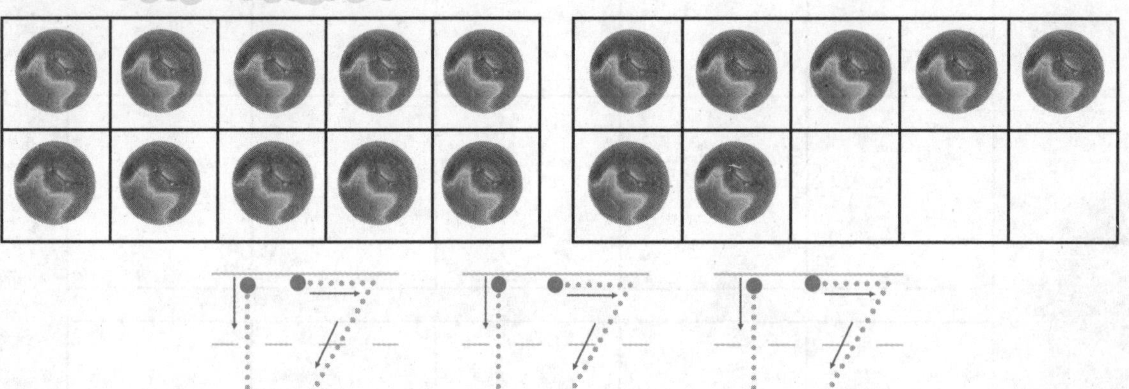

3

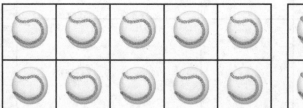

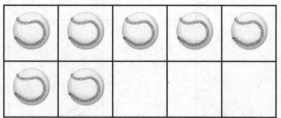

**Directions: 1-2.** Count the objects. Say how many. Use Work Mat 4 and counters to show the number. Trace the number three times. **3.** Count the objects. Say how many. Use Work Mat 4 and counters to show the number. Write the number three times.

Name

## Independent Practice

4

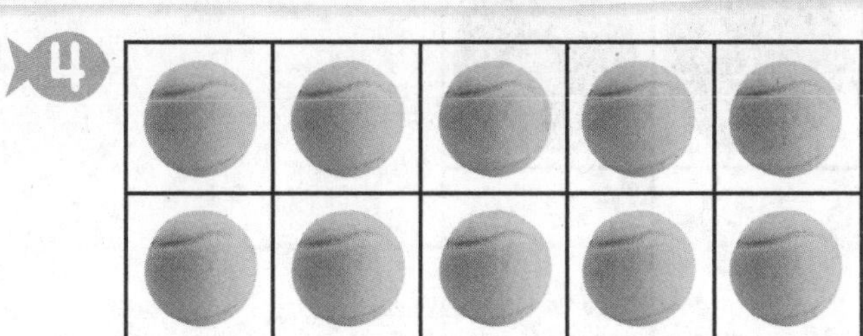
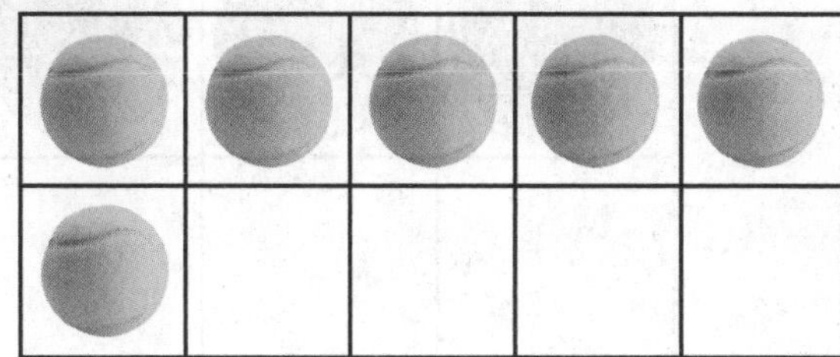

5

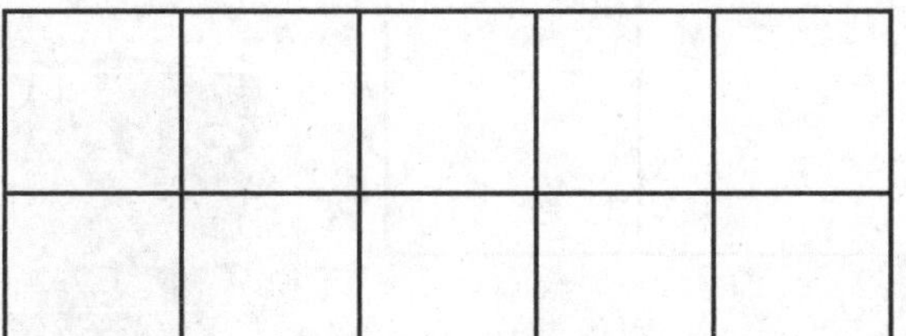
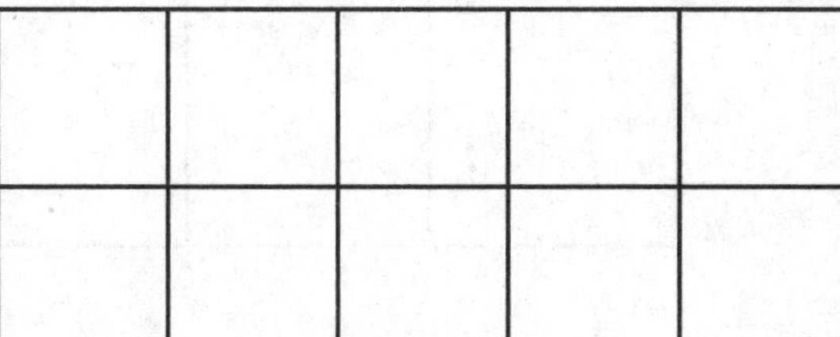

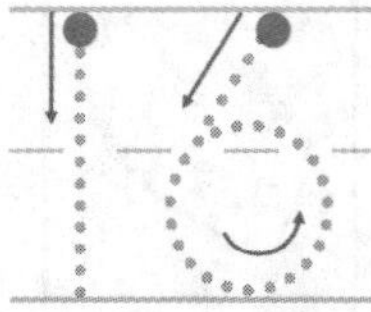
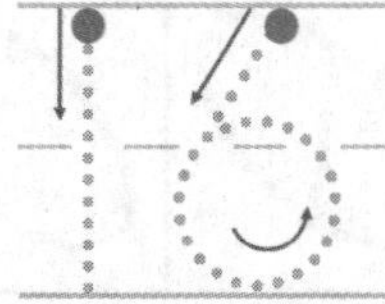
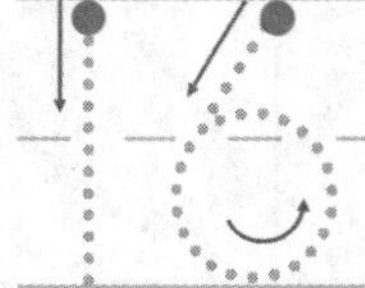

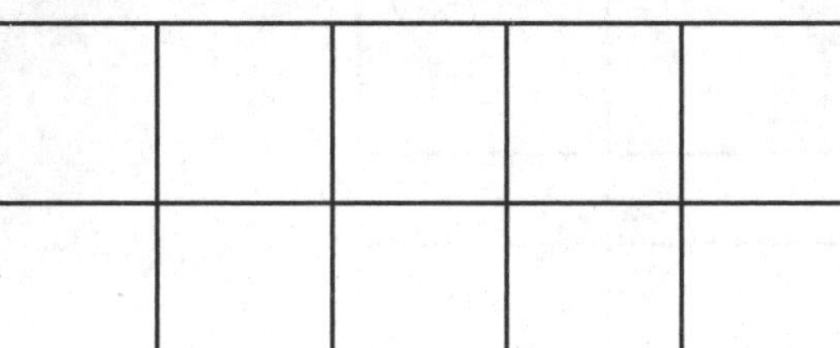
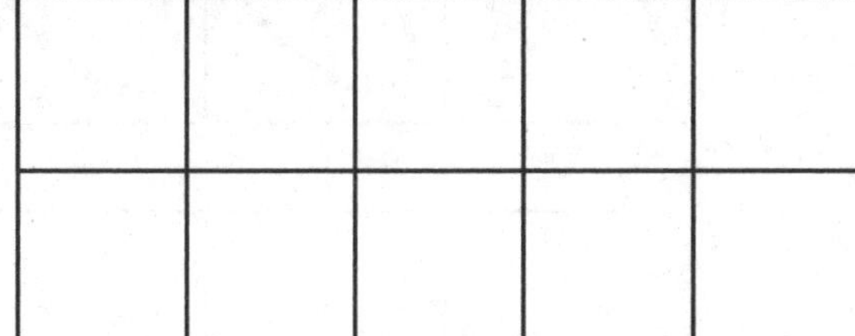

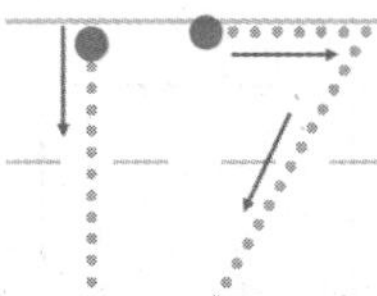
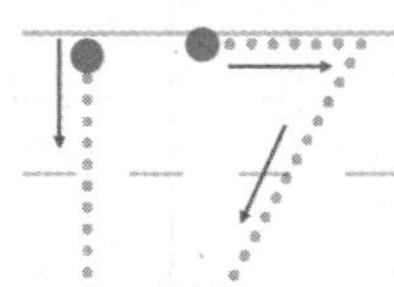
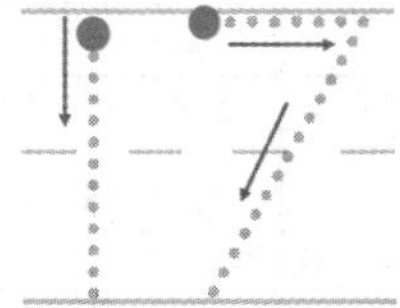

**Directions: 4.** Count the objects. Say how many. Use Work Mat 4 and counters to show the number. Write the number three times. **5–6.** Trace the number three times. Say how many. Use Work Mat 4 and counters to count the number. Draw circles for counters to show the number.

# Brain Builders

7

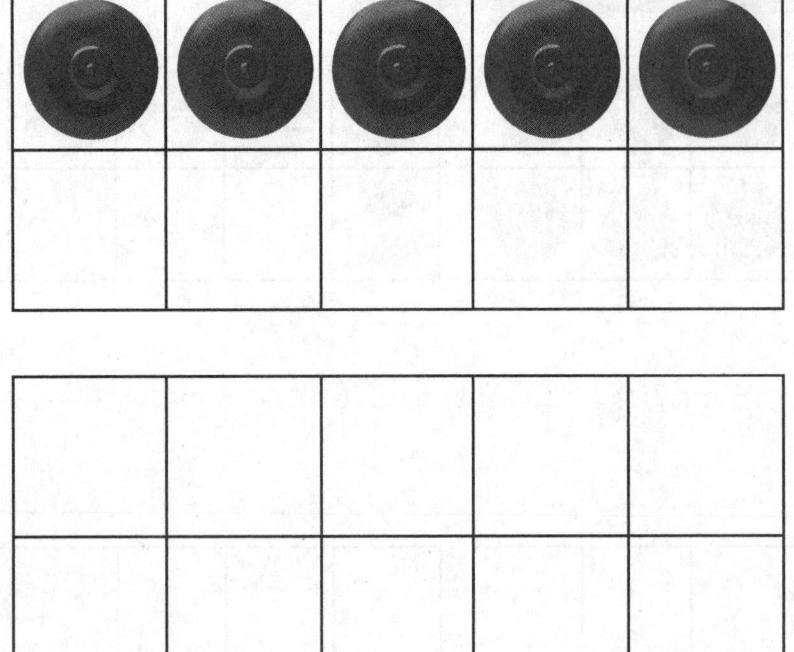

8

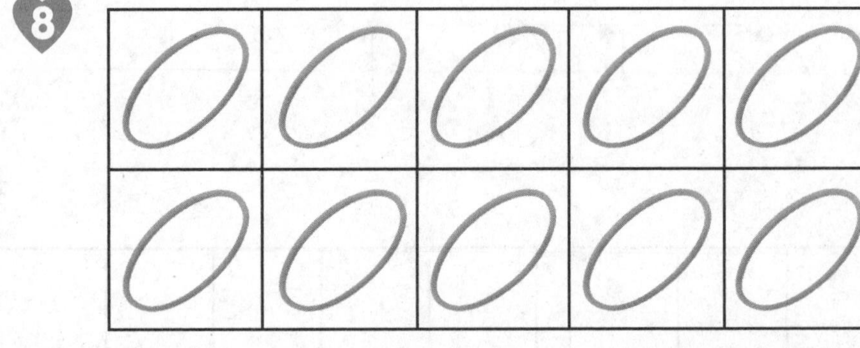

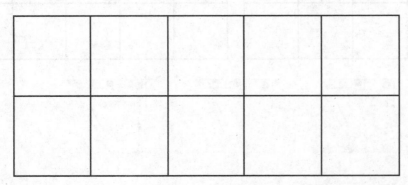

**Directions: 7.** Count the objects. Say how many. Draw more to show sixteen. Count with a friend to check the number of objects you drew. **8.** Count the objects. Say how many. Draw more to show seventeen. Count with a friend to check the number of objects you drew.

Name

# My Homework

Lesson 4

Numbers 16 and 17

## Homework Helper

Need help? connectED.mcgraw-hill.com

1

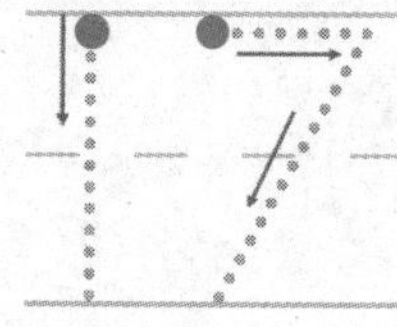

17 17

2

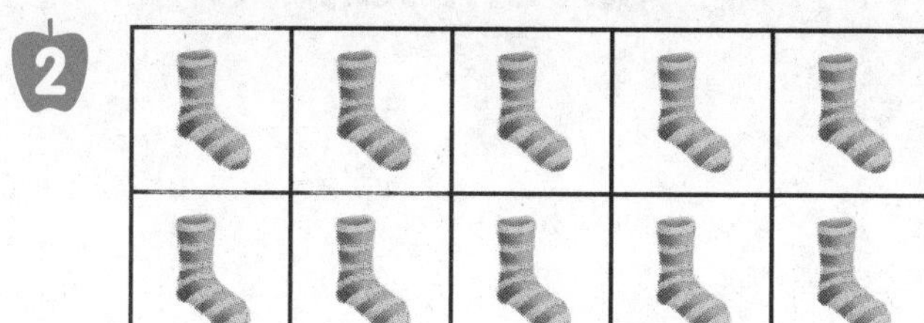

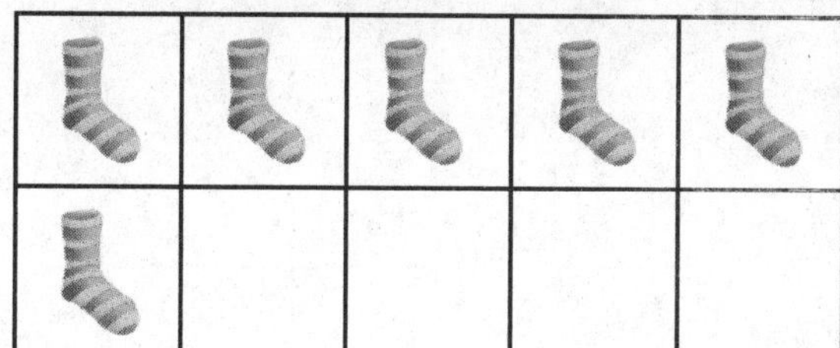

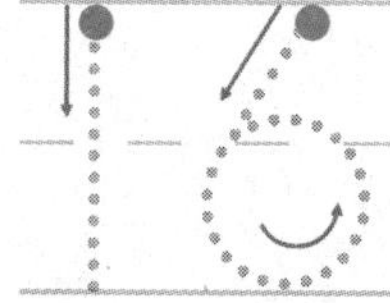

**Directions: 1–2.** Count the objects. Say how many. Trace the number. Write the number two times.

3

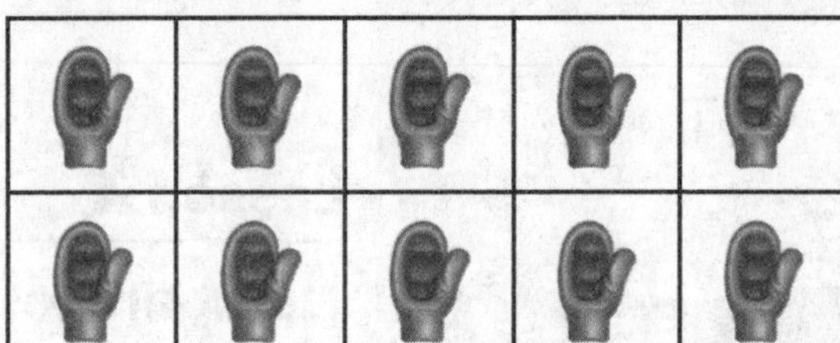

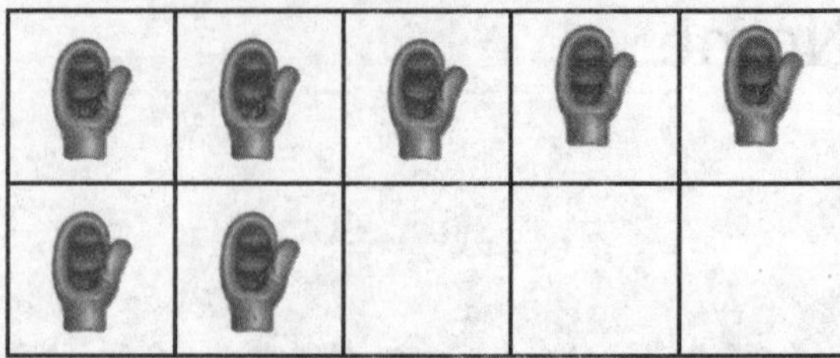

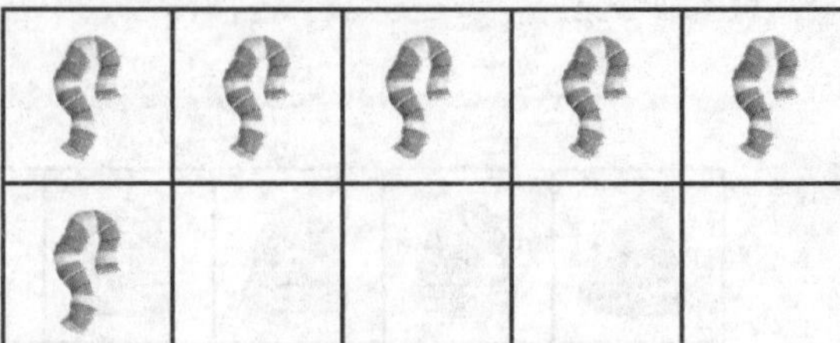

## Vocabulary Check

5 **sixteen 16**

**seventeen 17**

**Directions: 3–4.** Count the objects. Say how many. Write the number three times.
**5–6.** Draw stars to show the number.

**Math at Home** Ask your child to create groups of 16 and 17 using dry pasta. Then have him or her write 16 or 17 next to the corresponding group.

Name

# Check My Progress

## Vocabulary Check

2 twelve 12

## Concept Check

3

**Directions:** **1.** Draw more to show eleven. **2.** Draw more to show twelve. **3.** Count the objects. Say how many. Write the number three times.

4

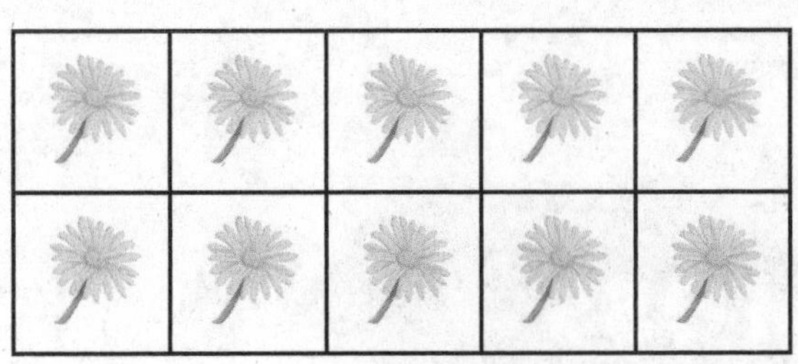

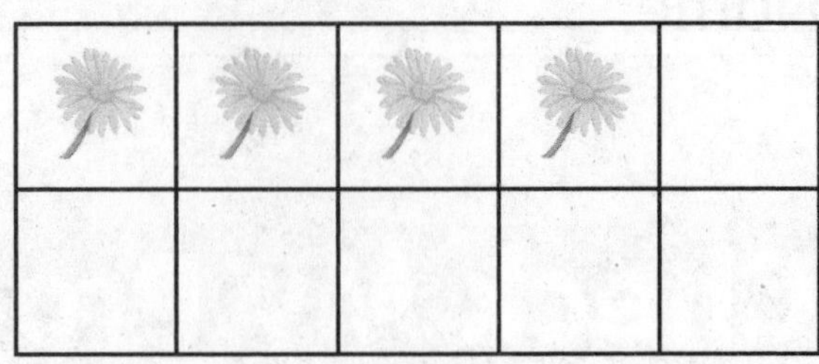

5

6

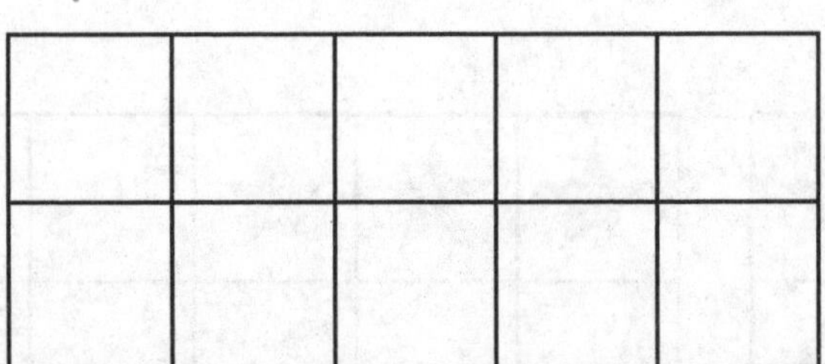

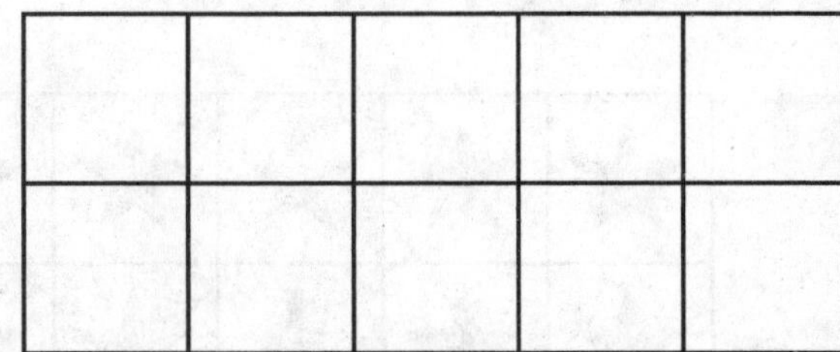

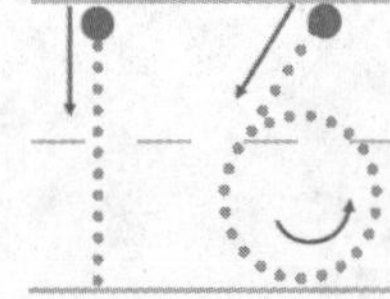

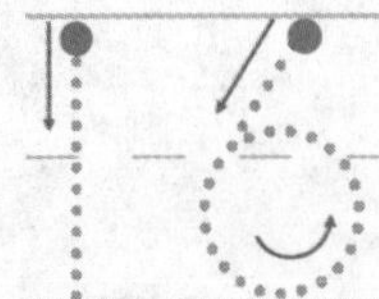

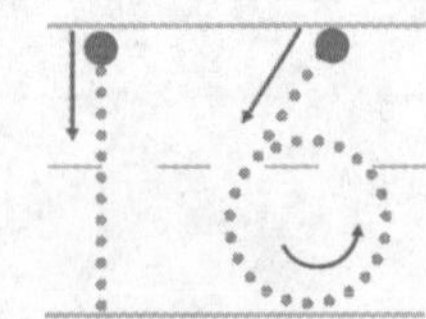

**Directions: 4–5.** Count the objects. Say how many. Write the number three times.
**6.** Trace the number. Say how many. Draw circles for counters to show the number.

Name

Lesson 5

# Numbers 18 and 19

**ESSENTIAL QUESTION**
How can I show numbers beyond 10?

## Math in My World

|  |  |  |  |  |
|---|---|---|---|---|
|  |  |  |  |  |
|  |  |  |  |  |

|  |  |  |  |  |
|---|---|---|---|---|
|  |  |  |  |  |
|  |  |  |  |  |

19

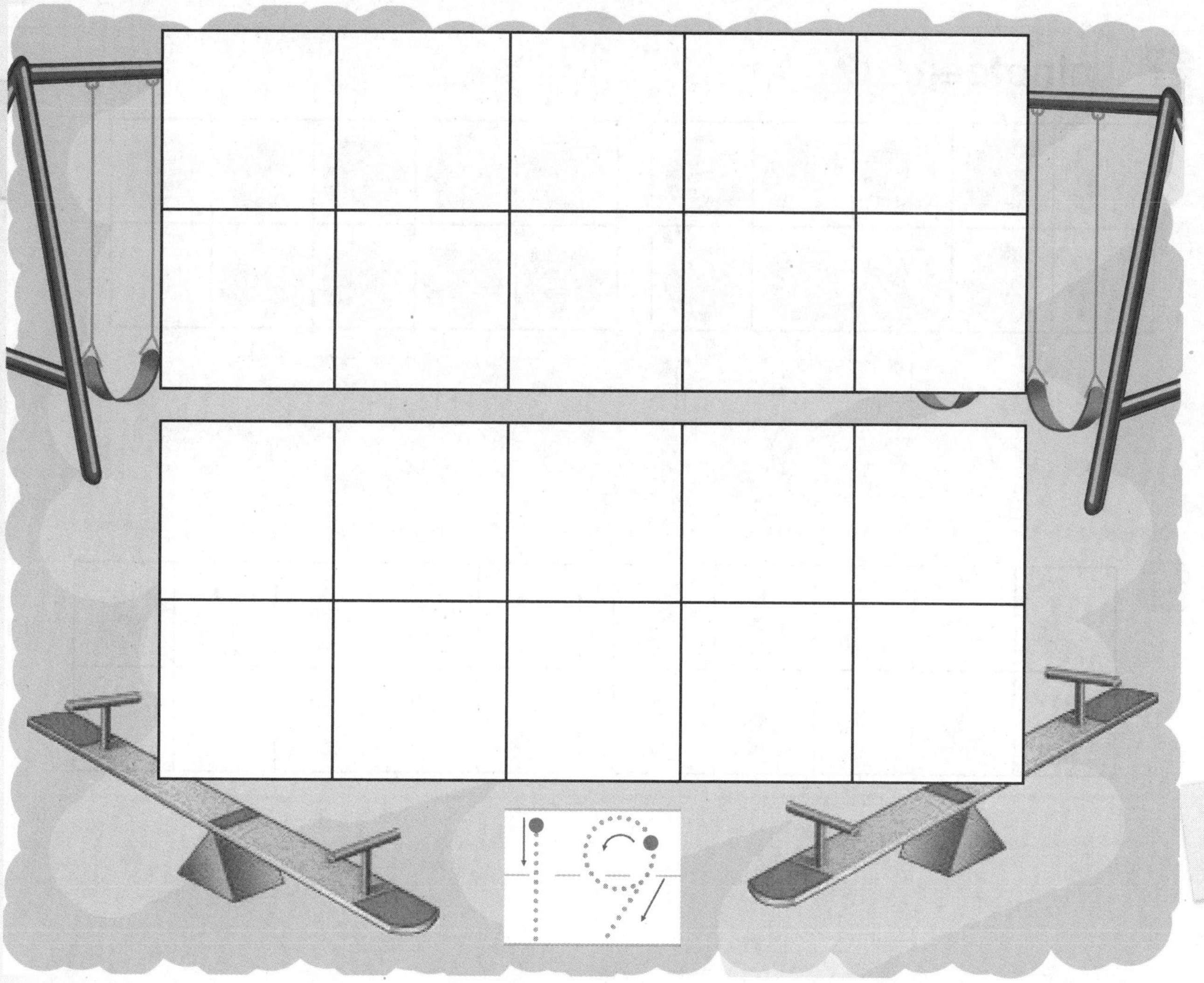

**Teacher Directions:** Place counters in the ten-frames to show numbers 18 and 19. Say how many. Trace the number. Trace the counters to show 19.

Processes & Practices

**1** eighteen 18

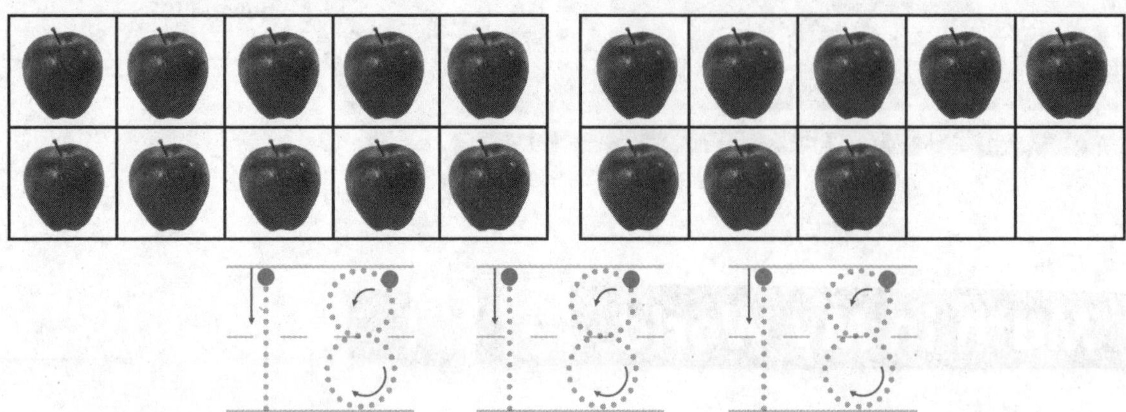

**2** nineteen 19

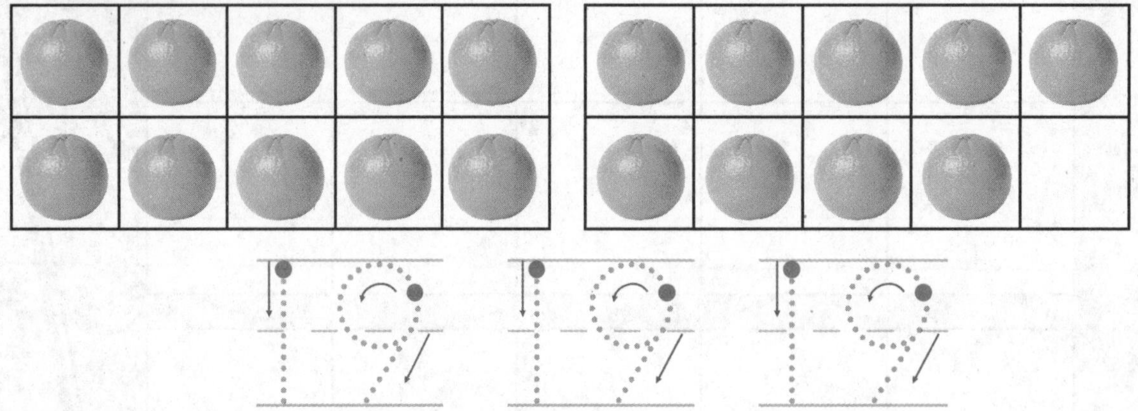

**3**

**Directions: 1–2.** Count the objects. Say how many. Use Work Mat 4 and counters to show the number. Trace the number three times. **3.** Count the objects. Say how many. Use Work Mat 4 and counters to show the number. Write the number three times.

Name

## Independent Practice

4

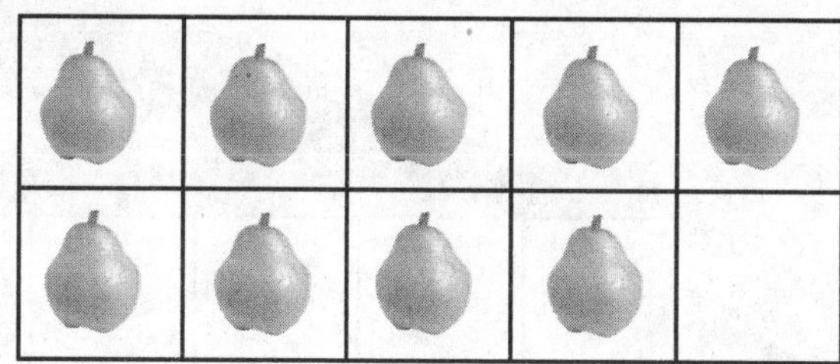

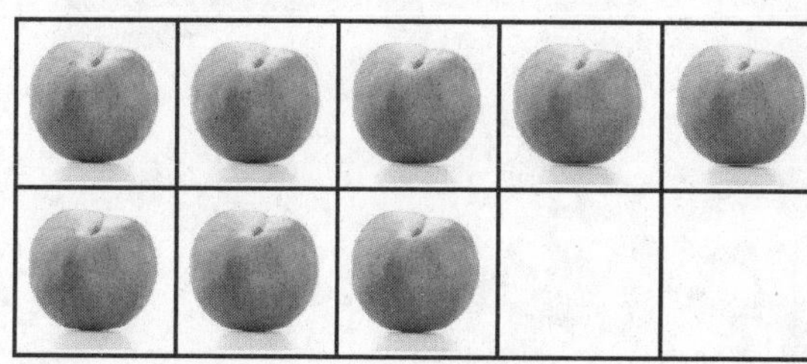

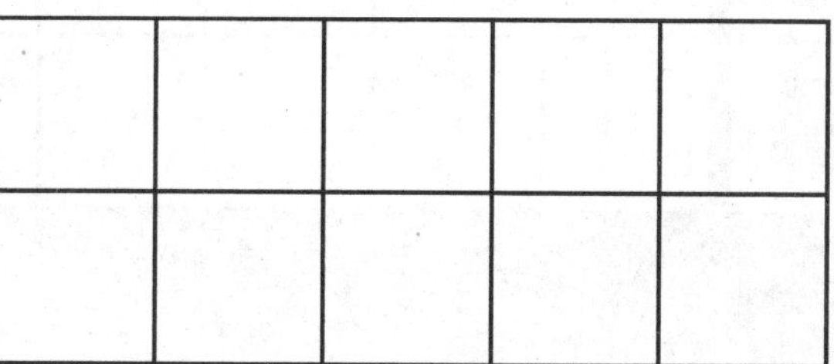

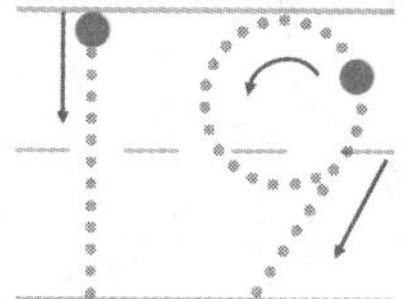
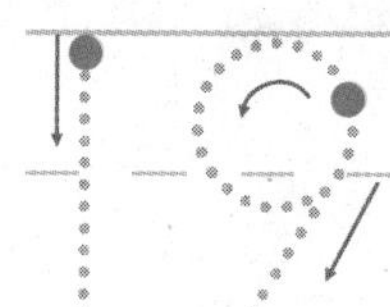
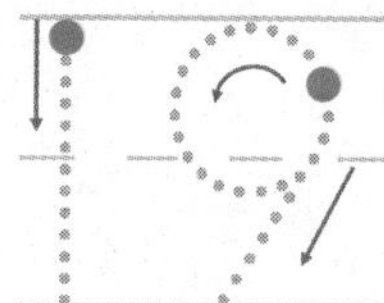

**Directions: 4–5.** Count the objects. Say how many. Use Work Mat 4 and counters to show the number. Write the number three times. **6.** Trace the number three times. Say how many. Use Work Mat 4 and counters to show the number. Draw circles for counters to show the number.

7

19

8

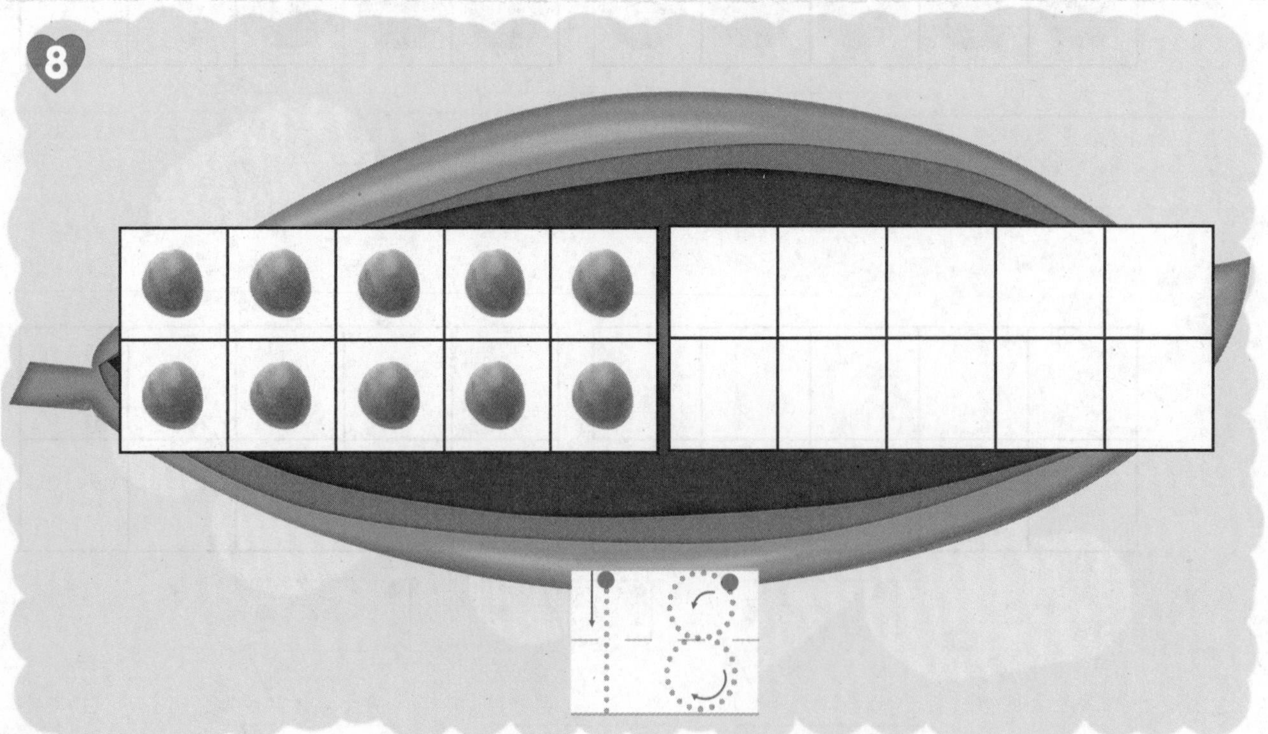

18

**Directions: 7.** Trace the number. Draw more in the ten-frame to show the number. **8.** Trace the number. Draw more in the ten-frame to show the number. Tell a friend if there are more pieces of corn or more pieces of peas in the ten-frames. Explain how you know.

Name ..............................

# My Homework

**Lesson 5**

**Numbers 18 and 19**

## Homework Helper

Need help? connectED.mcgraw-hill.com

1

19 19 19

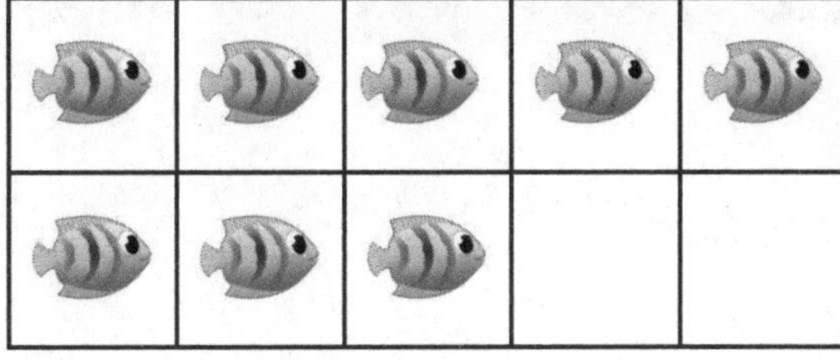

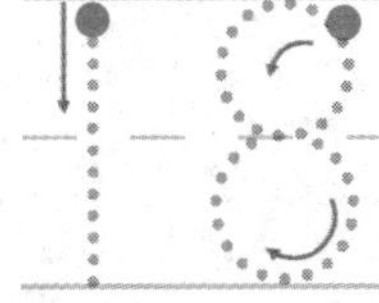

**Directions: 1–2.** Count the objects. Say how many. Trace the number. Write the number two times.

3

## Vocabulary Check

5 **eighteen 18**

6 **nineteen 19**

**Directions: 3–4.** Count the objects. Say how many. Write the number three times. **5–6.** Draw fish to show the number.

**Math at Home** Draw a piggy bank on a piece of paper. Count pennies in groups of 18 and 19. Place pennies on the piggy bank. Write the number. Repeat for other numbers.

Name

Lesson 6
# Number 20

**ESSENTIAL QUESTION**
**How can I show numbers beyond 10?**

## Math in My World

| | | | | |
|---|---|---|---|---|
| | | | | |
| | | | | |

| | | | | |
|---|---|---|---|---|
| | | | | |
| | | | | |

CITY PARK

20

**Teacher Directions:** Count the ladybugs going to the party. Use ● in the first ten-frame to show this number. Count the cupcakes on the table. Fill in the second ten-frame to show this number. Trace the number. Color the boxes to show 20.

## Guided Practice



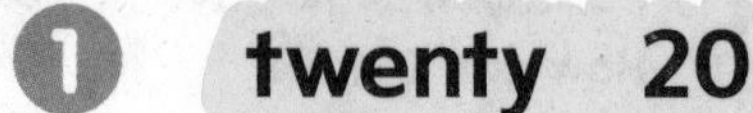

1 twenty 20

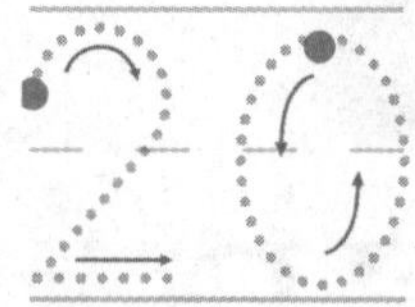
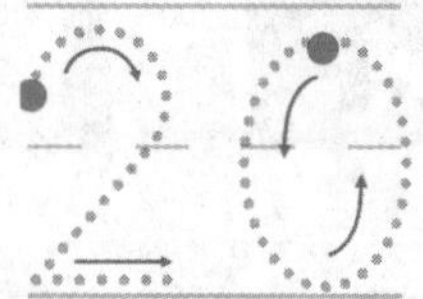
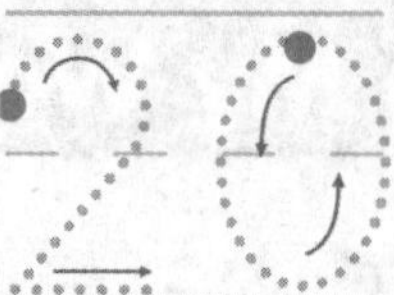

2

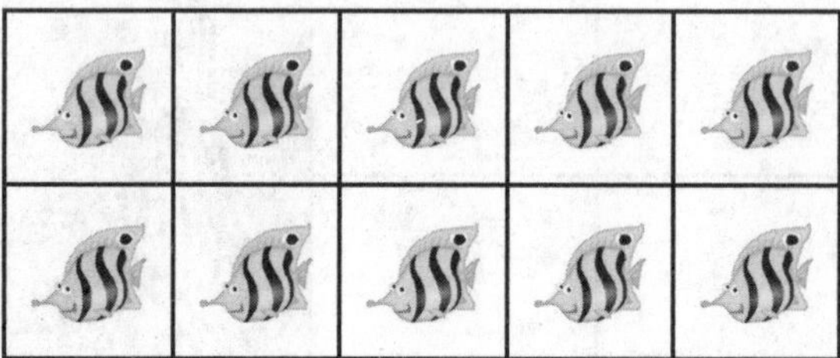
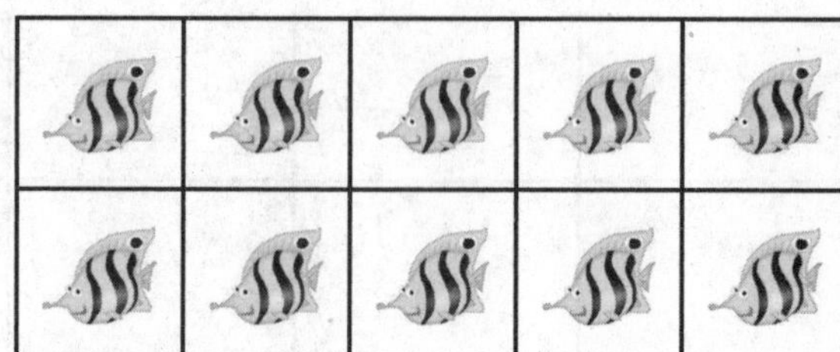

3

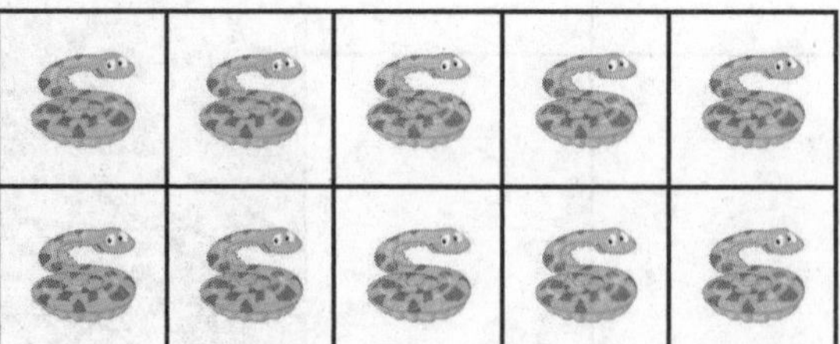
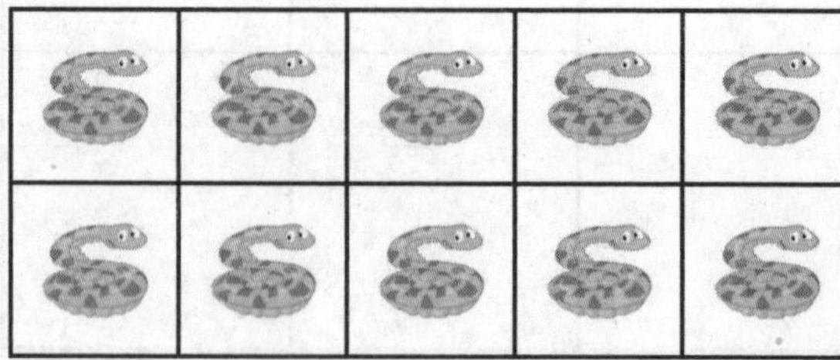

**Directions: 1.** Count the plants. Say how many. Use Work Mat 4 and counters to show the number. Trace the number three times. **2–3.** Count the animals. Say how many. Use Work Mat 4 and counters to show the number. Write the number three times.

Name

## Independent Practice

4

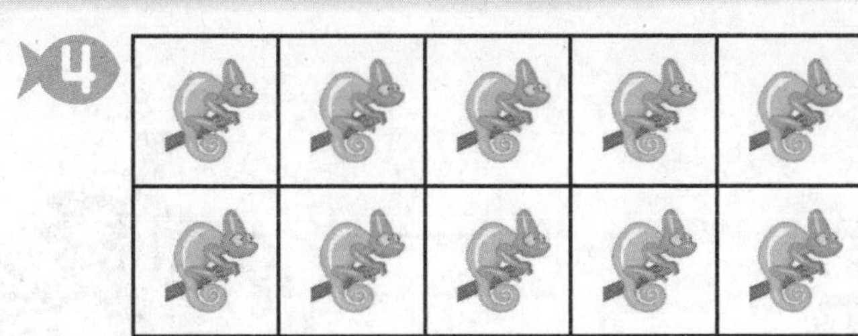
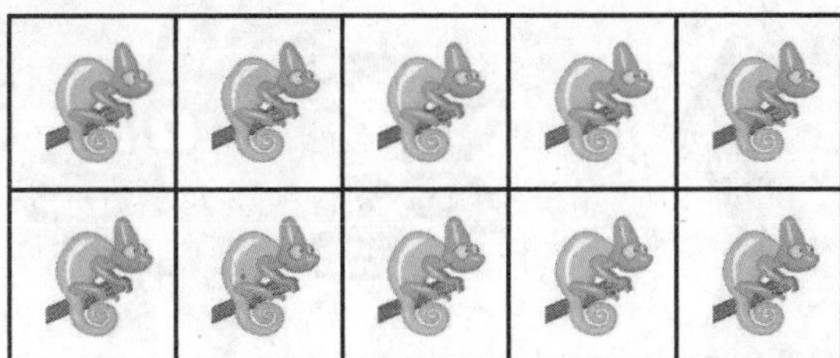

5

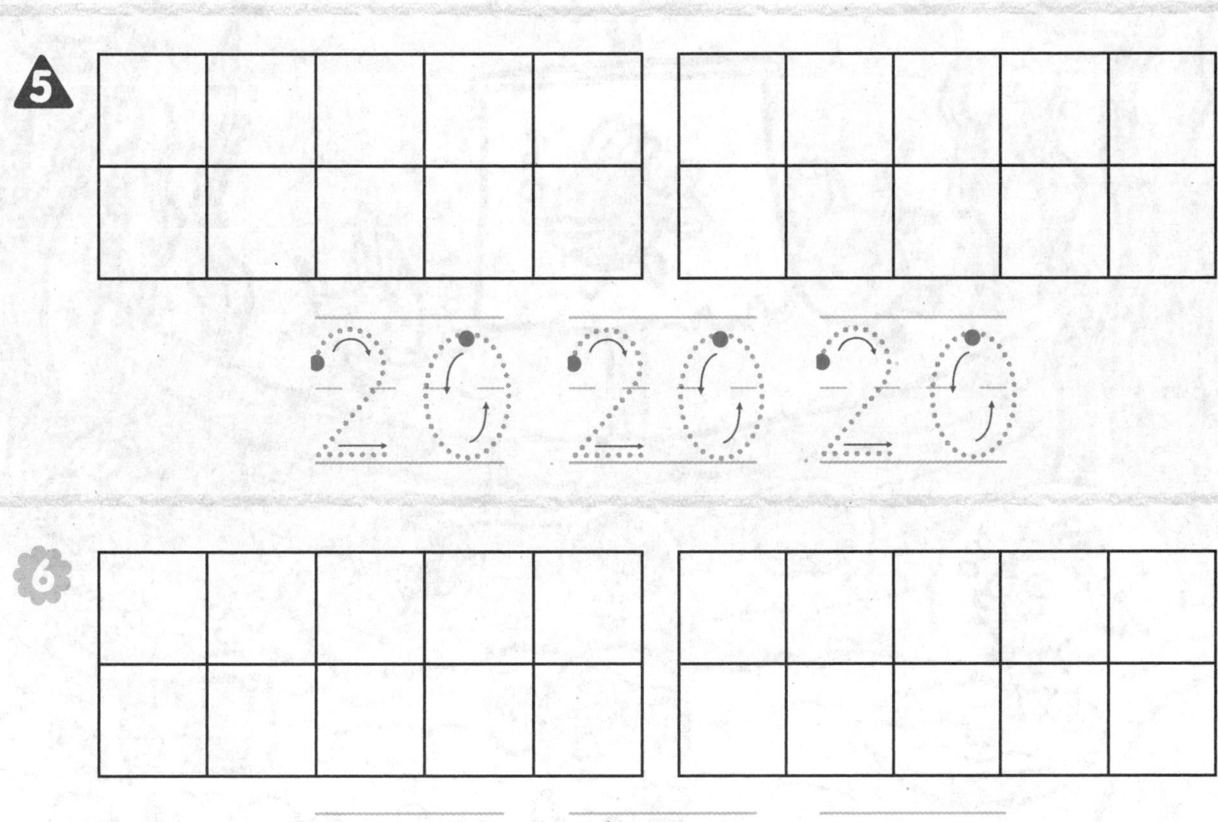

6

**Directions: 4.** Count the objects. Say how many. Use Work Mat 4 and counters to show the number. Write the number three times. **5–6.** Trace the number three times. Say how many. Use Work Mat 4 and counters to count the number. Draw circles for counters to show the number.

## Brain Builders

7

**Directions:** 7. Count the fish. Say how many. Draw more fish to show twenty. Write the number. Tell a friend if 20 is more or less than 10. Tell how you know.

Name

# My Homework

Lesson 6

Number 20

## Homework Helper

Need help? connectED.mcgraw-hill.com

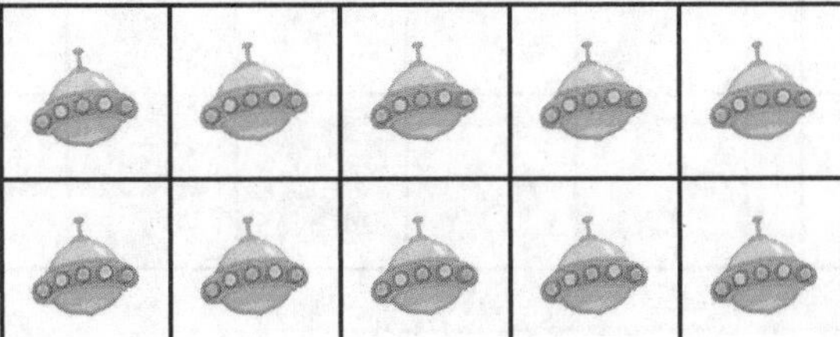
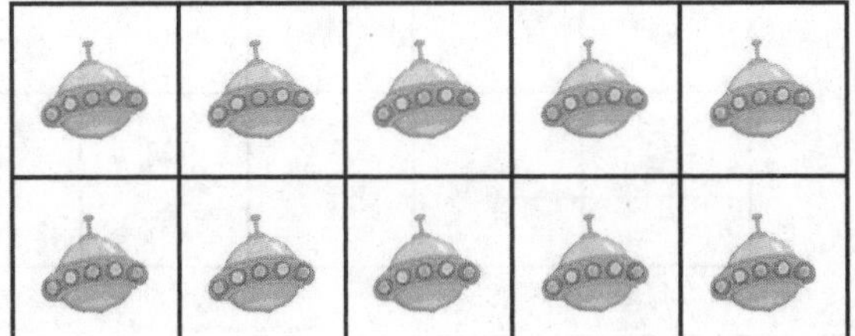

20 20 20

2

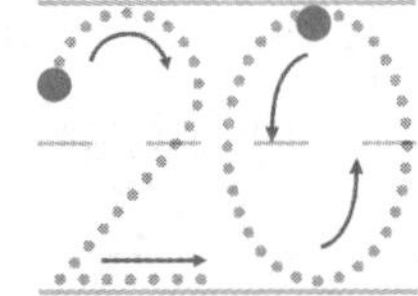

**Directions: 1–2.** Count the objects. Say how many. Trace the number. Write the number two times.

3

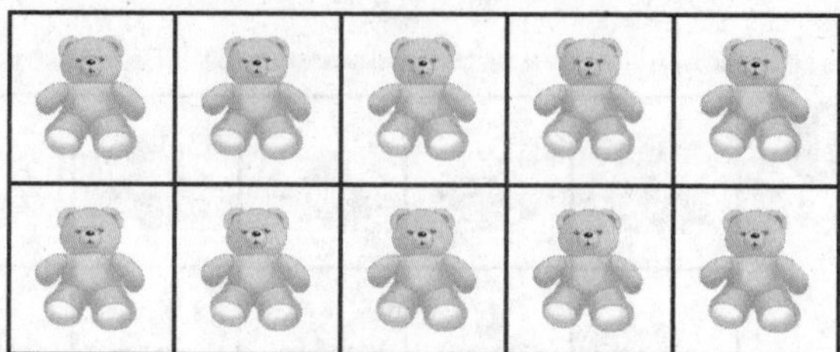

## Vocabulary Check

twenty 20

**Directions: 3–4.** Count the objects. Say how many. Write the number three times. **5.** Draw baseballs to show twenty.

**Math at Home** Have your child gather 10 items as you gather 10 items. Put the objects together to create a group of 20. Have your child count the objects.

Name

Lesson 7

# Problem Solving

**STRATEGY: Draw a Diagram**

**ESSENTIAL QUESTION**
How can I show numbers beyond 10?

## How many gumballs?

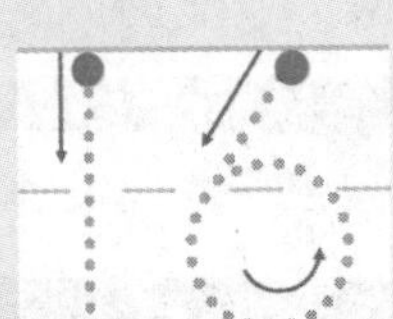


## Draw to Solve

**Directions:** Say and trace the number. Count the gumballs. Use cubes to show the number. At the bottom, draw an X for every gumball counted. Write the number.

## How many fish?

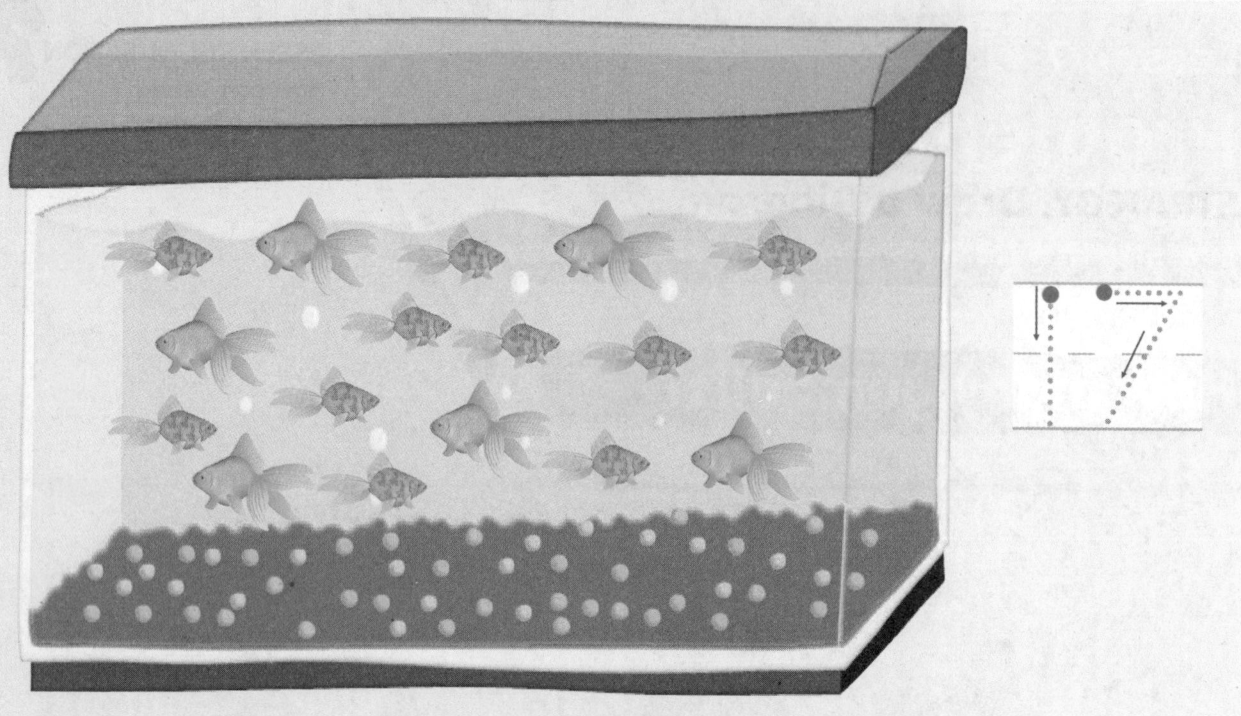

## Draw to Solve

**Directions:** Say and trace the number. Count the fish. Use cubes to show the number. At the bottom, draw an X for every fish counted. Write the number.

Name

## How many spots?

## Draw to Solve

**Directions:** Say and trace the number. Count the spots. Use cubes to show the number. At the bottom, draw an X for every spot counted. Write the number. Cross out 4 of the spots. Tell a friend how many spots there are now.

## How many bees?

## Draw to Solve

**Directions:** Say and trace the number. Count the bees. Use cubes to show the number. At the bottom, draw an X for every bee counted. Write the number.

Name

# My Homework

Lesson 7

Problem Solving: Draw a Diagram

## How many balls?

## Draw to Solve

**Directions:** Say and trace the number. Count the balls. At the bottom, draw that number of balls. Write the number.

## How many magnets?

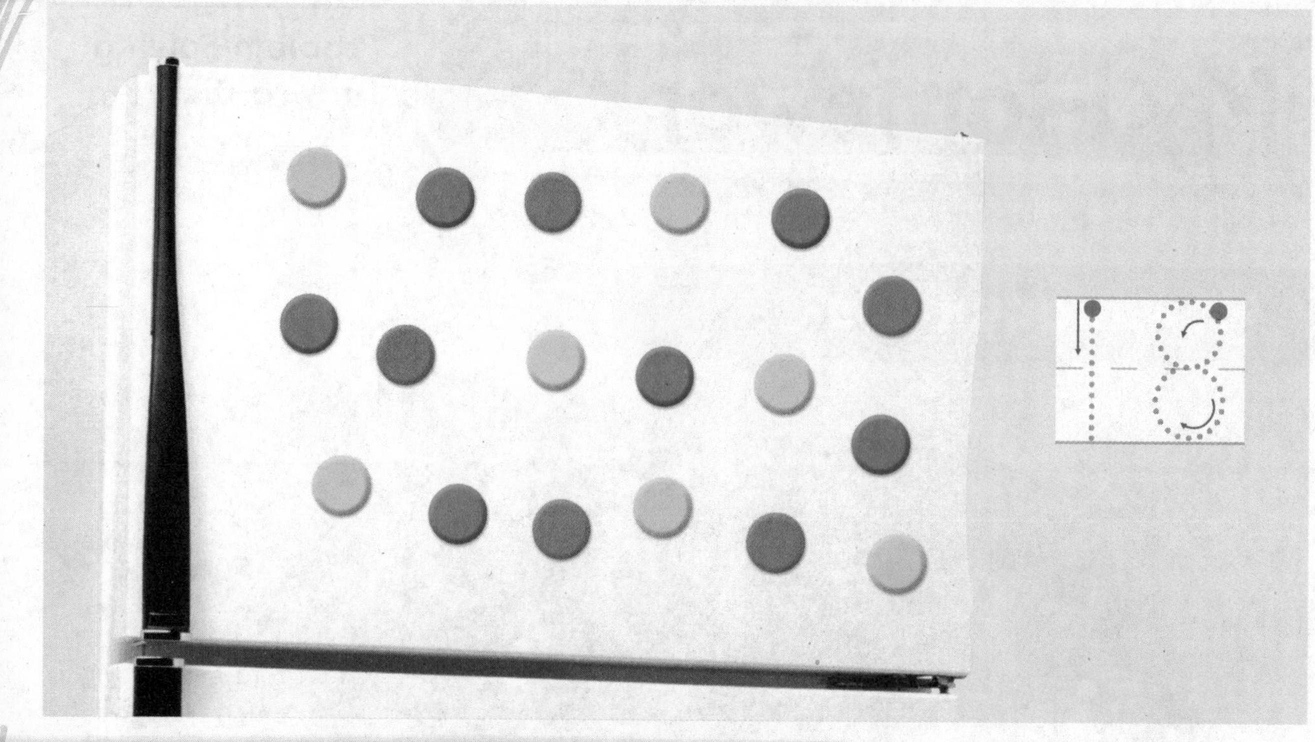


## Draw to Solve

**Directions:** Say and trace the number. Count the magnets. At the bottom, draw that number of magnets. Write the number.

**Math at Home** Take advantage of problem-solving opportunities during daily routines such as going to the grocery store. Have your child help you make a grocery list by drawing pictures of the grocery items needed. Have him or her count the items.

Name

# Check My Progress

## Vocabulary Check 

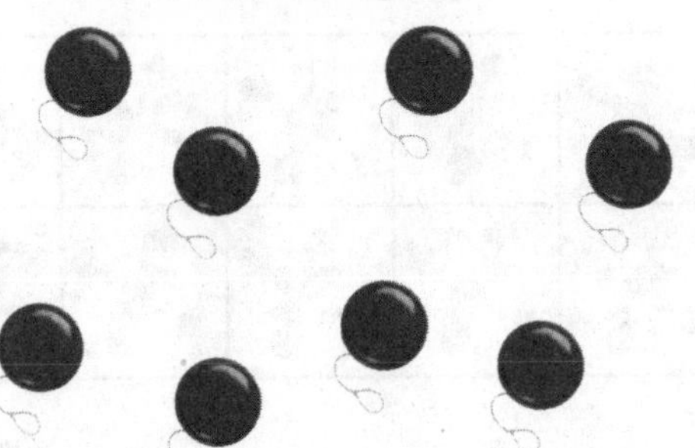

2 twenty 20

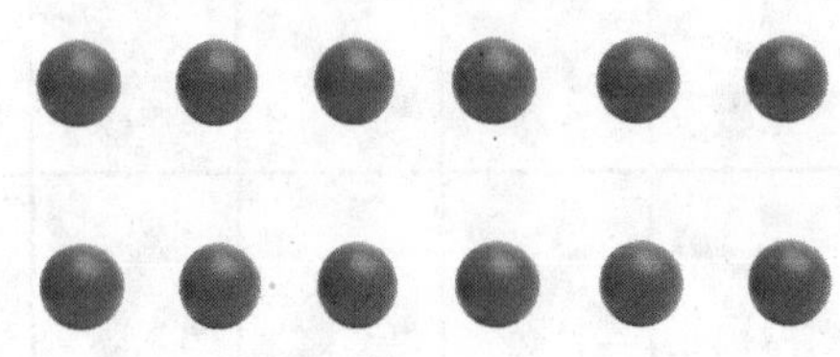

## Concept Check

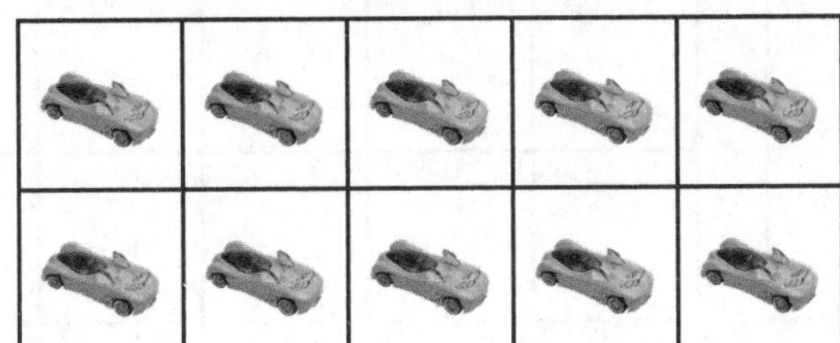

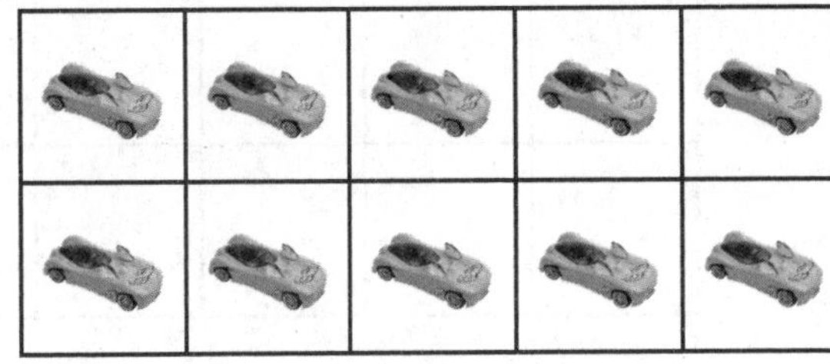

**Directions: 1.** Draw more to show nineteen. **2.** Draw more to show twenty. **3.** Count the objects. Say how many. Write the number three times.

4

5

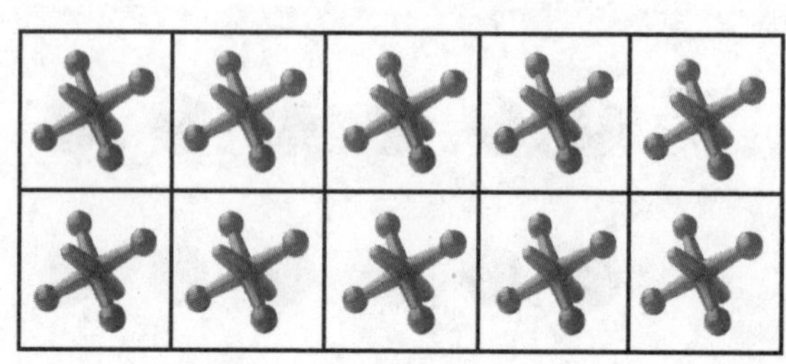

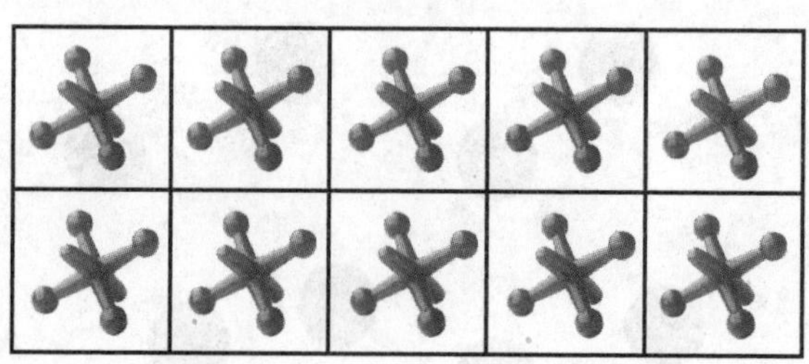

6

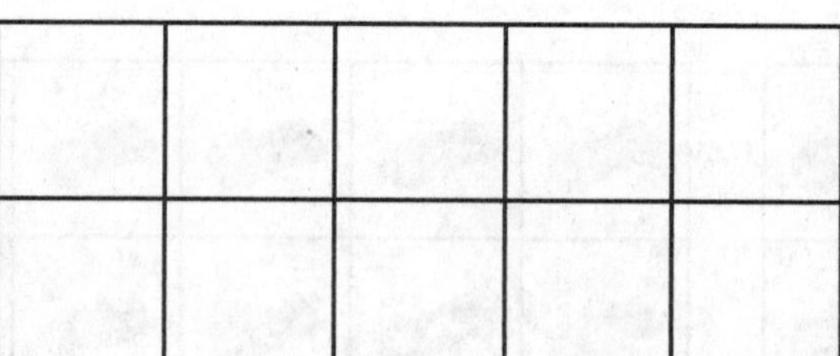

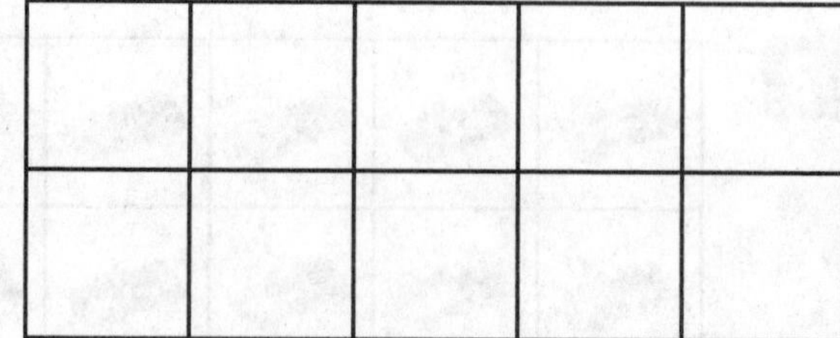

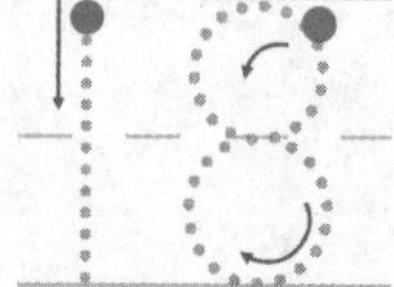

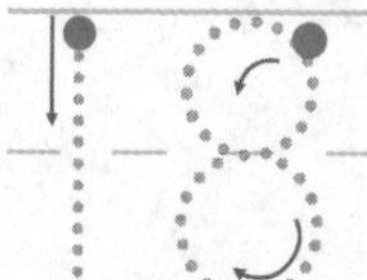

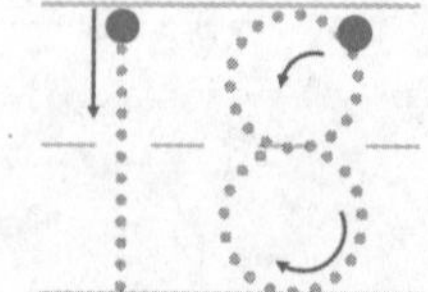

**Directions: 4–5.** Count the objects. Say how many. Write the number three times.
**6.** Trace the number three times. Say how many. Draw circles for counters to show the number.

Name

Lesson 8

# Count to 50 by Ones

**ESSENTIAL QUESTION**

How can I show numbers beyond 10?

## Math in My World

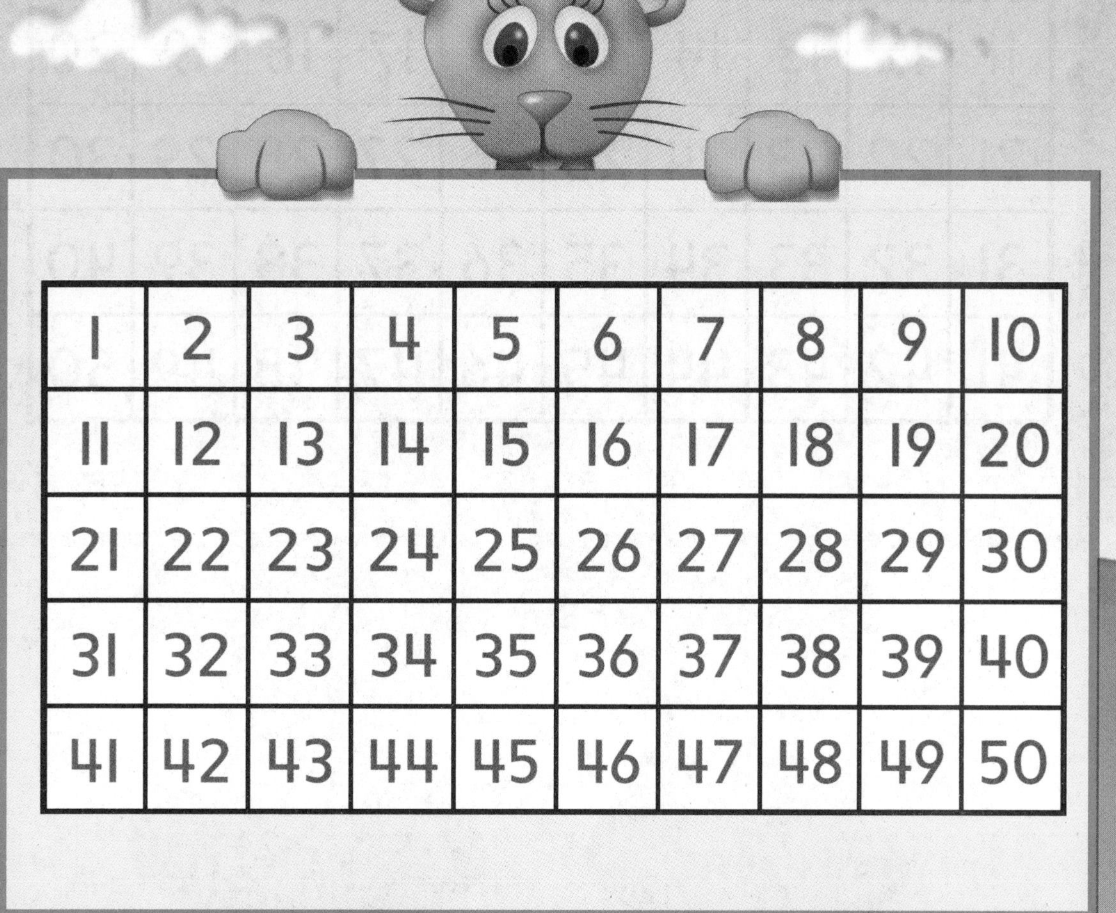

| 1 | 2 | 3 | 4 | 5 | 6 | 7 | 8 | 9 | 10 |
|---|---|---|---|---|---|---|---|---|---|
| 11 | 12 | 13 | 14 | 15 | 16 | 17 | 18 | 19 | 20 |
| 21 | 22 | 23 | 24 | 25 | 26 | 27 | 28 | 29 | 30 |
| 31 | 32 | 33 | 34 | 35 | 36 | 37 | 38 | 39 | 40 |
| 41 | 42 | 43 | 44 | 45 | 46 | 47 | 48 | 49 | 50 |

**Teacher Directions:** Touch and count from 1 to 25. Color those numbers blue. Then touch and count from 26 to 50. Color those numbers yellow.

# Guided Practice

Processes & Practices

| 1 | 2 | 3 | 4 | 5 | 6 | 7 | 8 | 9 | 10 |
|---|---|---|---|---|---|---|---|---|---|
| 11 | 12 | 13 | 14 | 15 | 16 | 17 | 18 | 19 | 20 |
| 21 | 22 | 23 | 24 | 25 | 26 | 27 | 28 | 29 | 30 |
| 31 | 32 | 33 | 34 | 35 | 36 | 37 | 38 | 39 | 40 |
| 41 | 42 | 43 | 44 | 45 | 46 | 47 | 48 | 49 | 50 |

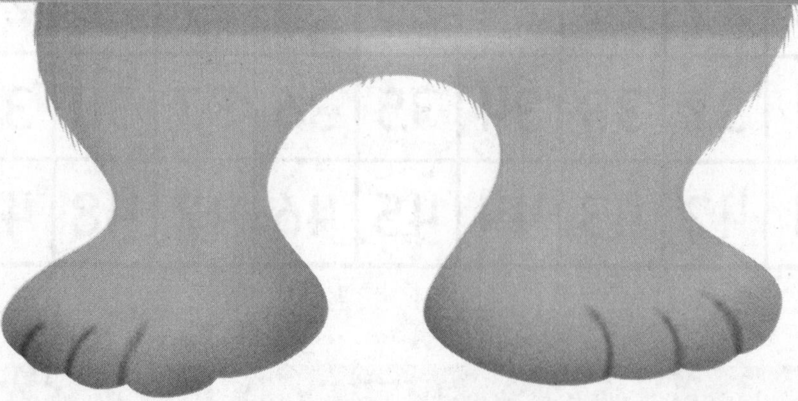

**Directions:** Touch, count, and circle 1 through 10 yellow. Touch, count, and circle 11 through 25 blue. Touch, count, and circle 26 through 40 orange. Touch, count, and circle 41 through 50 black.

Name

## Independent Practice

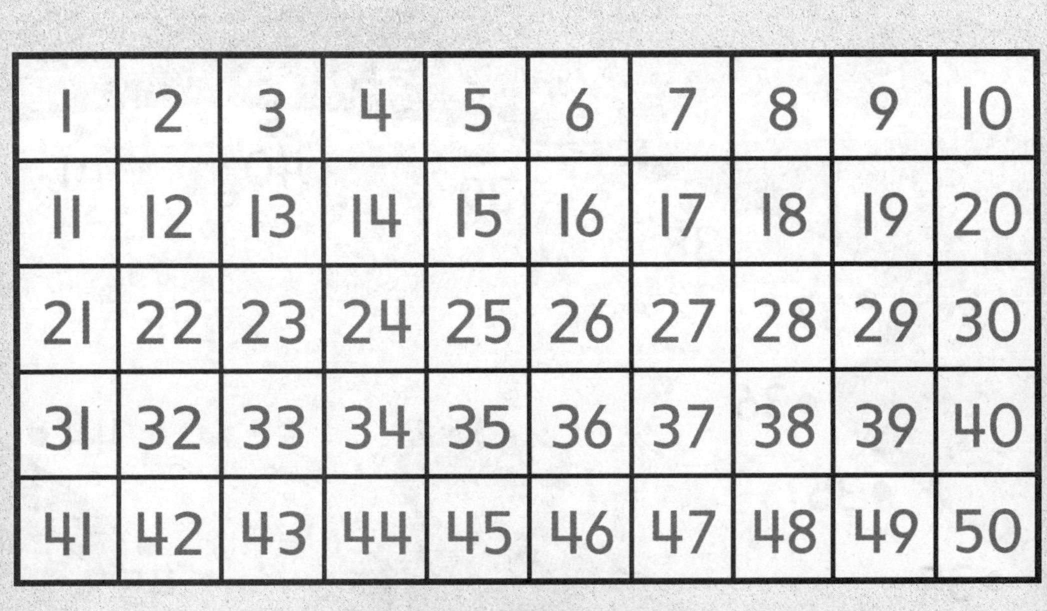

| 1 | 2 | 3 | 4 | 5 | 6 | 7 | 8 | 9 | 10 |
|---|---|---|---|---|---|---|---|---|---|
| 11 | 12 | 13 | 14 | 15 | 16 | 17 | 18 | 19 | 20 |
| 21 | 22 | 23 | 24 | 25 | 26 | 27 | 28 | 29 | 30 |
| 31 | 32 | 33 | 34 | 35 | 36 | 37 | 38 | 39 | 40 |
| 41 | 42 | 43 | 44 | 45 | 46 | 47 | 48 | 49 | 50 |

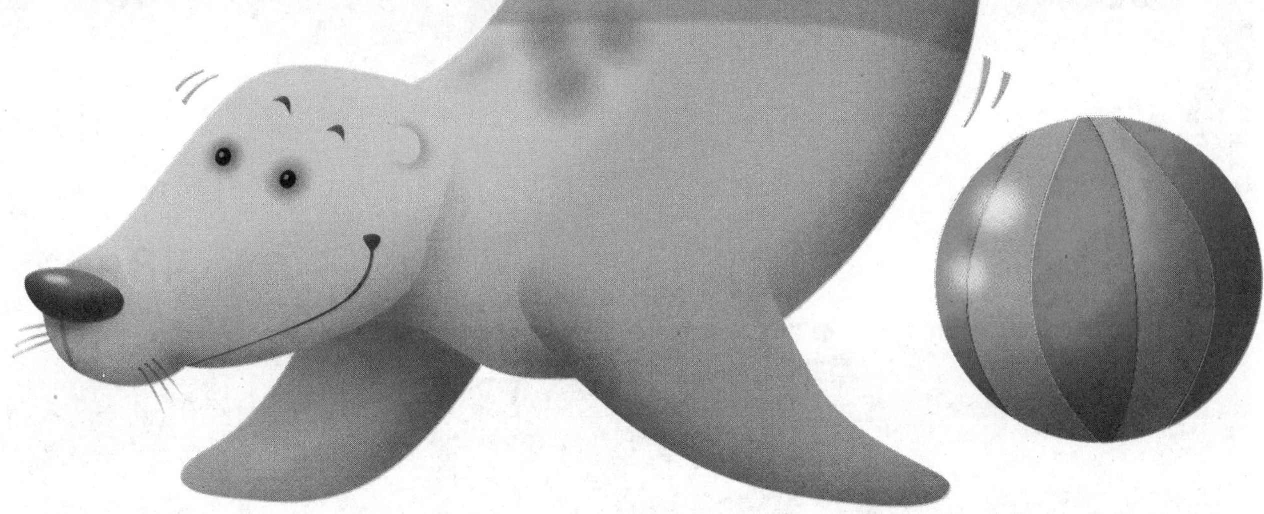

**Directions:** Touch, count, and color numbers 1 through 10 red. Touch, count, and color numbers 11 through 20 blue. Touch, count, and color numbers 21 through 30 green. Touch, count, and color numbers 31 through 50 orange.

Processes &Practices

## Brain Builders

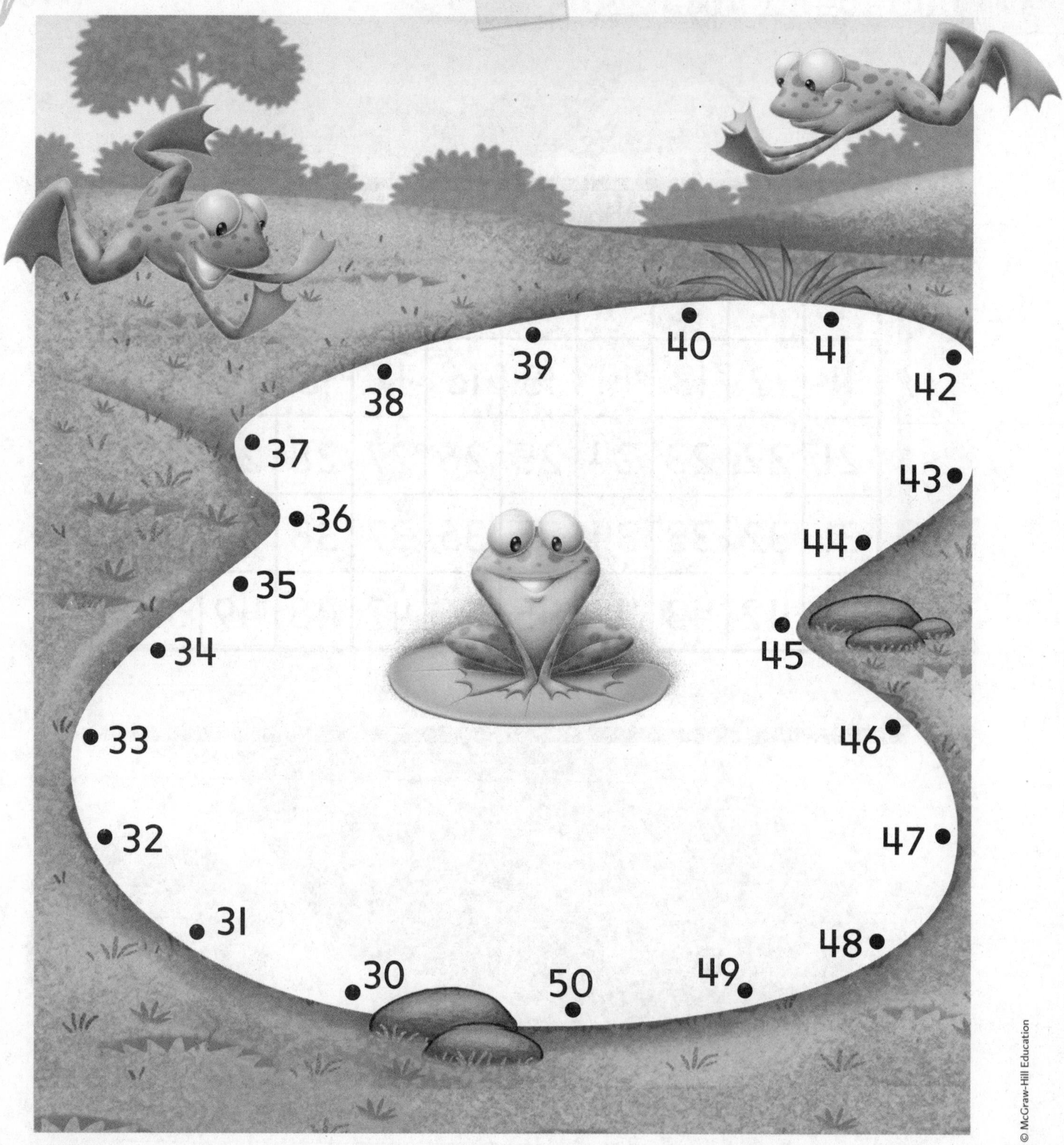

30 31 32 33 34 35 36 37 38 39 40 41 42 43 44 45 46 47 48 49 50

**Directions:** Start at 30. Say each number as you connect the dots by counting from 30–50. Tell what you made. Color the picture. Tell a friend what number would have come before 30. Explain how you know.

Name

Lesson 8

Count to 50 by Ones

# My Homework

## Homework Helper

Need help? connectED.mcgraw-hill.com

| 1 | 2 | 3 | 4 | 5 | 6 | 7 | 8 | 9 | 10 |
|---|---|---|---|---|---|---|---|---|---|
| 11 | 12 | 13 | 14 | 15 | 16 | 17 | 18 | 19 | 20 |
| 21 | 22 | 23 | 24 | 25 | 26 | 27 | 28 | 29 | 30 |
| 31 | 32 | 33 | 34 | 35 | 36 | 37 | 38 | 39 | 40 |
| 41 | 42 | 43 | 44 | 45 | 46 | 47 | 48 | 49 | 50 |

**Directions:** Touch, count, and color numbers 1 through 9 red. Touch, count, and color numbers 10 through 19 blue. Touch, count, and color numbers 20 through 39 green. Touch, count, and color numbers 40 through 50 yellow.

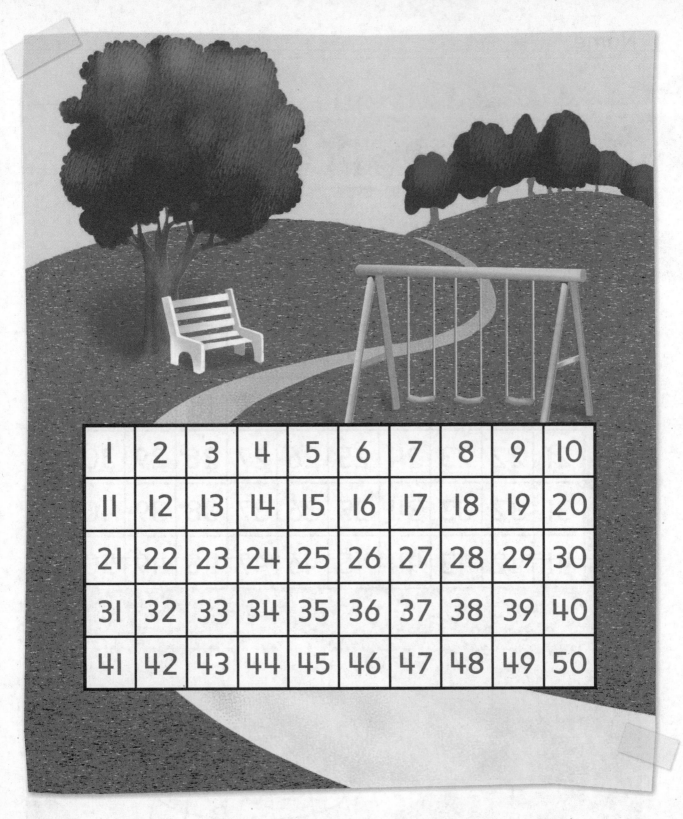

| 1 | 2 | 3 | 4 | 5 | 6 | 7 | 8 | 9 | 10 |
|---|---|---|---|---|---|---|---|---|---|
| 11 | 12 | 13 | 14 | 15 | 16 | 17 | 18 | 19 | 20 |
| 21 | 22 | 23 | 24 | 25 | 26 | 27 | 28 | 29 | 30 |
| 31 | 32 | 33 | 34 | 35 | 36 | 37 | 38 | 39 | 40 |
| 41 | 42 | 43 | 44 | 45 | 46 | 47 | 48 | 49 | 50 |

**Directions:** Touch, count, and color numbers 1 through 20 purple. Touch, count, and color the numbers 21 through 32 yellow. Touch, count, and color the numbers 33 through 50 orange.

**Math at Home** Have your child count 50 pieces of cereal or 50 blocks. Have your child lay the cereal or blocks on this chart as he or she counts them.

Name

# Lesson 9
# Count to 100 by Ones

**ESSENTIAL QUESTION**
How can I show numbers beyond 10?

## Math in My World

| 1 | 2 | 3 | 4 | 5 | 6 | 7 | 8 | 9 | 10 |
|---|---|---|---|---|---|---|---|---|---|
| 11 | 12 | 13 | 14 | 15 | 16 | 17 | 18 | 19 | 20 |
| 21 | 22 | 23 | 24 | 25 | 26 | 27 | 28 | 29 | 30 |
| 31 | 32 | 33 | 34 | 35 | 36 | 37 | 38 | 39 | 40 |
| 41 | 42 | 43 | 44 | 45 | 46 | 47 | 48 | 49 | 50 |
| 51 | 52 | 53 | 54 | 55 | 56 | 57 | 58 | 59 | 60 |
| 61 | 62 | 63 | 64 | 65 | 66 | 67 | 68 | 69 | 70 |
| 71 | 72 | 73 | 74 | 75 | 76 | 77 | 78 | 79 | 80 |
| 81 | 82 | 83 | 84 | 85 | 86 | 87 | 88 | 89 | 90 |
| 91 | 92 | 93 | 94 | 95 | 96 | 97 | 98 | 99 | 100 |

**Teacher Directions:** Touch and count from 1 to 25. Color those numbers blue. Touch and count from 26 to 68. Color those numbers orange. Touch and count from 69 to 100. Color those numbers yellow.

## Guided Practice

| 1 | 2 | 3 | 4 | 5 | 6 | 7 | 8 | 9 | 10 |
|---|---|---|---|---|---|---|---|---|---|
| 11 | 12 | 13 | 14 | 15 | 16 | 17 | 18 | 19 | 20 |
| 21 | 22 | 23 | 24 | 25 | 26 | 27 | 28 | 29 | 30 |
| 31 | 32 | 33 | 34 | 35 | 36 | 37 | 38 | 39 | 40 |
| 41 | 42 | 43 | 44 | 45 | 46 | 47 | 48 | 49 | 50 |
| 51 | 52 | 53 | 54 | 55 | 56 | 57 | 58 | 59 | 60 |
| 61 | 62 | 63 | 64 | 65 | 66 | 67 | 68 | 69 | 70 |
| 71 | 72 | 73 | 74 | 75 | 76 | 77 | 78 | 79 | 80 |
| 81 | 82 | 83 | 84 | 85 | 86 | 87 | 88 | 89 | 90 |
| 91 | 92 | 93 | 94 | 95 | 96 | 97 | 98 | 99 | 100 |

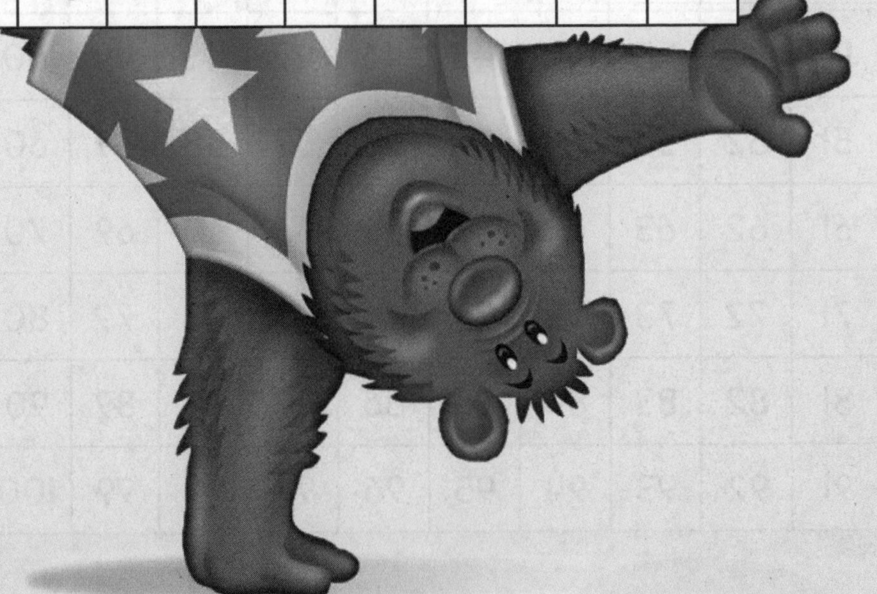

**Directions:** Touch, count, and circle 1 to 10 green. Touch, count, and circle 11 to 35 purple. Touch, count, and circle 36 to 62 orange. Touch, count, and circle 63 to 100 brown.

Name ..........................................

## Independent Practice

| 1 | 2 | 3 | 4 | 5 | 6 | 7 | 8 | 9 | 10 |
|---|---|---|---|---|---|---|---|---|---|
| 11 | 12 | 13 | 14 | 15 | 16 | 17 | 18 | 19 | 20 |
| 21 | 22 | 23 | 24 | 25 | 26 | 27 | 28 | 29 | 30 |
| 31 | 32 | 33 | 34 | 35 | 36 | 37 | 38 | 39 | 40 |
| 41 | 42 | 43 | 44 | 45 | 46 | 47 | 48 | 49 | 50 |
| 51 | 52 | 53 | 54 | 55 | 56 | 57 | 58 | 59 | 60 |
| 61 | 62 | 63 | 64 | 65 | 66 | 67 | 68 | 69 | 70 |
| 71 | 72 | 73 | 74 | 75 | 76 | 77 | 78 | 79 | 80 |
| 81 | 82 | 83 | 84 | 85 | 86 | 87 | 88 | 89 | 90 |
| 91 | 92 | 93 | 94 | 95 | 96 | 97 | 98 | 99 | 100 |

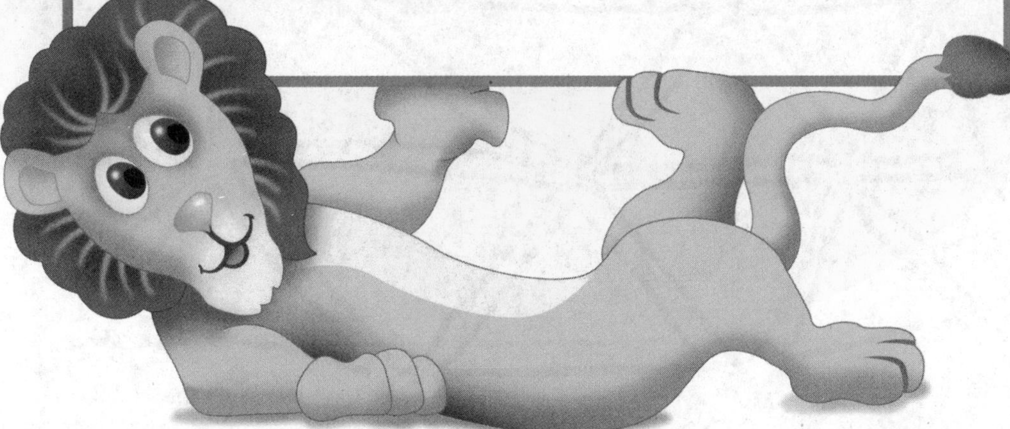

**Directions:** Touch, count, and circle numbers 1 to 15. Touch, count, and color the numbers 16 to 36 red. Touch, count, and circle numbers 37 to 47. Touch, count, and color the numbers 48 to 68 blue. Touch, count, and circle the numbers 69 to 79. Touch, count, and color the numbers 80 to 100 yellow.

# Brain Builders

**Directions:** Start at 72. Say each number as you connect the dots by counting from 72–90. What did you make? Tell a friend what would be the next 5 numbers that come after 90.

Name ..........................................

# My Homework

**Lesson 9**

Count to 100 by Ones

## Homework Helper

Need help? connectED.mcgraw-hill.com

1

| | | | | | | | | | |
|---|---|---|---|---|---|---|---|---|---|
| 1 | 2 | 3 | 4 | 5 | 6 | 7 | 8 | 9 | 10 |
| 11 | 12 | 13 | 14 | 15 | 16 | 17 | 18 | 19 | 20 |
| 21 | 22 | 23 | 24 | 25 | 26 | 27 | 28 | 29 | 30 |
| 31 | 32 | 33 | 34 | 35 | 36 | 37 | 38 | 39 | 40 |
| 41 | 42 | 43 | 44 | 45 | 46 | 47 | 48 | 49 | 50 |
| 51 | 52 | 53 | 54 | 55 | 56 | 57 | 58 | 59 | 60 |
| 61 | 62 | 63 | 64 | 65 | 66 | 67 | 68 | 69 | 70 |
| 71 | 72 | 73 | 74 | 75 | 76 | 77 | 78 | 79 | 80 |
| 81 | 82 | 83 | 84 | 85 | 86 | 87 | 88 | 89 | 90 |
| 91 | 92 | 93 | 94 | 95 | 96 | 97 | 98 | 99 | 100 |

**Directions:** Touch, count, and color the numbers 1 to 15 red. Touch, count, and color the numbers 16 to 33 yellow. Touch, count, and color 34 to 75 purple. Touch, count, and color 76 to 100 green.

2

| 1 | 2 | 3 | 4 | 5 | 6 | 7 | 8 | 9 | 10 |
|---|---|---|---|---|---|---|---|---|---|
| 11 | 12 | 13 | 14 | 15 | 16 | 17 | 18 | 19 | 20 |
| 21 | 22 | 23 | 24 | 25 | 26 | 27 | 28 | 29 | 30 |
| 31 | 32 | 33 | 34 | 35 | 36 | 37 | 38 | 39 | 40 |
| 41 | 42 | 43 | 44 | 45 | 46 | 47 | 48 | 49 | 50 |
| 51 | 52 | 53 | 54 | 55 | 56 | 57 | 58 | 59 | 60 |
| 61 | 62 | 63 | 64 | 65 | 66 | 67 | 68 | 69 | 70 |
| 71 | 72 | 73 | 74 | 75 | 76 | 77 | 78 | 79 | 80 |
| 81 | 82 | 83 | 84 | 85 | 86 | 87 | 88 | 89 | 90 |
| 91 | 92 | 93 | 94 | 95 | 96 | 97 | 98 | 99 | 100 |

**Directions:** Touch, count, and color 1 to 24 blue. Touch, count, and color 25 to 59 orange. Touch, count, and color 60 to 100 yellow.

**Math at Home** Using this page, ask your child to read aloud the numbers from 1 to 100. Talk about any numbers that might be special to your family such as the number of people in your family, numbers in your home address, or numbers that tell your age.

Name

Lesson 10

# Count to 100 by Tens

**ESSENTIAL QUESTION**
How can I show numbers beyond 10?

## Math in My World

| 1 | 2 | 3 | 4 | 5 | 6 | 7 | 8 | 9 | 10 |
|---|---|---|---|---|---|---|---|---|---|
| 11 | 12 | 13 | 14 | 15 | 16 | 17 | 18 | 19 | 20 |
| 21 | 22 | 23 | 24 | 25 | 26 | 27 | 28 | 29 | 30 |
| 31 | 32 | 33 | 34 | 35 | 36 | 37 | 38 | 39 | 40 |
| 41 | 42 | 43 | 44 | 45 | 46 | 47 | 48 | 49 | 50 |
| 51 | 52 | 53 | 54 | 55 | 56 | 57 | 58 | 59 | 60 |
| 61 | 62 | 63 | 64 | 65 | 66 | 67 | 68 | 69 | 70 |
| 71 | 72 | 73 | 74 | 75 | 76 | 77 | 78 | 79 | 80 |
| 81 | 82 | 83 | 84 | 85 | 86 | 87 | 88 | 89 | 90 |
| 91 | 92 | 93 | 94 | 95 | 96 | 97 | 98 | 99 | 100 |

**Teacher Directions:** Touch and count by 10s from 10 to 100. Color 10, 20, and 30 green. Color 40, 50, and 60 yellow. Color 70, 80, 90, and 100 red.

# Guided Practice

1

10 20 30

2

40 ___ 60

3

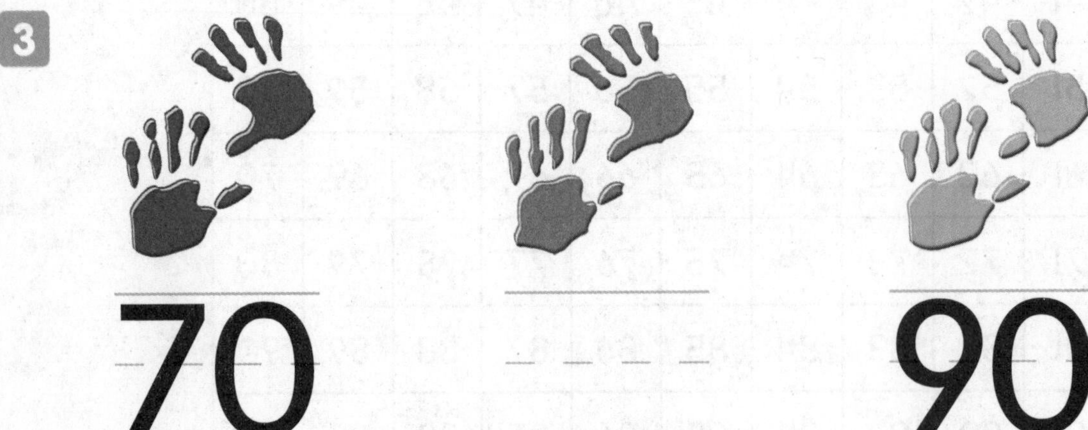

70 ___ 90

**Directions: 1–3.** Count by 10s. Use a hundred chart to find the missing numbers. Write the missing numbers.

Name

## Independent Practice

4

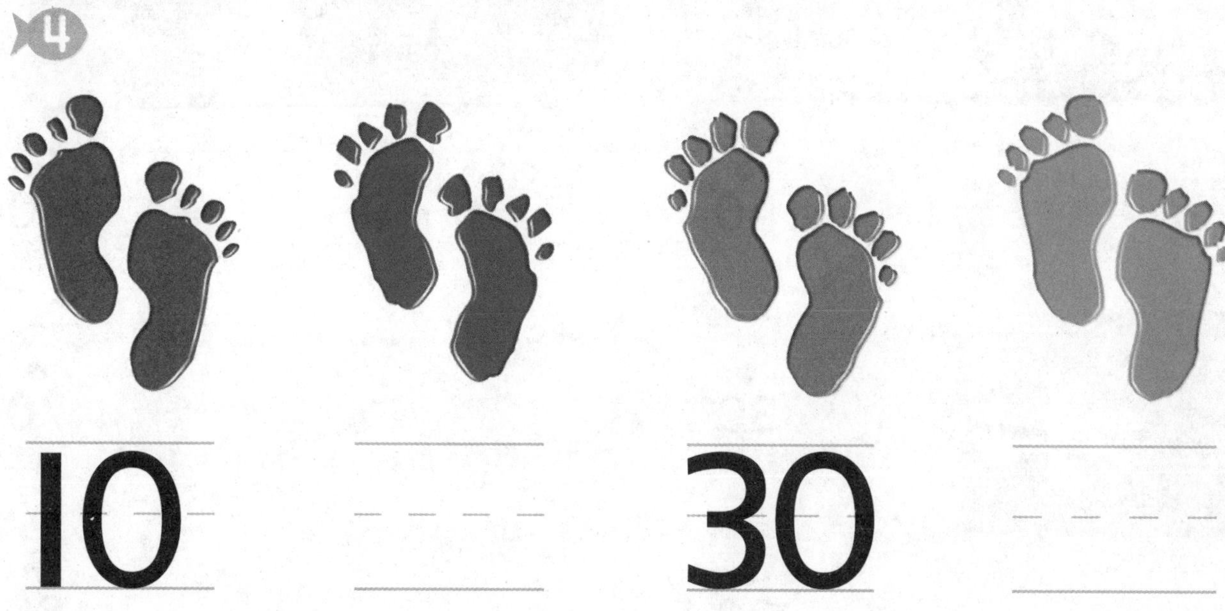

10 ____ 30 ____

5

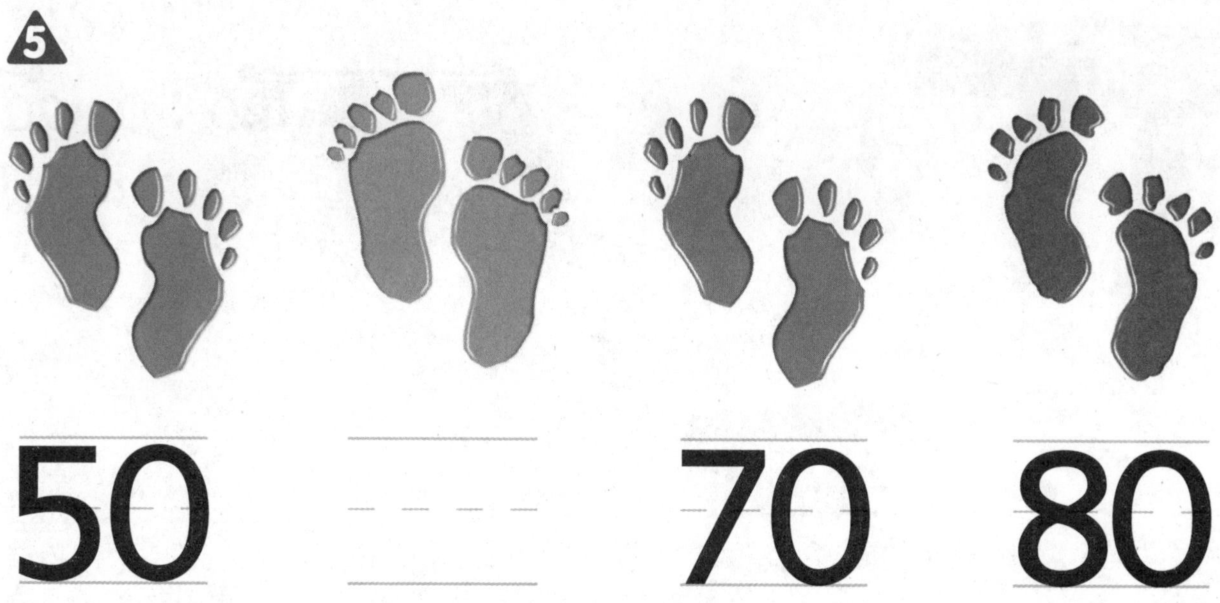

50 ____ 70 80

**Directions:** 4–5. Count by 10s. Use a hundred chart to find the missing numbers. Write the missing numbers.

Processes &Practices

# Brain Builders

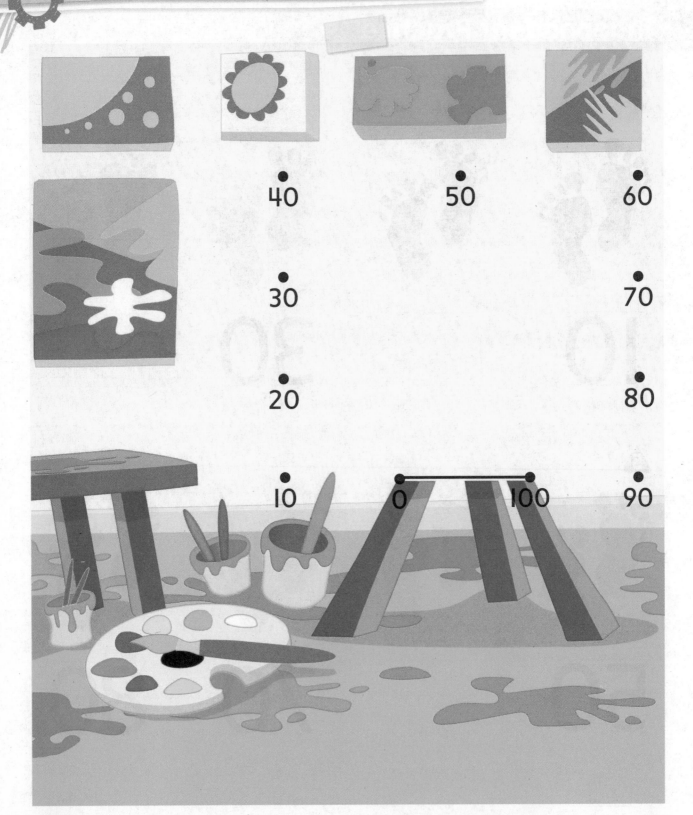

40 50 60

30 70

20 80

10 0 100 90

**Directions:** Start at 0. Count by tens to connect the dots. Say the numbers. Draw a picture on the easel. Count backward from 100 to 0 by tens with a friend. Discuss with each other how the picture could help you.

Name

# My Homework

Lesson 10

Count to 100 by Tens

## Homework Helper

eHelp

Need help? connectED.mcgraw-hill.com

1

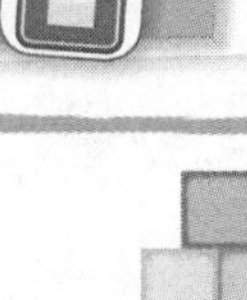

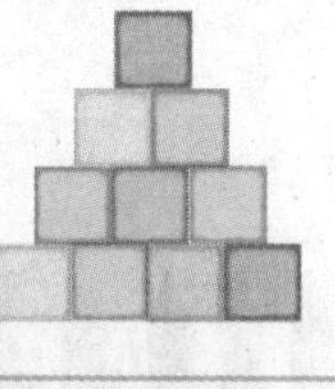

10 20 30

2

40

3

70

**Directions: 1–3.** Count by 10s. Write the missing numbers.

4

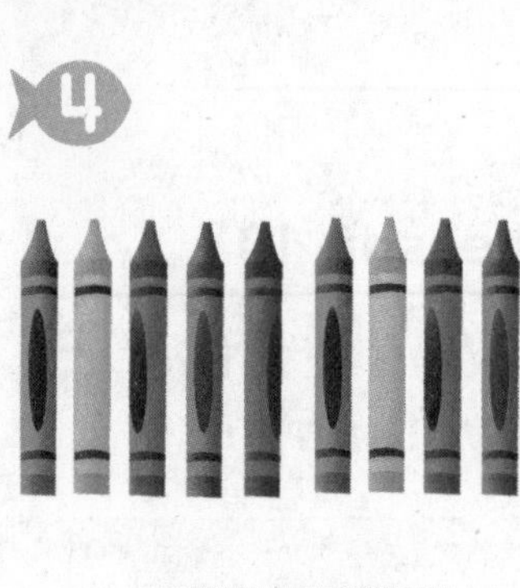

10

5

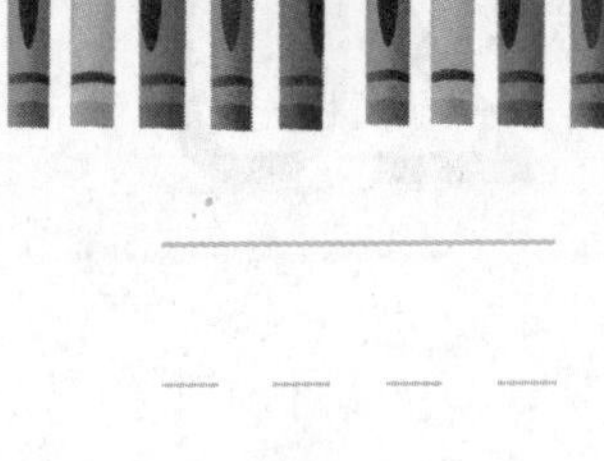

40

6

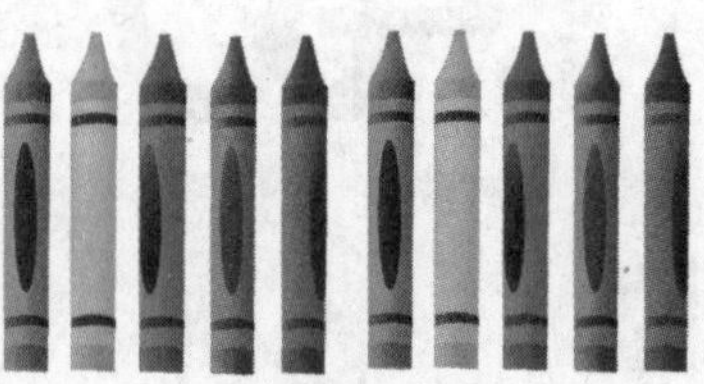

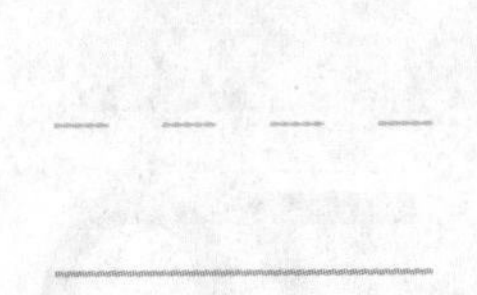

70

**Directions: 4–6.** Count by 10s. Write the missing numbers.

**Math at Home** Help your child make ten groups of 10 using cereal. Have him or her count to 100 by tens.

Name

# Fluency Practice

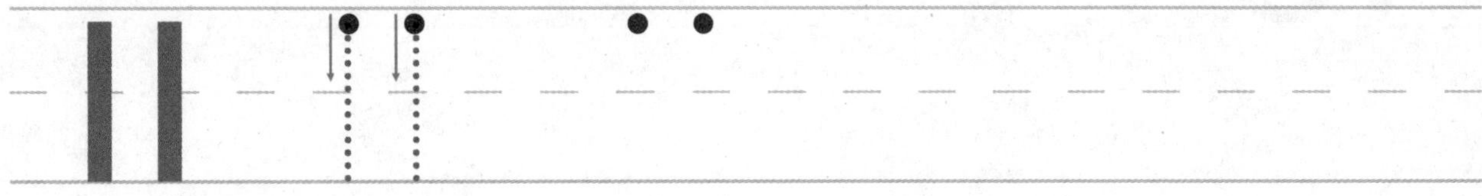

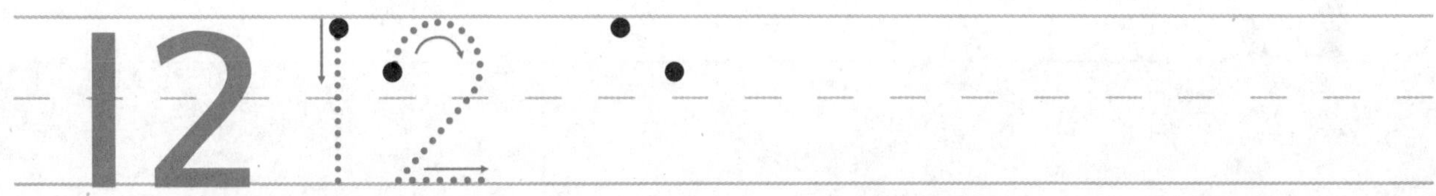

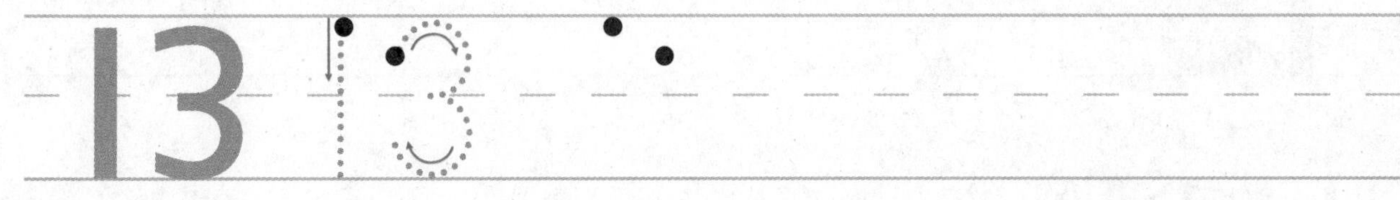

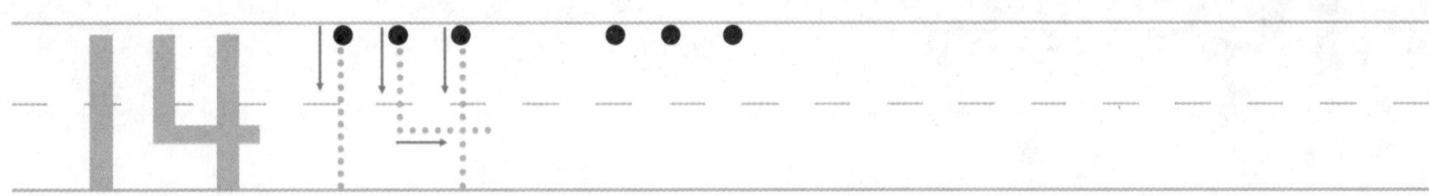

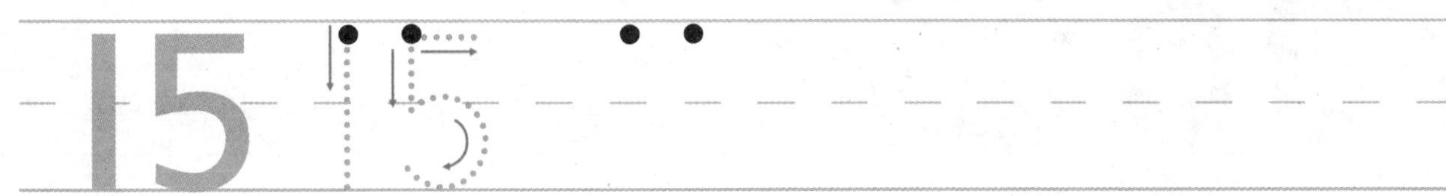



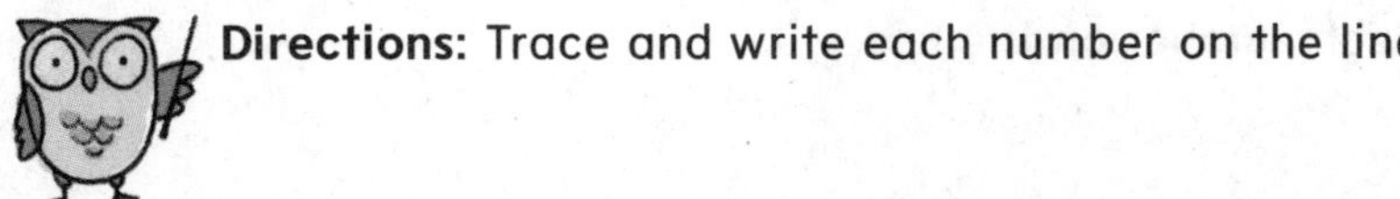

**Directions:** Trace and write each number on the line.

# Fluency Practice

16

17

18

19

20

**Directions:** Trace and write each number on the line.

Name ..............................

# My Review

Chapter 3

Numbers Beyond 10

## Vocabulary Check 

eighteen

fifteen

nineteen

twenty

seventeen

sixteen

**Directions:** **1.** Draw a branch in the mouth of the giraffe with twenty spots. **2.** Draw fifteen spots on the giraffe whose spots are missing. **3.** Draw a necktie on the giraffe with seventeen spots. **4.** Give the giraffe with eighteen spots a flower by her ear. **5.** Draw a circle around the giraffe with sixteen spots. **6.** Put an X on the giraffe with nineteen spots.

# Concept Check

**1**

| 1 | 2 | 3 | 4 | 5 | 6 | 7 | 8 | 9 | 10 |
|---|---|---|---|---|---|---|---|---|---|
| 11 | 12 | 13 | 14 | 15 | 16 | 17 | 18 | 19 | 20 |

**2**

| 1 | 2 | 3 | 4 | 5 | 6 | 7 | 8 | 9 | 10 |
|---|---|---|---|---|---|---|---|---|---|
| 11 | 12 | 13 | 14 | 15 | 16 | 17 | 18 | 19 | 20 |
| 21 | 22 | 23 | 24 | 25 | 26 | 27 | 28 | 29 | 30 |
| 31 | 32 | 33 | 34 | 35 | 36 | 37 | 38 | 39 | 40 |
| 41 | 42 | 43 | 44 | 45 | 46 | 47 | 48 | 49 | 50 |

**3**

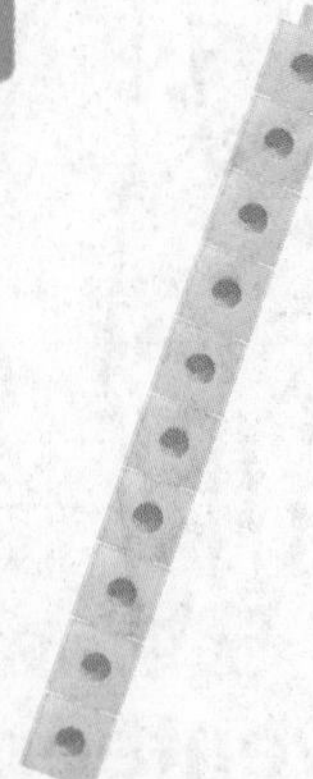

10

____

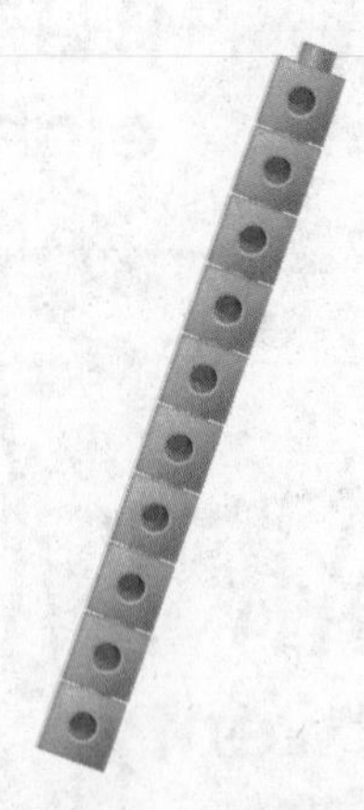

30

____

**Directions: 1.** Touch and color 1 to 9 blue. Touch and color 10 to 16 yellow. Touch and color 17 to 20 red. **2.** Touch and color 1 to 15 green. Touch and color 16 to 32 orange. Touch and color 33 to 50 blue. **3.** Count by 10s. Write the missing numbers.

Name

## Brain Builders

4

**Directions: 4.** Color 10 circles purple. Color 10 circles green. Count all the circles. Say how many. Write the number.

# Reflect

**Chapter 3**

**ESSENTIAL QUESTION**

**How can I show numbers beyond 10?**

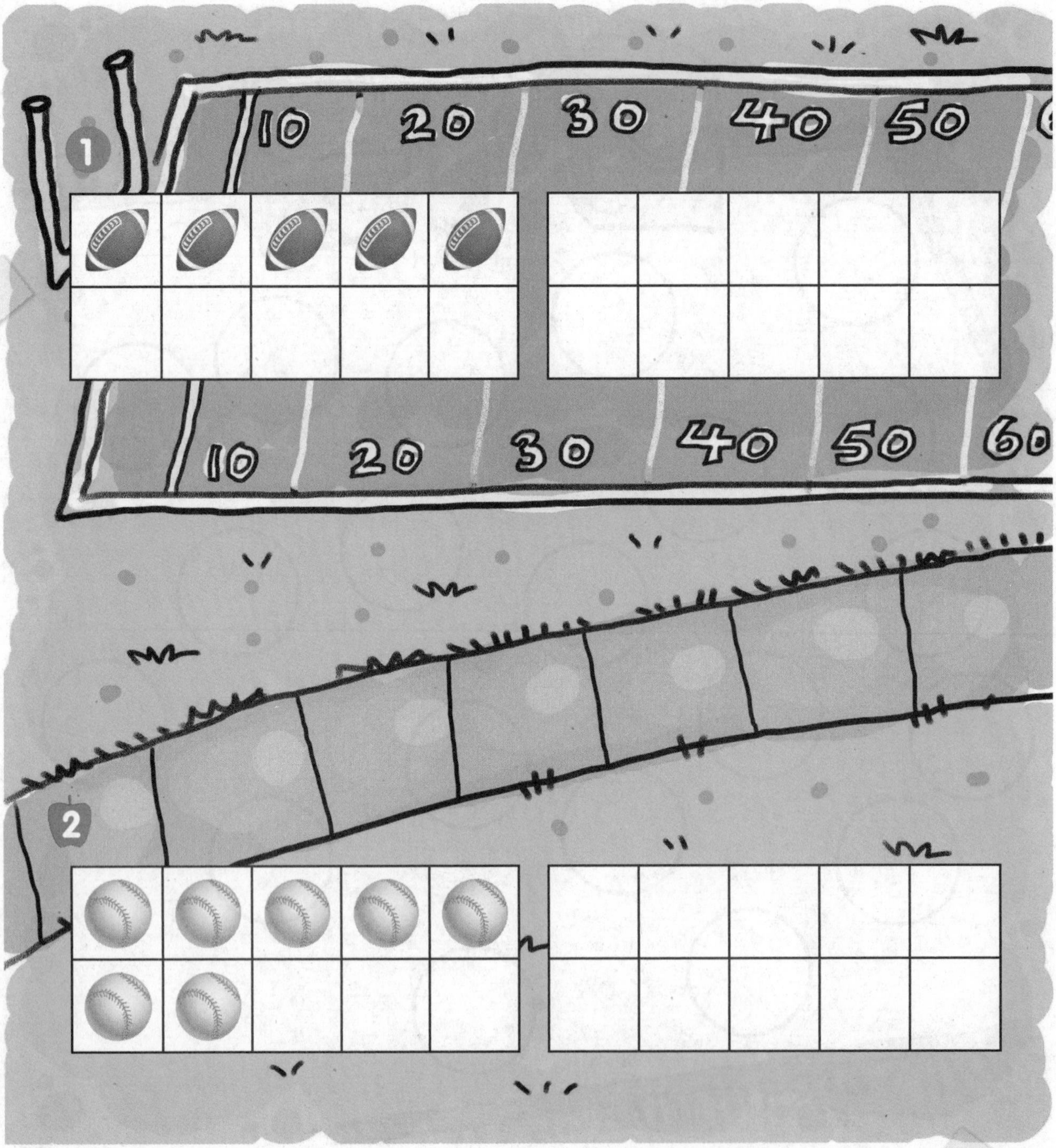

1

2

**Directions: 1.** Count the objects. Say how many. Draw more to the group to make seventeen. **2.** Count the objects. Say how many. Draw more to the group to make twenty.

Name

Date

Score

# Performance Task

## At the Sporting Goods Store

You are visiting a sporting goods store.

**Show all of your work to receive full credit.**

### Part A

Count the baseball gloves. Write the number.

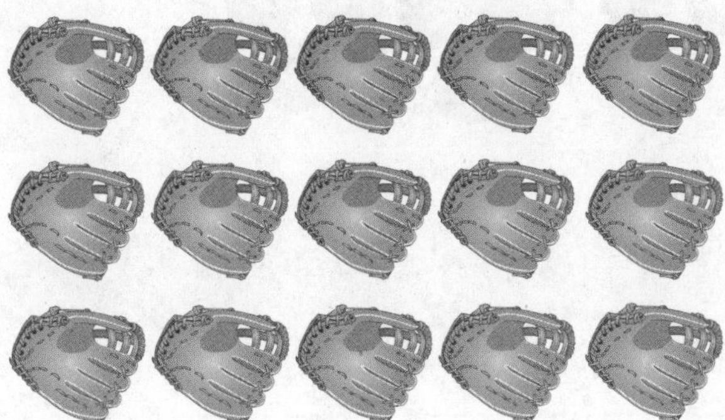

### Part B

Count the baseballs. Write the number.

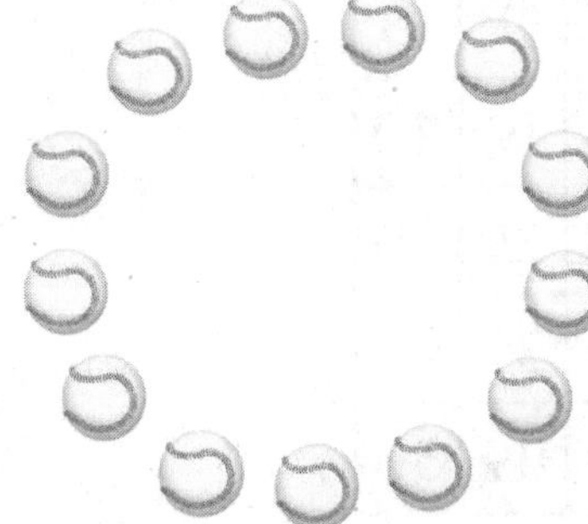

## Part C

There are 20 marbles. Color the boxes in the ten-frames to show that number of marbles.

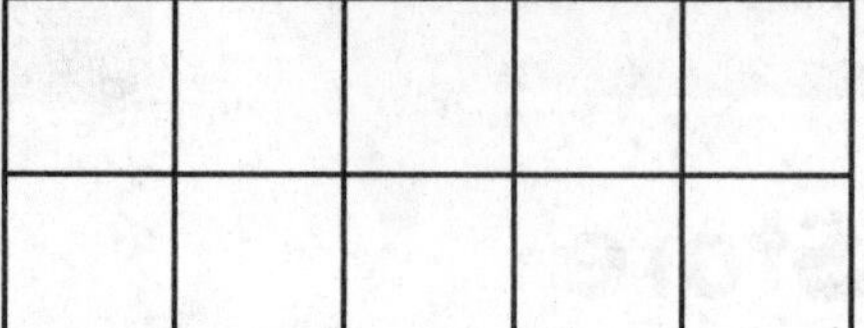

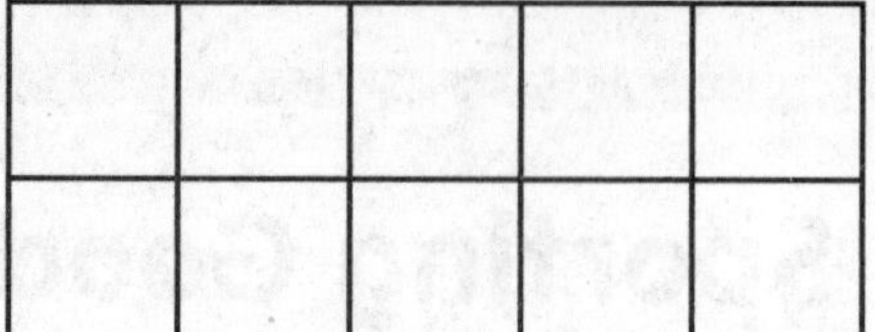

## Part D

Count by 10s. Write the numbers.

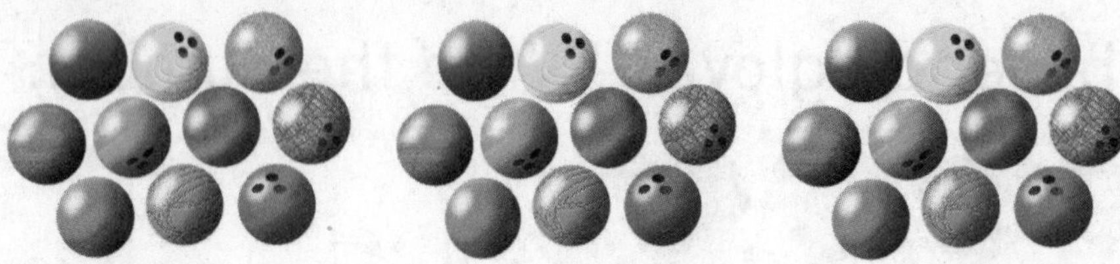

10 ____ ____

## Part E

Count to 50 by 10s. Color each number you count yellow.

| 1 | 2 | 3 | 4 | 5 | 6 | 7 | 8 | 9 | 10 |
|---|---|---|---|---|---|---|---|---|---|
| 11 | 12 | 13 | 14 | 15 | 16 | 17 | 18 | 19 | 20 |
| 21 | 22 | 23 | 24 | 25 | 26 | 27 | 28 | 29 | 30 |
| 31 | 32 | 33 | 34 | 35 | 36 | 37 | 38 | 39 | 40 |
| 41 | 42 | 43 | 44 | 45 | 46 | 47 | 48 | 49 | 50 |

Chapter 4

# Compose and Decompose Numbers to 10

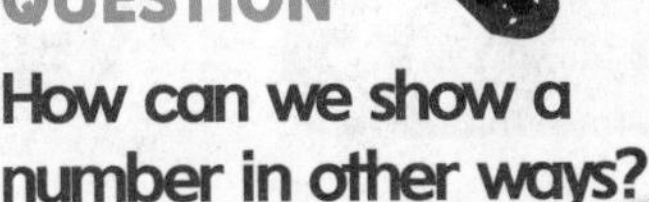

ESSENTIAL QUESTION

How can we show a number in other ways?

I'm Going to the City!

Watch a video!

Name

# Chapter 4 Project

## Our "Ways to Make Numbers" Chart

1. Choose a number between 5–10. Write that number on the line below.
2. Show 4 different ways to make the number below.
3. Draw groups of objects to make that number. Use two different colored crayons to show the different ways you make your number in each of the boxes below.

| | |
|---|---|
| | |
| | |

Name

# Am I Ready?

1

2

3

**Directions: 1–3.** Count the objects. Write the number.

Name

# My Math Words

## Review Vocabulary

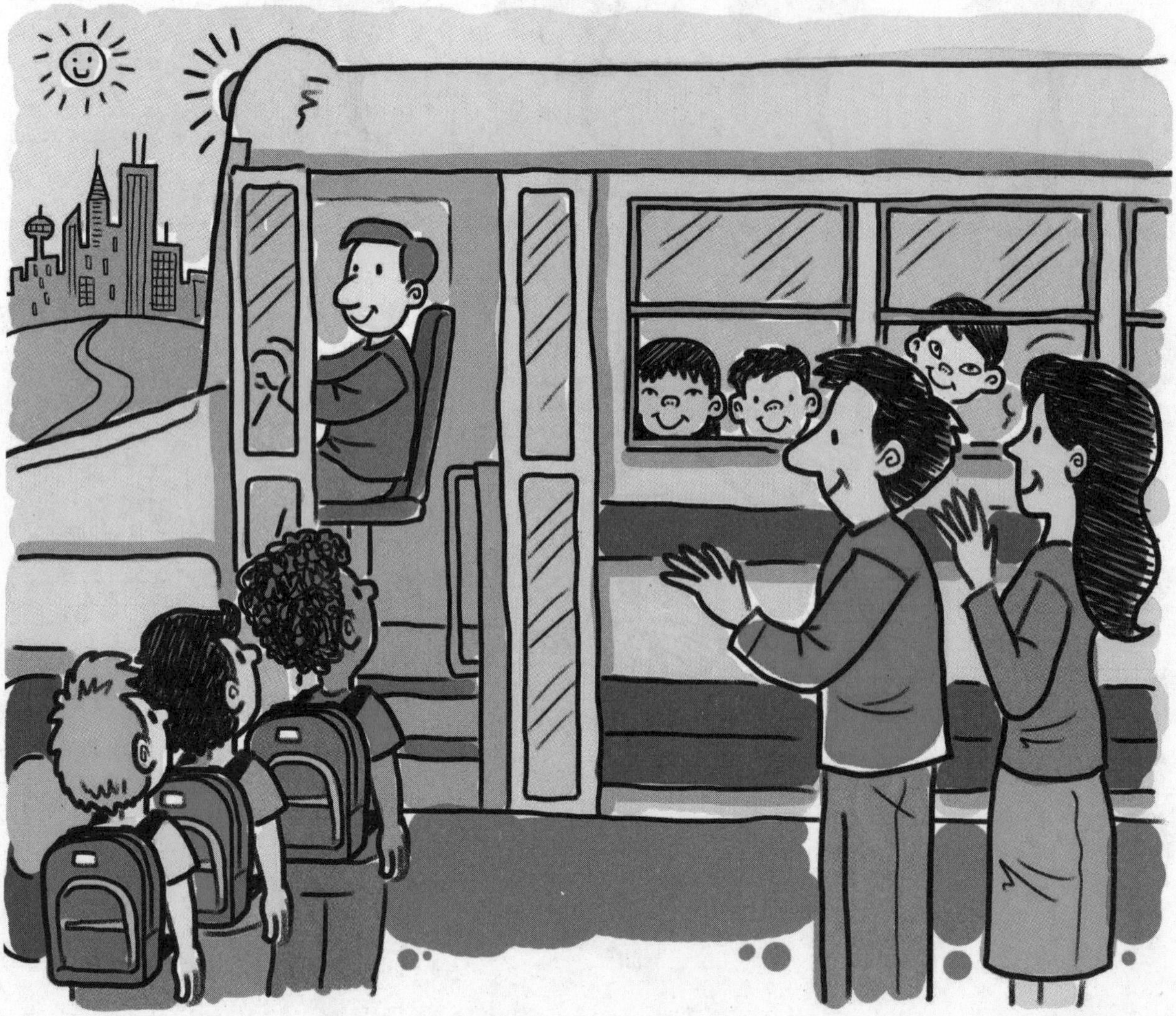

four 4

five 5

**Directions:** Trace each word and number. Draw an X on each person in the bus. Count how many Xs. Draw an X on the number that tells how many. Circle each person outside the bus. Count how many circles. Draw a circle around the number that tells how many.

# My Vocabulary Cards

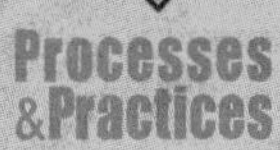


eight 8

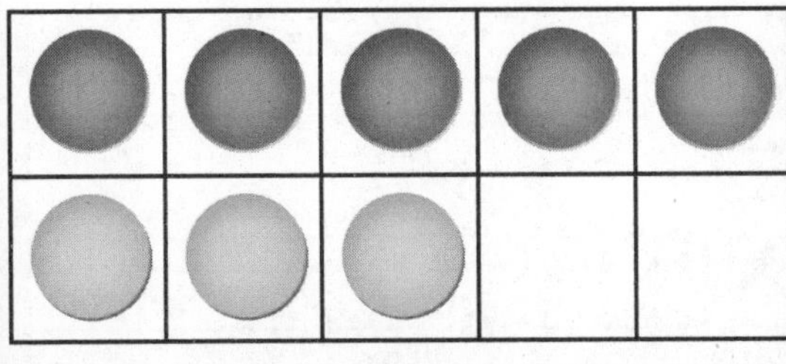

five 5

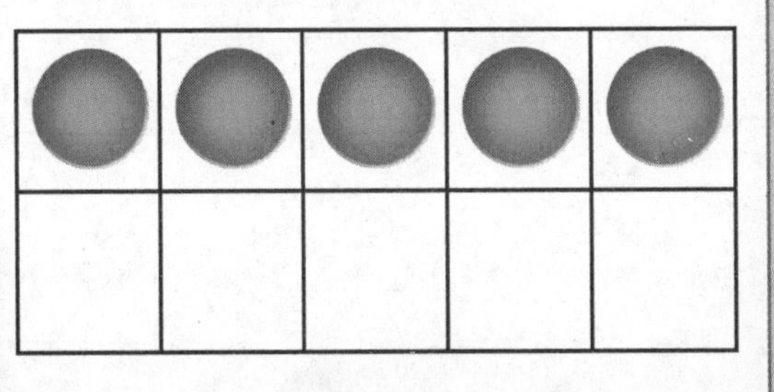

four 4

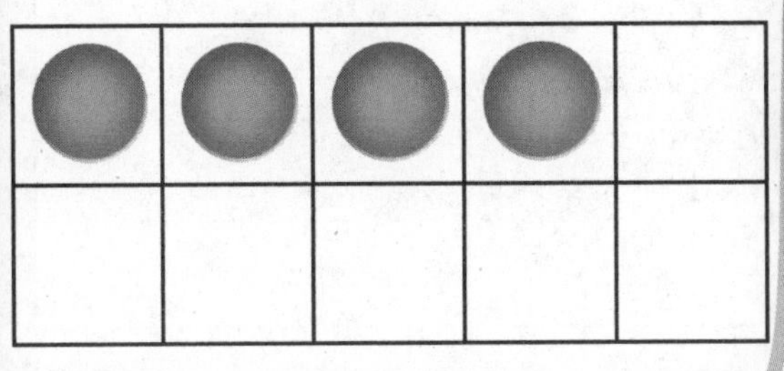

nine 9

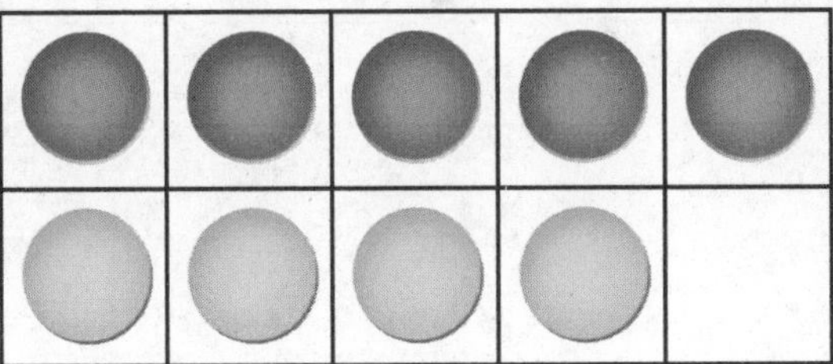

seven 7

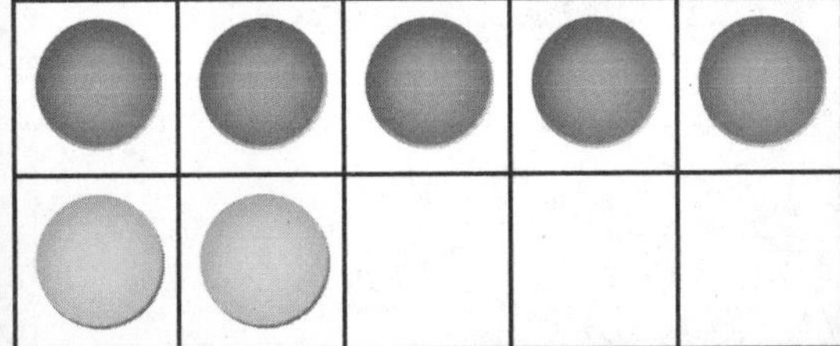

six 6

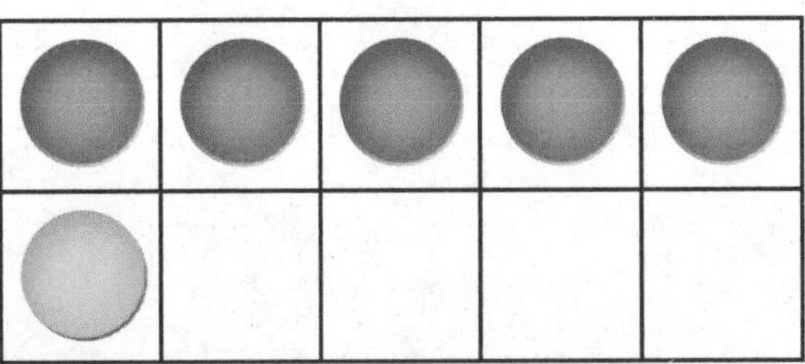

**Teacher Directions:**
**Ideas for Use**

- Show students cards six through nine. Have them discuss how many red and how many yellow counters are on the number six card. Repeat with number cards seven through nine.
- Have students select a card and show the number using counters and a ten-frame.

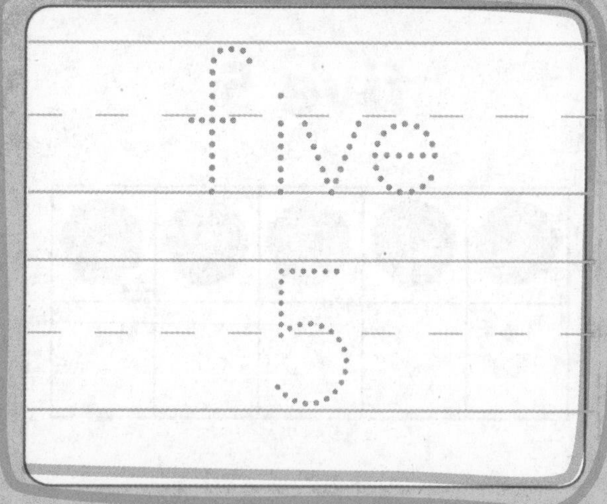


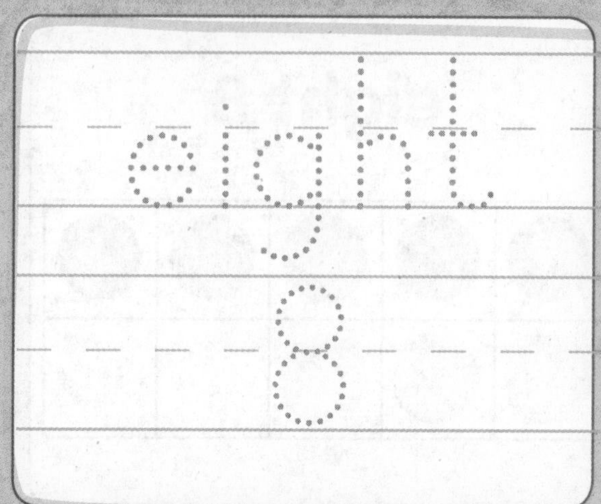

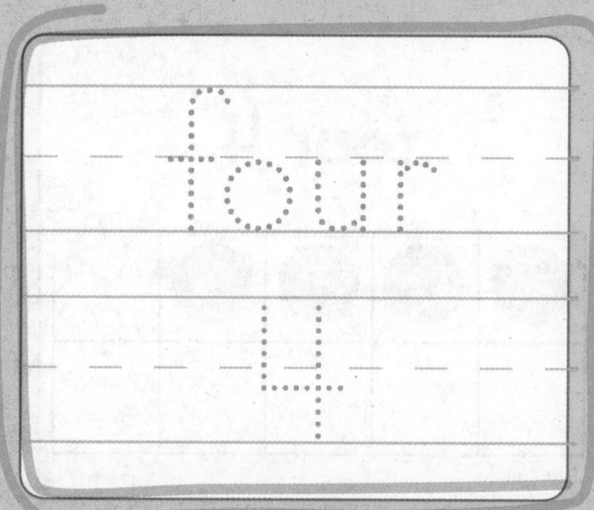


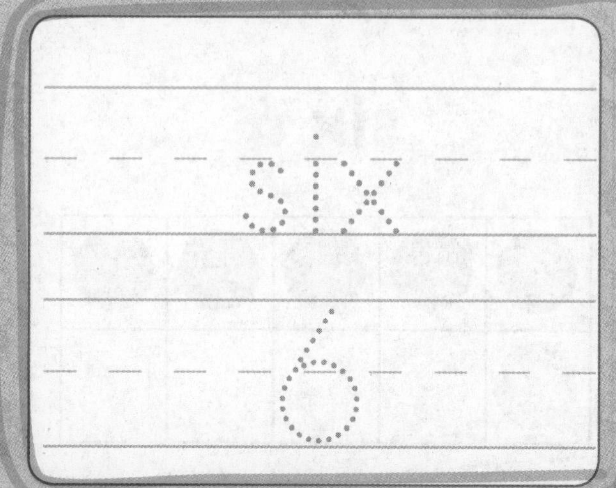

# My Foldable

FOLDABLES® Follow the steps on the back to make your Foldable.

4

and

9

and

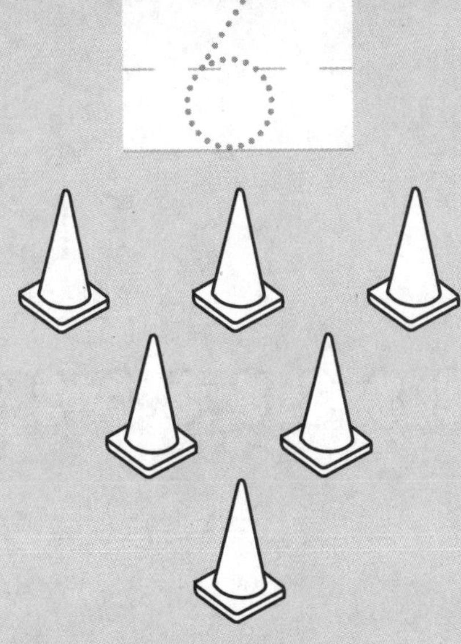

6

and

7

and

FOLDABLES®
Study Organizer

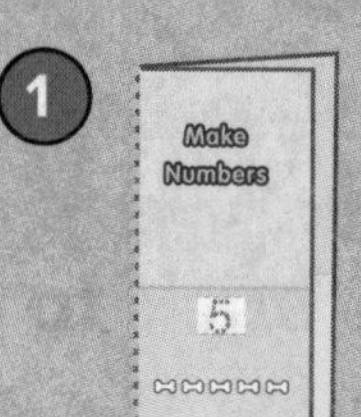
Make
Numbers
5
and

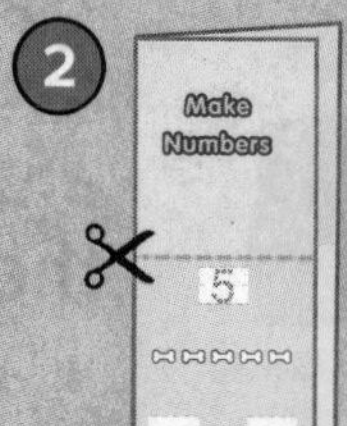
Make
Numbers
5
and

Make
Numbers
5
and

Make
Numbers

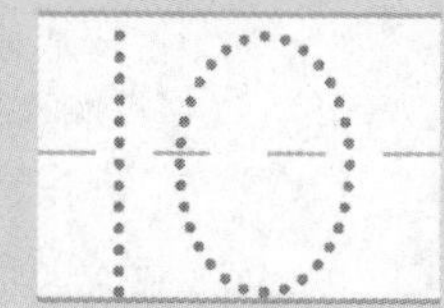
10

and

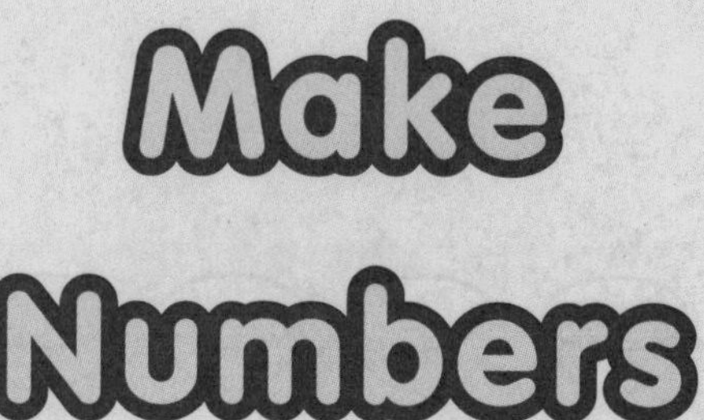
Make
Numbers

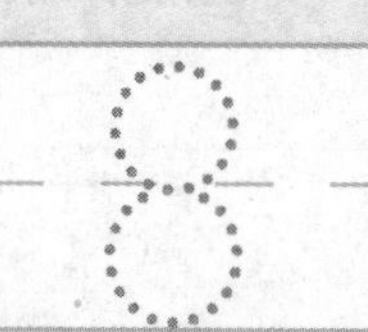
8

and

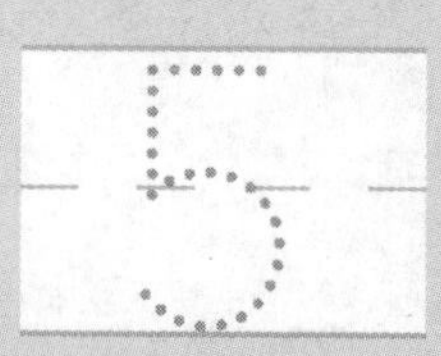
5

and

Name

Lesson 1
# Make 4 and 5

**ESSENTIAL QUESTION**
How can we show a number in other ways?

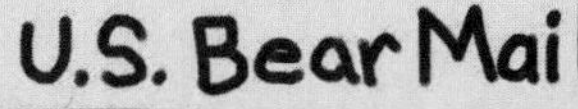

1

2

**Teacher Directions: 1.** Use and to show ways to make four. Color the row of four bears red and yellow to show one way. **2.** Repeat to show ways to make five. Color the row of bears red and yellow to show one way.

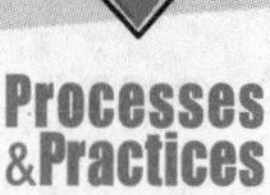


## Guided Practice

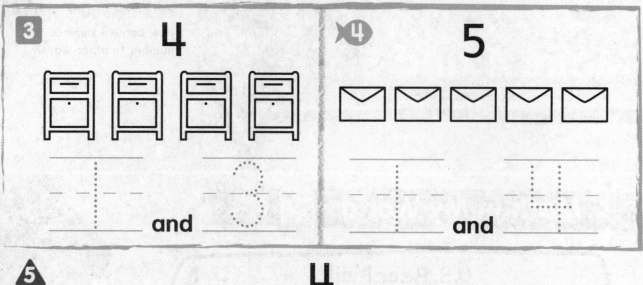

3 4

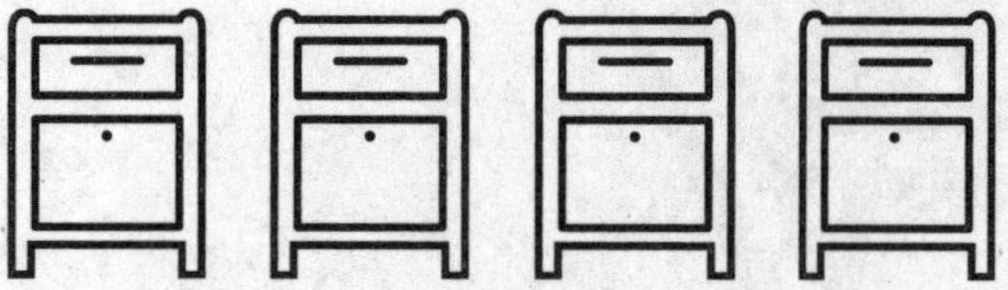

1 and 3

4 5

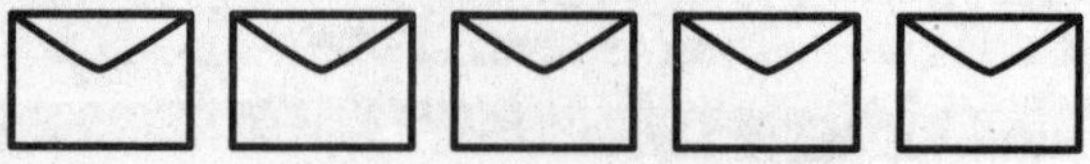

1 and 4

5 4

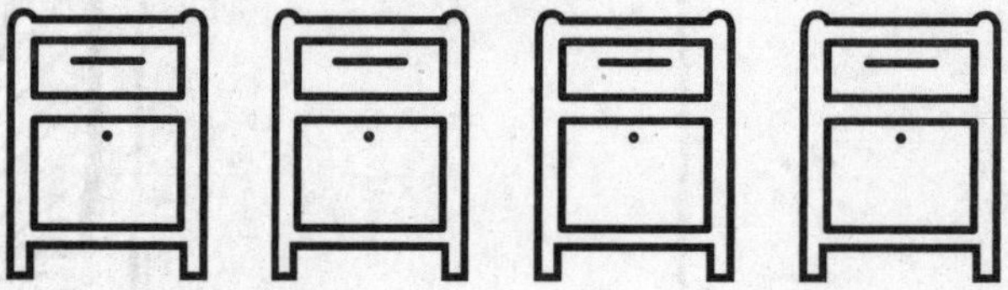

and

6 5

and

**Directions: 3–4.** Trace the numbers to show a way to make four and five. Use red and yellow to color the objects to show this way. **5–6.** Use red and yellow to color the objects to show a way to make four and five. Write the numbers.

Name

## Independent Practice

7

4

______ ______

______ and ______

8

5

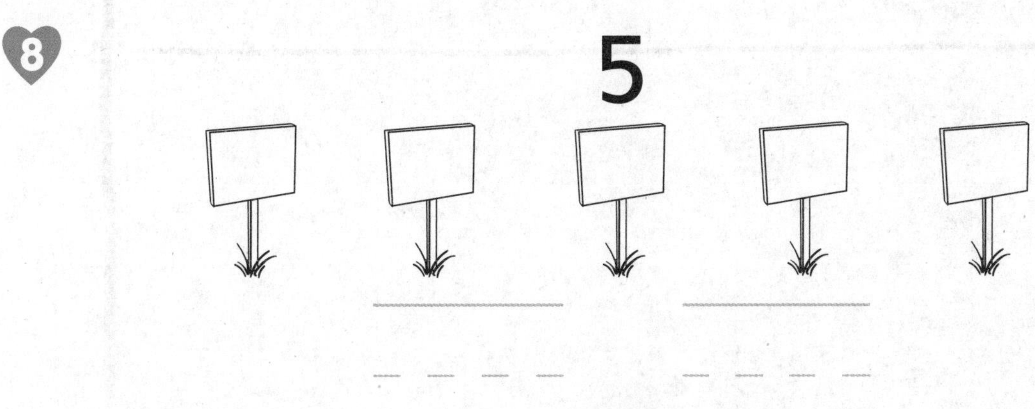

______ ______

______ and ______

9

5

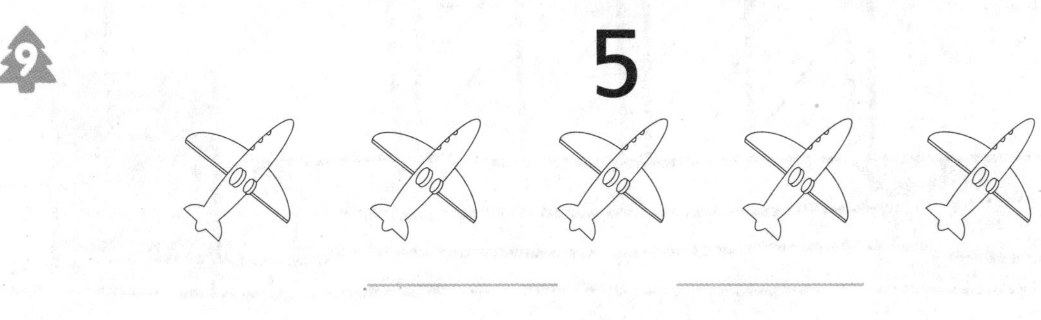

______ ______

______ and ______

**Directions: 7–9.** Use red and yellow to color the objects to show a way to make four and five. Write the numbers.

# Brain Builders

Processes & Practices

PUBLIC LIBRARY

10

11

**Directions: 10.** Draw four windows. Color the windows yellow and blue to show a way to make four. **11.** Draw five windows. Color the windows yellow and blue to show a way to make five. Explain all of the ways that you could make 5 to a friend.

Name

# My Homework

Lesson 1

Make 4 and 5

## Homework Helper

eHelp

Need help? connectED.mcgraw-hill.com

1

4

2 and 2

2

5

1 and 4

3

4

and

**Directions: 1–2.** Use red and blue to color the objects to show a way to make four and five. Write the numbers. **3.** Use red and blue to color the objects to show a way to make four. Write the numbers.

4

_______ and _______

5

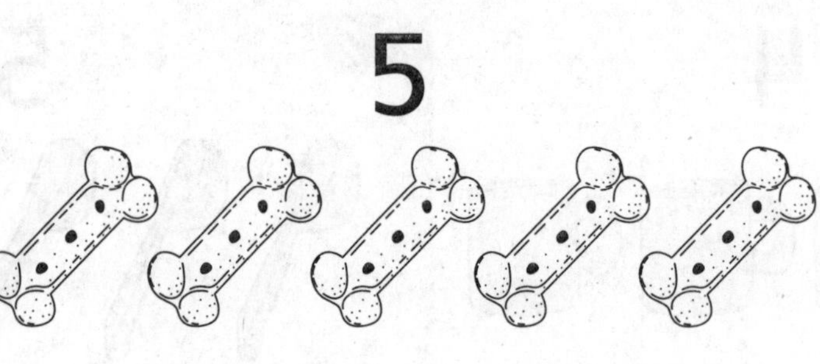

_______ and _______

5

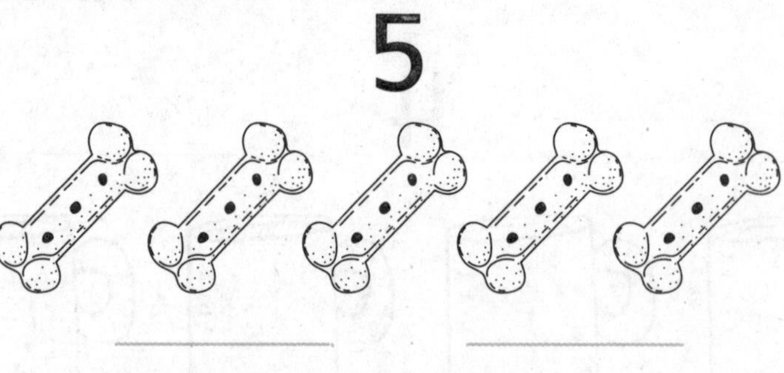

_______ and _______

**Directions: 4–6.** Use red and blue to color the objects to show a way to make four and five. Write the numbers.

**Math at Home** Draw a picture of four objects such as four cups or four plates. Have your child color the objects to show a way to make four. Repeat with five objects having your child color to show a way to make five.

Name

Lesson 2

# Take Apart 4 and 5

**ESSENTIAL QUESTION**
How can we show a number in other ways?

##  Math in My World 

TAXI

1

2

**Teacher Directions: 1.** Use [bear] to show four. Break apart the counting bears into two groups to take apart four. Circle the bears to show one way to take apart four. **2.** Repeat with ways to take apart five. Circle the bears to show one way to take apart five.

Processes &Practices

## Guided Practice

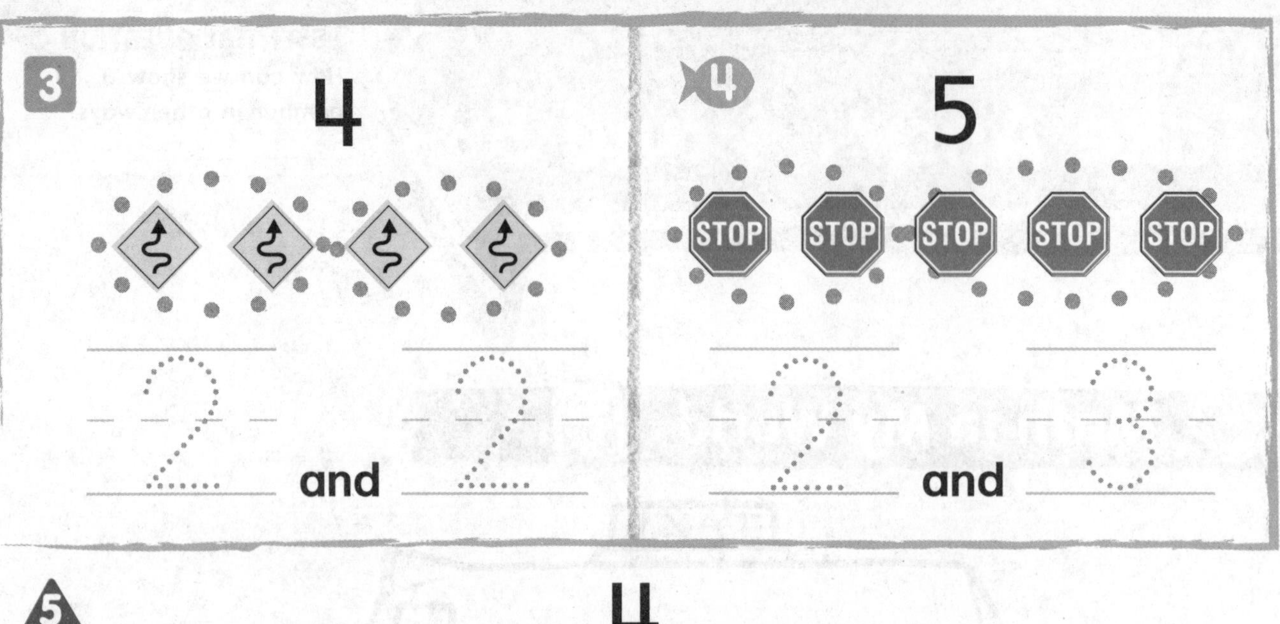


5 4

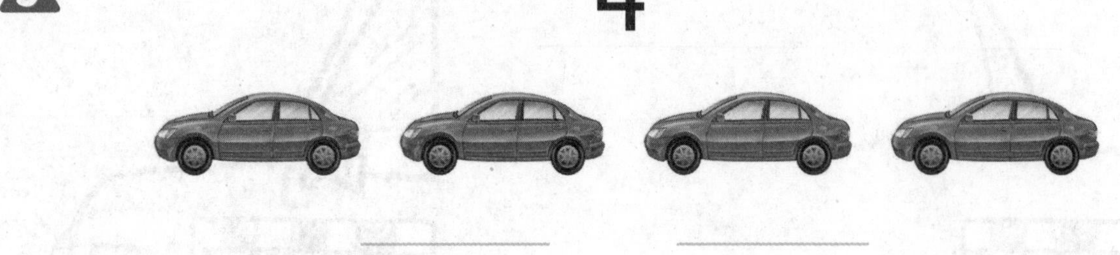

and

6 5

and

**Directions: 3–4.** Look at the number. Count the objects. Trace the dashed circles to show a way to take apart the number. Trace the numbers. **5–6.** Look at the number. Count the objects. Circle the objects to show a way to take apart the number. Write the numbers.

Name

## Independent Practice

5

_____ and _____

4

_____ and _____

5

_____ and _____

**Directions: 7–9.** Look at the number. Count the objects. Circle the objects to show a way to take apart the number. Write the numbers.

Processes &Practices

## Brain Builders

10

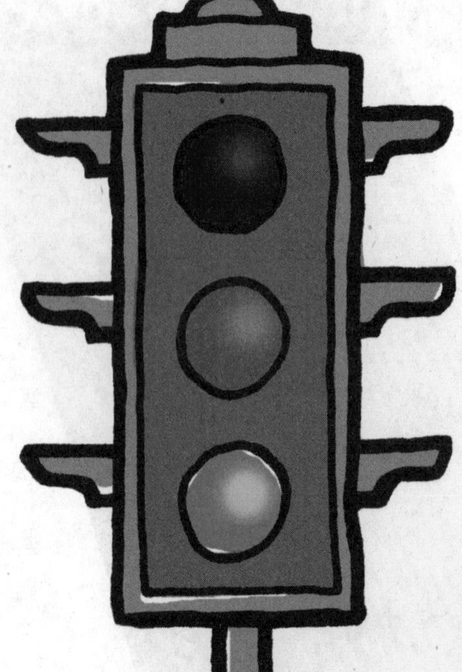

1 and

11

1 and

**Directions: 10–11.** Count how many counters. Write the number above the counters. Use counters to show the number. Take apart a group of one counter. Trace the number one. Count the counters in the other group. Write the number. Circle the counters to show each group. Discuss with a partner all of the ways that you could take apart 4 and 5 counters.

Name ..........................................

# My Homework

**Lesson 2**

Take Apart 4 and 5

## Homework Helper

Need help? connectED.mcgraw-hill.com

1

4

3 and 1

2

4

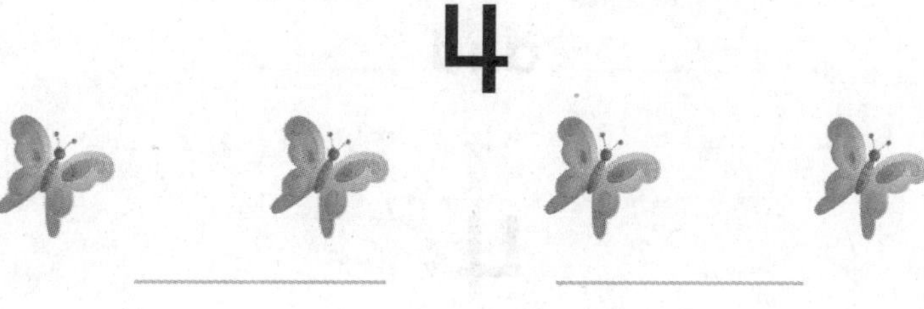

______ and ______

3

5

______ and ______

**Directions: 1–3.** Look at the number. Count the objects. Circle the objects to show a way to take apart the number. Write the numbers.

4

# 4

______ ______

______ and ______

# 5

______ ______

______ and ______

# 4

______ ______

______ and ______

**Directions: 4–6.** Look at the number. Count the objects. Circle the objects to show a way to take apart the number. Write the numbers.

**Math at Home** Show your child a group of four cups. Have your child break apart the group into two groups to show a way to take apart four. Guide your child in writing the numbers that tell how many are in each group. Repeat using five cups.

Name

# Lesson 3
# Make 6 and 7

**ESSENTIAL QUESTION**
How can we show a number in other ways?

## Math in My World

1 ○ ○ ○ ○ ○ ○

2 ○ ○ ○ ○ ○ ○ ○

**Teacher Directions: 1.** Use (red) and (yellow) to show ways to make six. Color the row of six counters red and yellow to show one way. **2.** Repeat to show ways to make seven. Color the row of counters red and yellow to show one way.

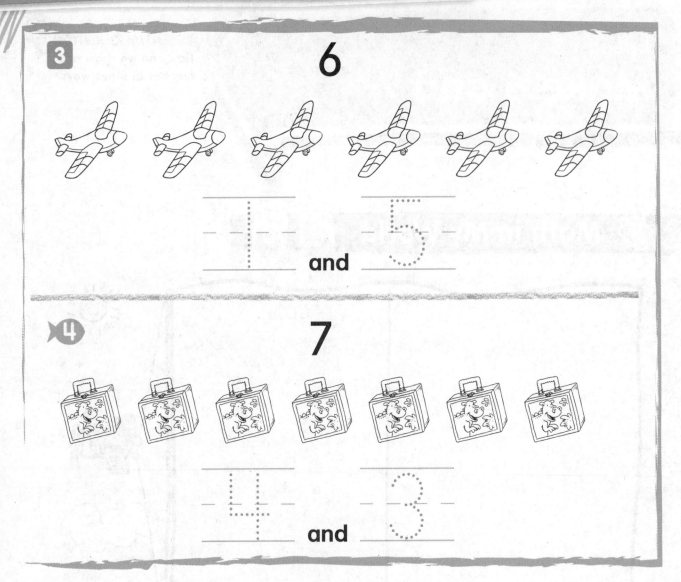

## Guided Practice

**3** 6

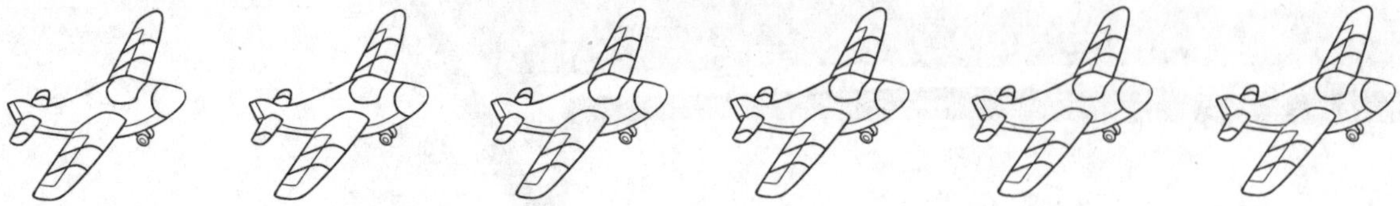

1 and 5

**4** 7

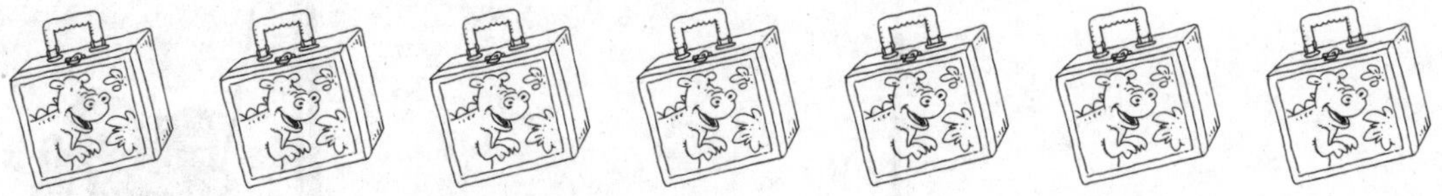

4 and 3

**5** 6

____ and ____

**Directions: 3–4.** Trace the numbers to show a way to make six and seven. Use red and blue to color the objects to show this way. **5.** Use red and blue to color the objects to show a way to make six. Write the numbers.

Name

## Independent Practice

6

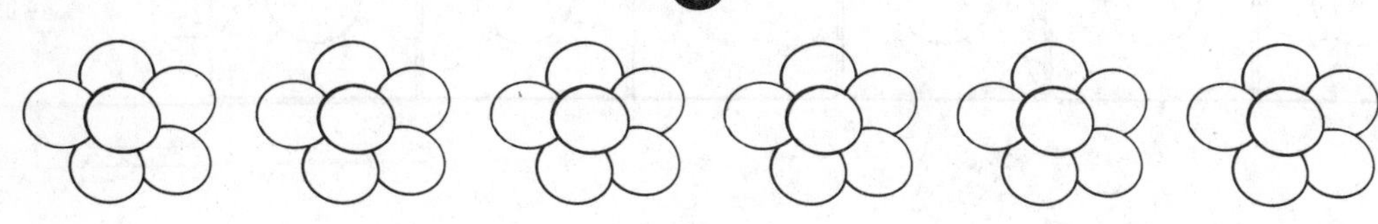

_____ and _____

7

_____ and _____

7

_____ and _____

**Directions: 6–8.** Use red and blue to color the objects to show a way to make six and seven. Write the numbers.

Processes & Practices

# Brain Builders

6

9 I and 5

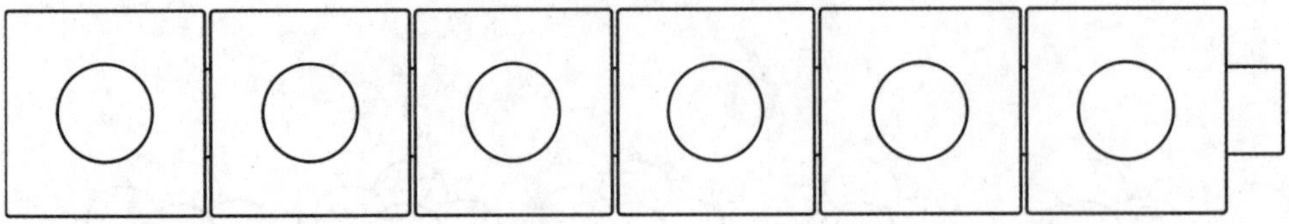

10 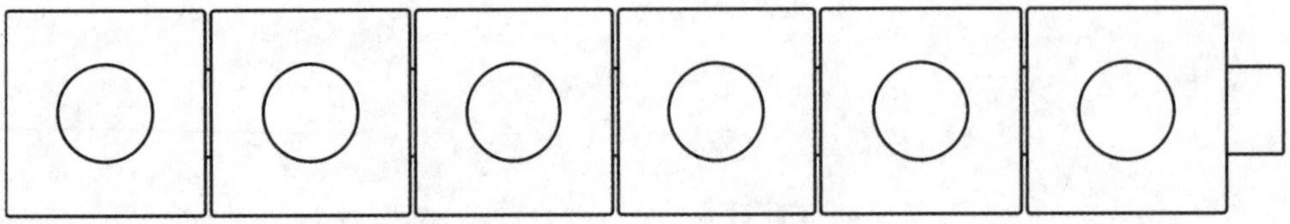 2 and 4

11 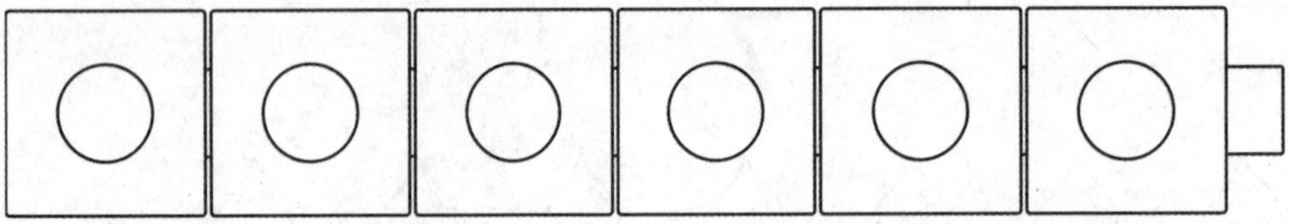 3 and 3

12 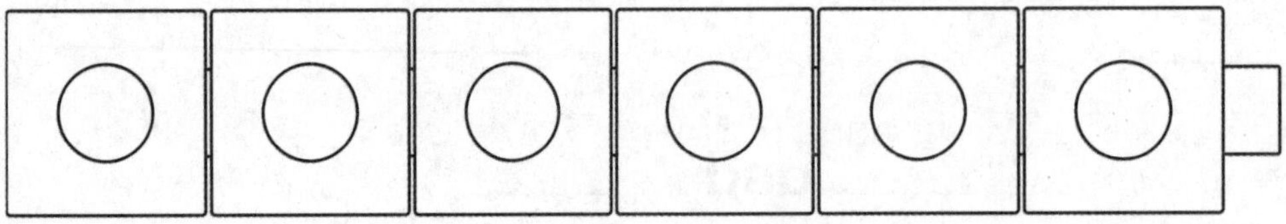 4 and 2

13 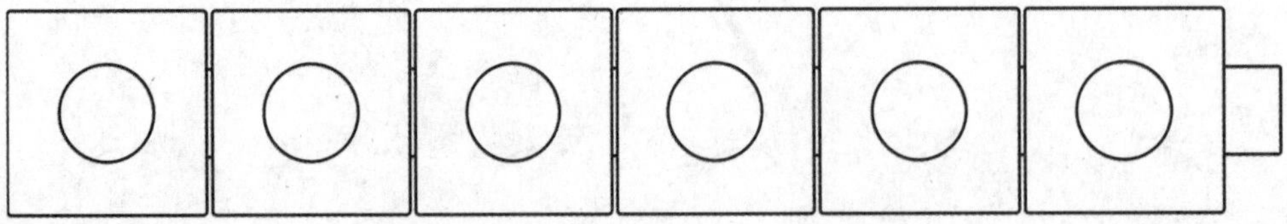 5 and I

14 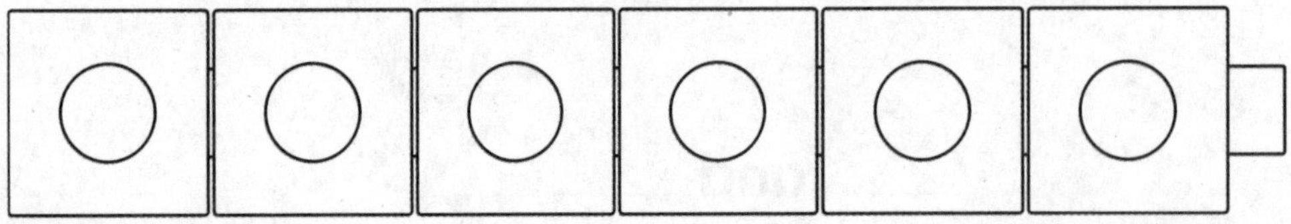 6 and 0

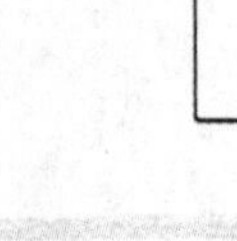

**Directions: 9–I4.** Say the numbers that show a way to make six. Color the cubes purple and green to show that way to make six. Discuss with a friend all of the ways to make seven using cubes.

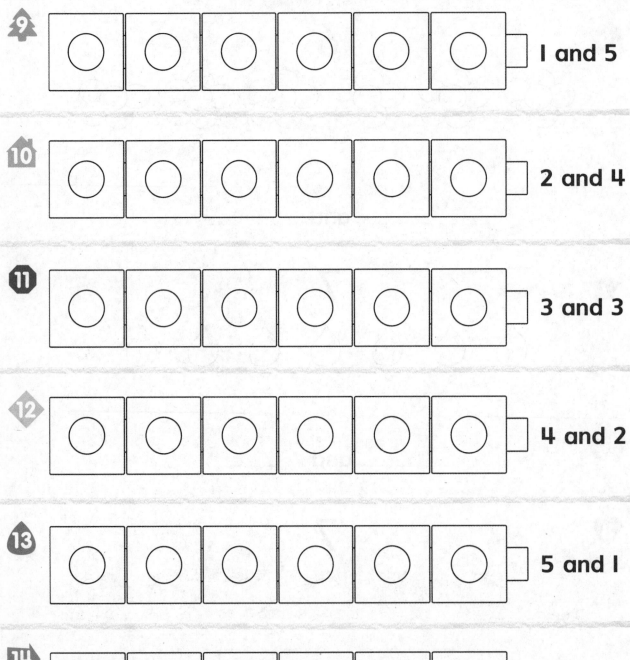

Name

Lesson 3

Make 6 and 7

# My Homework

## Homework Helper

Need help? connectED.mcgraw-hill.com

**1**

6

3 and 3

**2**

6

_____ and _____

**3**

7

_____ and _____

**Directions: I–3.** Use orange and purple to color the objects to show a way to make six and seven. Write the numbers.

4

6

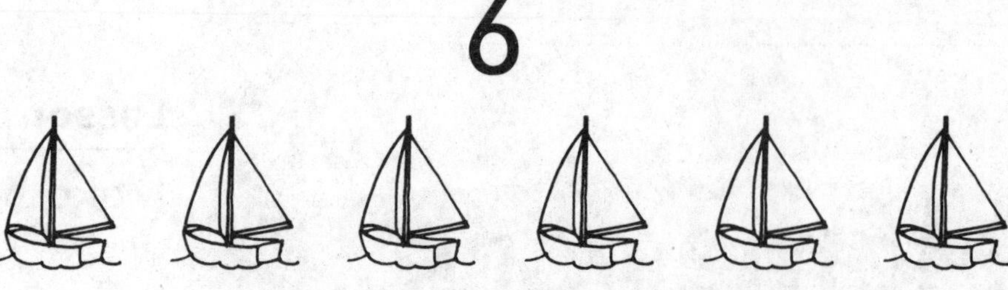

______ and ______

5

7

______ and ______

6

7

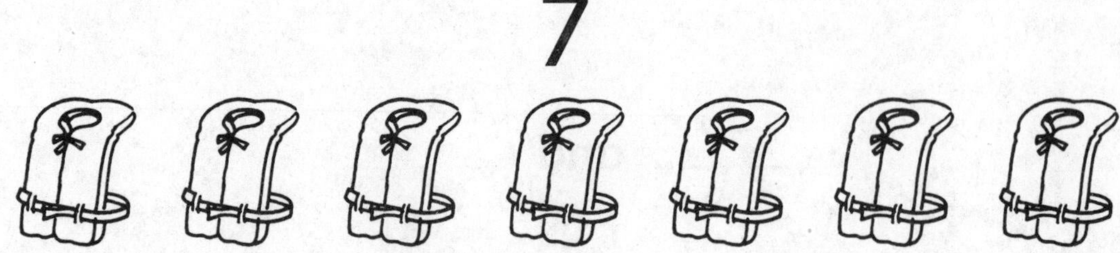

______ and ______

**Directions: 4–6.** Use orange and purple to color the objects to show a way to make six and seven. Write the numbers.

**Math at Home** Draw a picture of six objects such as six books or six crayons. Have your child color the objects to show a way to make six. Repeat by drawing seven objects and having your child show a way to make seven.

Name

Lesson 4

# Take Apart 6 and 7

**ESSENTIAL QUESTION**
How can we show a number in other ways?

## Math in My World

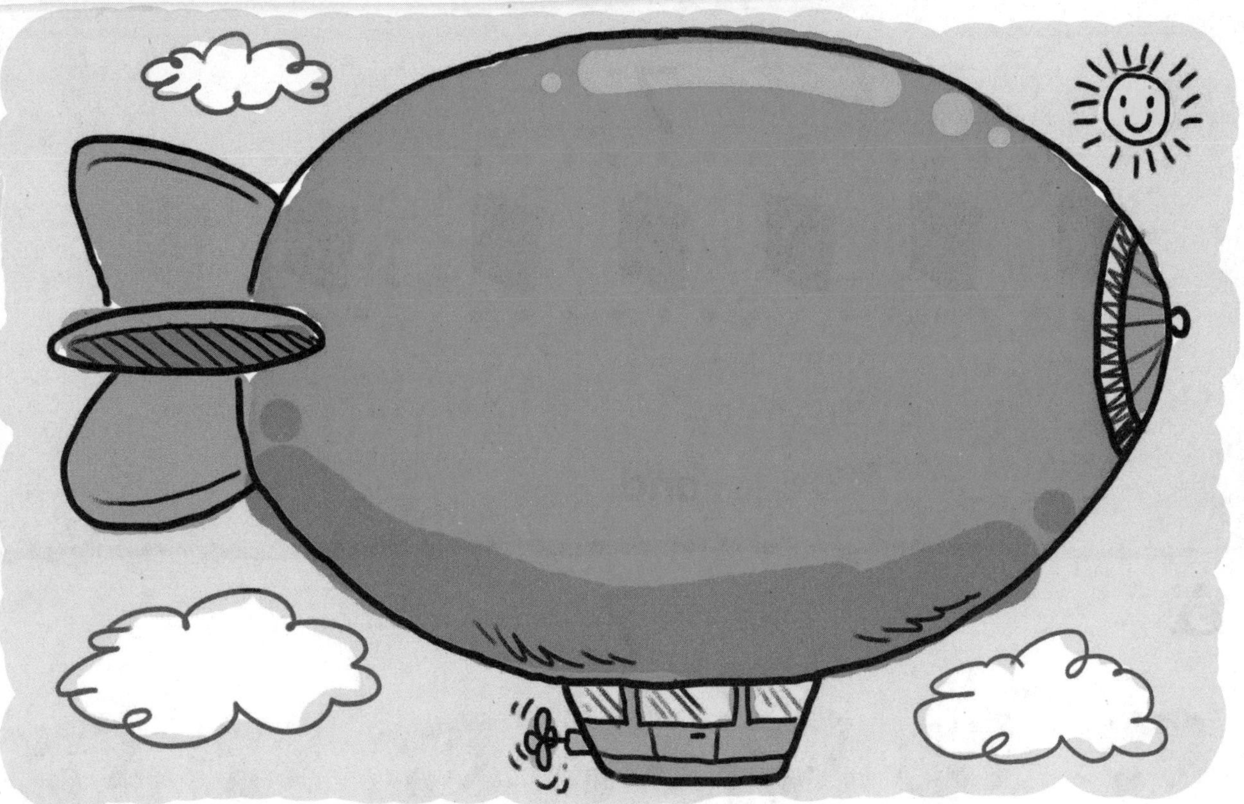

1

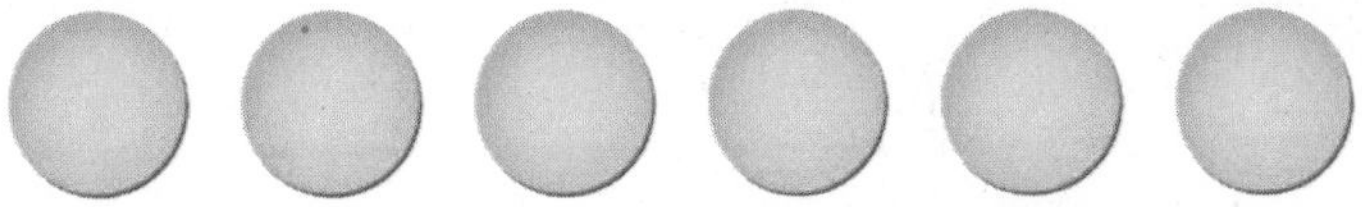

2

**Teacher Directions: 1.** Use ● to show six. Break apart the counters into two groups to take apart six. Circle the counters to show a way to take apart six. **2.** Repeat with ways to take apart seven. Circle the counters to show one way to take apart seven.

## Guided Practice

Processes &Practices

3

6

COME TO MY PARTY! COME TO MY PARTY! COME TO MY PARTY! COME TO MY PARTY! COME TO MY PARTY! COME TO MY PARTY!

3 and 3

4

7

5 and 2

5

7

____ and ____

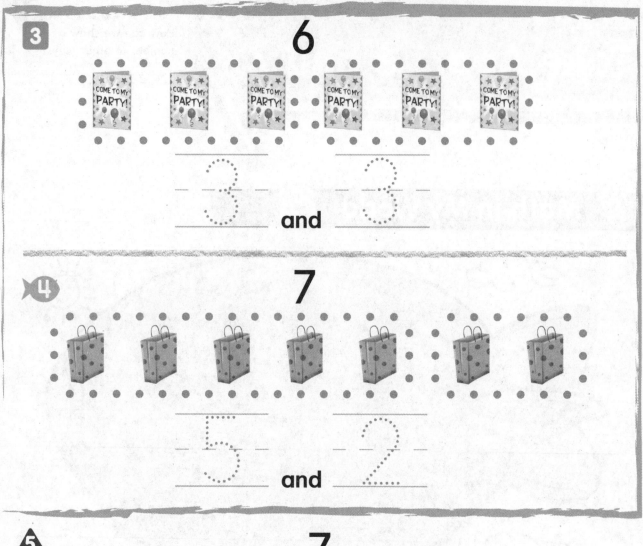

**Directions: 3–4.** Look at the number. Count the objects. Trace the dashed circles to show a way to take apart the number. Trace the numbers. **5.** Look at the number. Count the objects. Circle the objects to show a way to take apart the number. Write the numbers.

Name

## Independent Practice

6

6

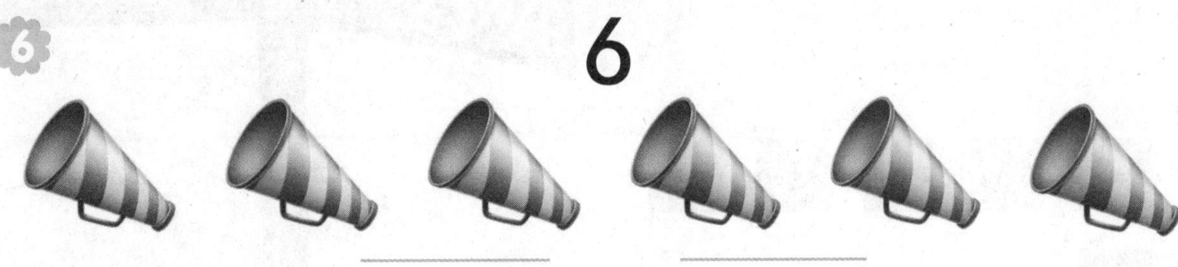

and

7

6

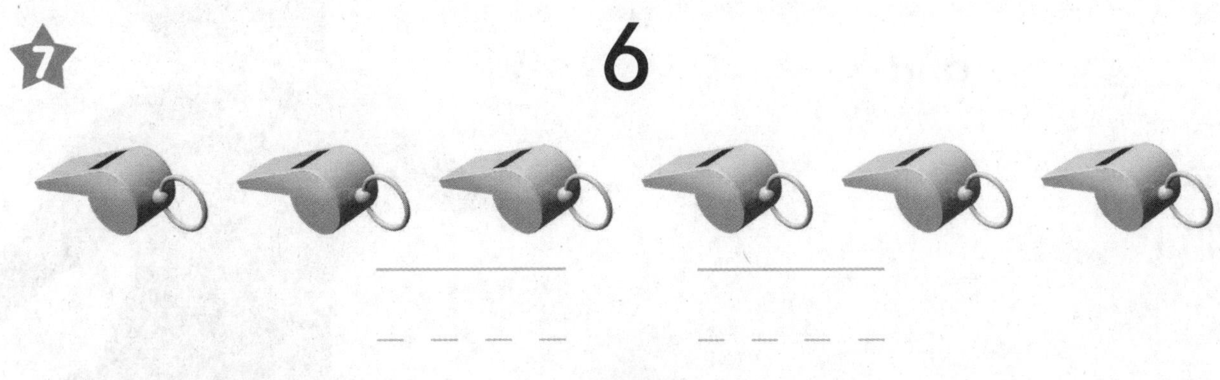

and

8

7

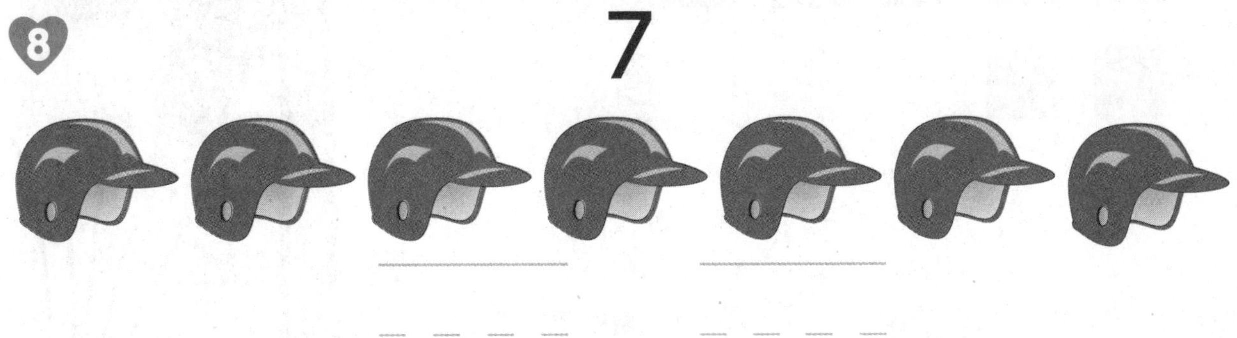

and

**Directions: 6–8.** Look at the number. Count the objects. Circle the objects to show a way to take apart the number. Write the numbers.

Processes &Practices

## Brain Builders

9

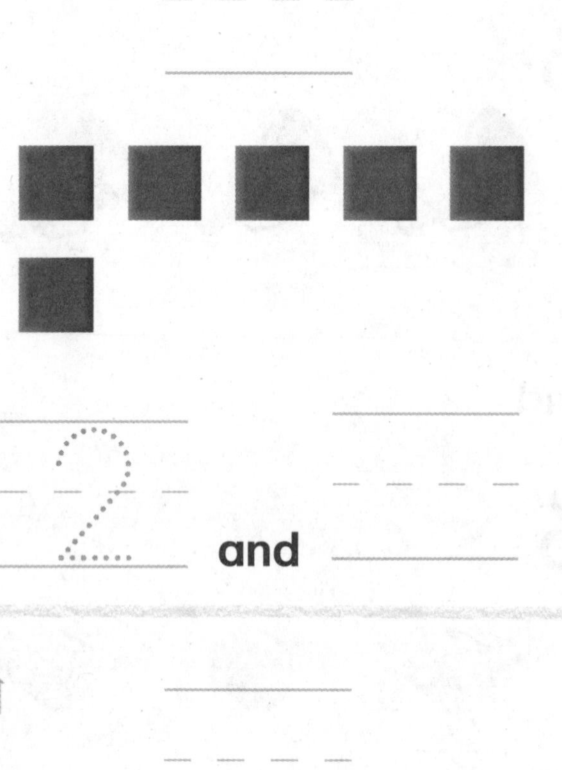

2 and

10

2 and

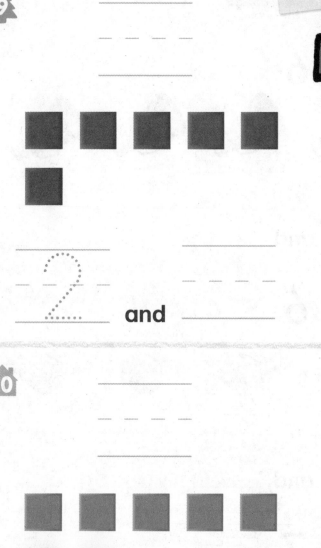

**Directions: 9–10.** Count how many tiles. Write the number above the tiles. Use tiles to show the number. Take apart a group of two tiles. Trace the number two. Count the tiles in the other group. Write the number. Circle the tiles to show each group. Explain to a friend another way that you can take apart the number. Use tiles to help you.

Name

# My Homework

Lesson 4

Take Apart 6 and 7

## Homework Helper

Need help? connectED.mcgraw-hill.com

1

6

5 and 1

2

7

______ and ______

3

6

______ and ______

**Directions: 1–3.** Look at the number. Count the objects. Circle the objects to show a way to take apart the number. Write the numbers.

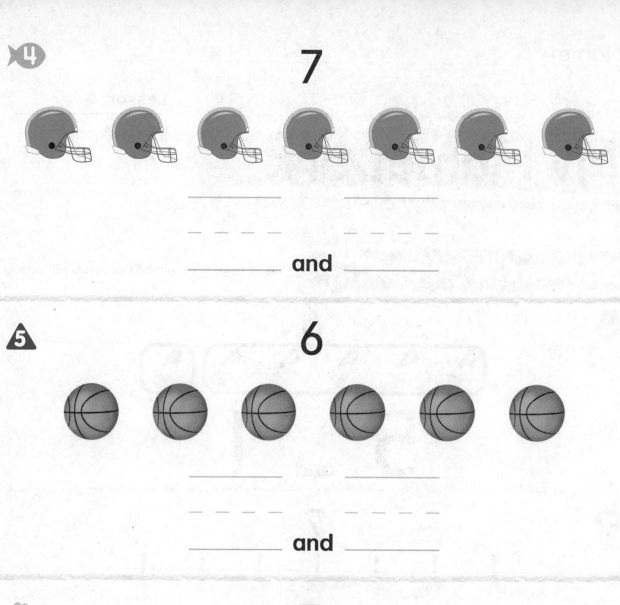

4

7

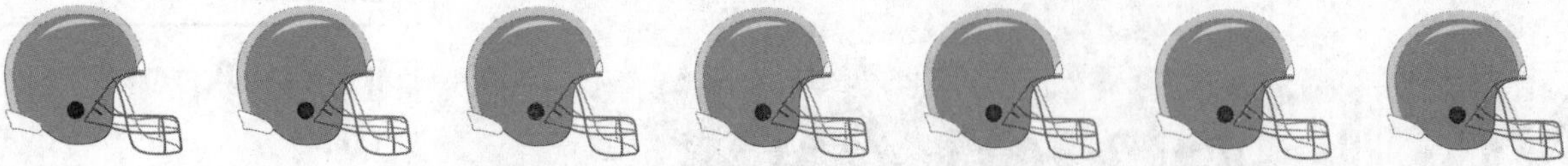

____ ____

____ and ____

5

6

____ ____

____ and ____

6

7

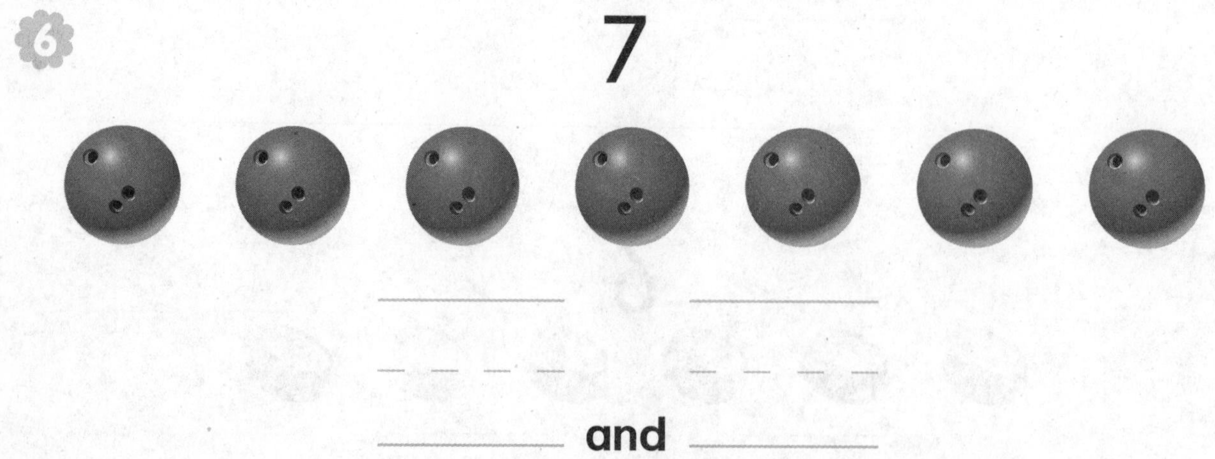

____ ____

____ and ____

**Directions: 4–6.** Look at the number. Count the objects. Circle the objects to show a way to take apart six and seven. Write the numbers.

**Math at Home** Show your child a group of six pennies. Have your child break apart the group into two groups to show a way to take apart six. Guide your child in writing the numbers that tell how many are in each group. Repeat using seven pennies.

Name

Lesson 5

# Problem Solving

**STRATEGY: Act It Out**

**ESSENTIAL QUESTION**

How can we show a number in other ways?

## How do you make the number?

## Act It Out

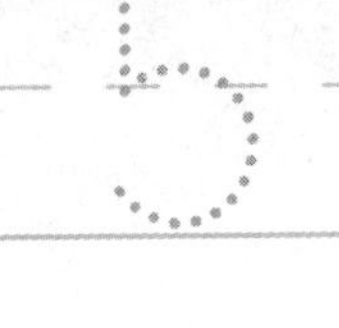

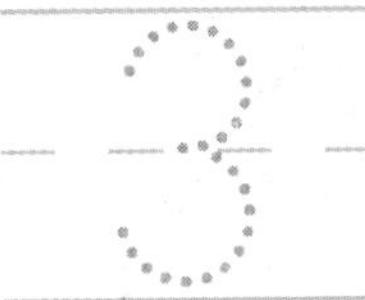

and

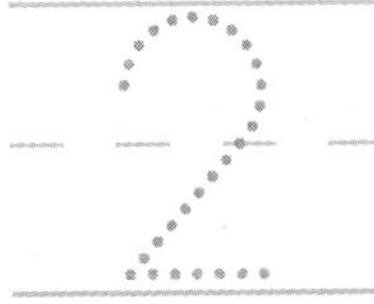

**Teacher Directions:** Look at the attribute buttons. Use ▲ and ▲ to show them. Count the buttons. Say how many. Trace the number on the top line. Trace the numbers that show a way to make that number.

## How do you make the number?

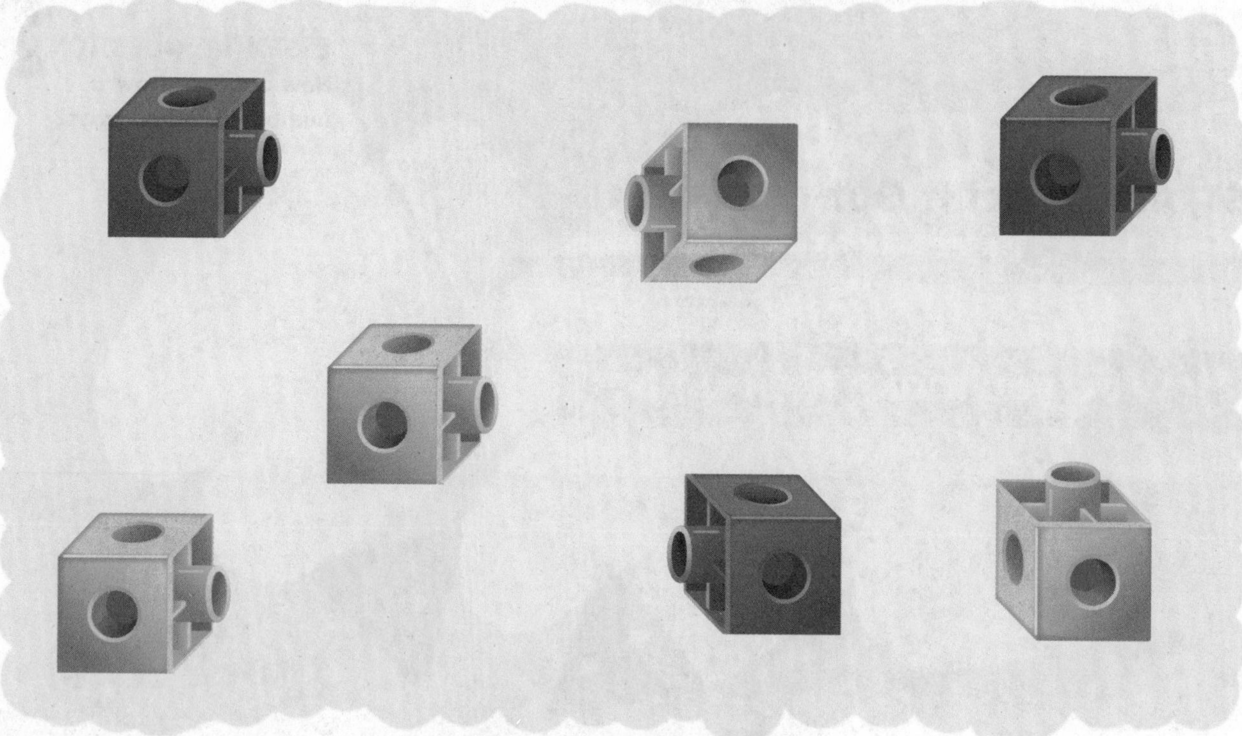

## Act It Out

and

**Directions:** Look at the connecting cubes. Use and to show them. Count the cubes. Say how many. Write the number on the top line. Write the numbers that show how many purple cubes and how many orange cubes as a way to make that number.

Name

## How do you make the number?

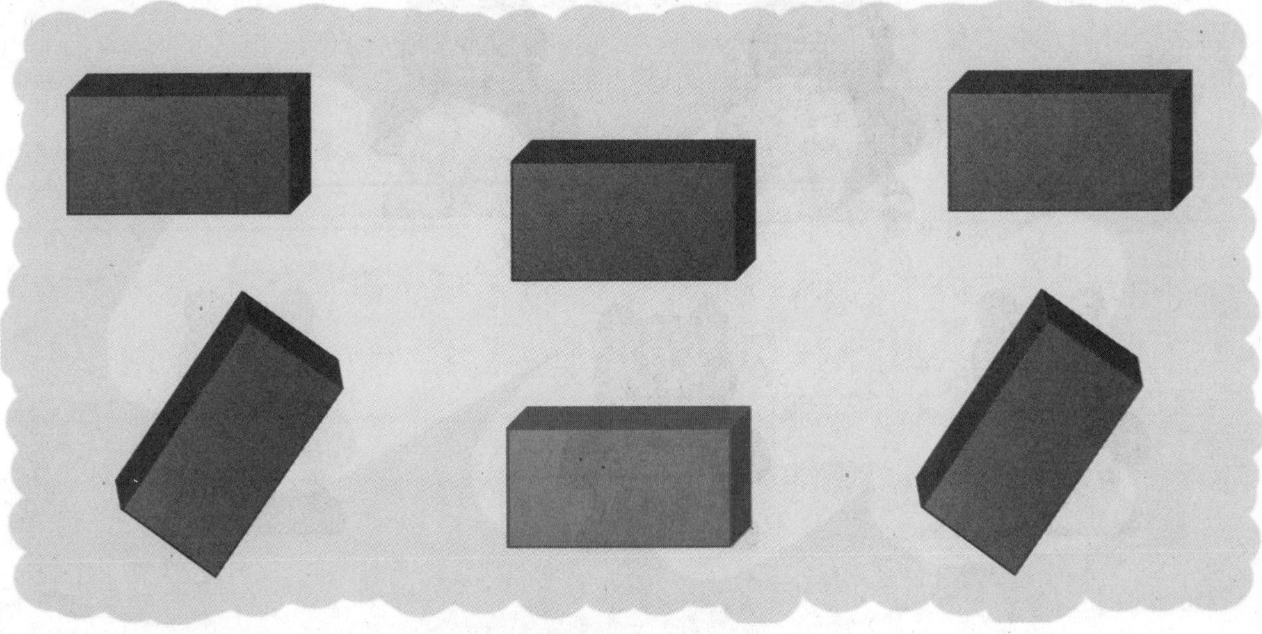

## Act It Out

and

**Directions:** Look at the attribute blocks. Use [blue block] and [red block] to show them. Count the blocks. Say how many. Write the number on the top line. Write the numbers that show how many blue blocks and how many red blocks as a way to make that number. Tell a friend how many blocks there would be if you would draw another red block. Tell the friend the number of blue blocks and the number of red blocks that make that number.

## How do you make the number?

## Act It Out

and

**Directions:** Look at the bears. Use and to show them. Count the bears. Say how many. Write the number on the top line. Write the numbers that show how many green bears and how many blue bears as a way to make that number.

Name



# My Homework

## How do you make the number?

## Act It Out

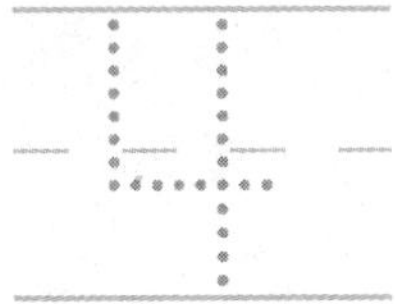

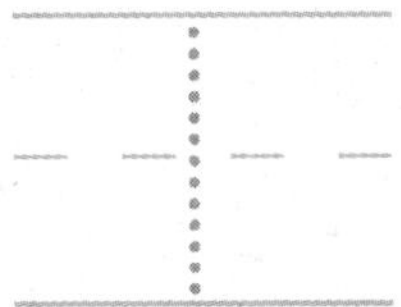

and

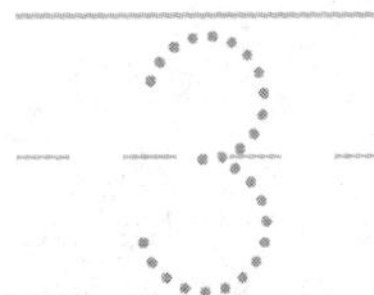

**Directions:** Look at the crayons. Use crayons to show them. Count the crayons. Say how many. Trace the number on the top line. Trace the numbers that show a way to make that number.

## How do you make the number?

## Act It Out

________

________ **and** ________

**Directions:** Look at the crayons. Use crayons to show them. Count the crayons. Say how many. Write the number on the top line. Write the numbers that show how many red crayons and how many blue crayons as a way to make that number.

**Math at Home** Take advantage of problem-solving opportunities during daily routines such as cooking. Use teaspoons and tablespoons to show ways to make a number.

Name ..............................

# Check My Progress

## Vocabulary Check 

**1** six 6

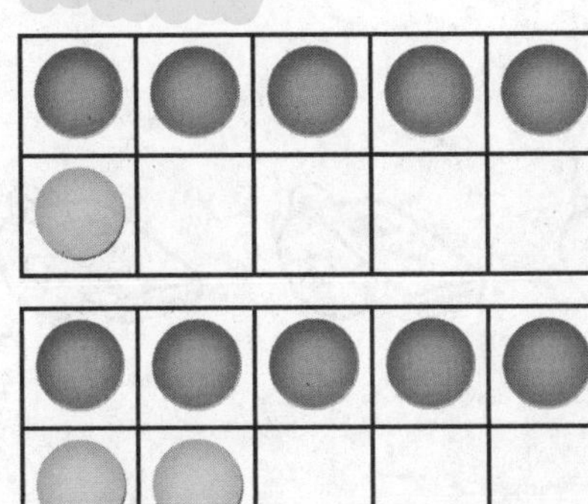

**2** seven 7

## Concept Check

**3**

4

_______ _______

_______ and _______

**Directions: 1.** Count the objects in each ten-frame. Circle the ten-frame that shows six. **2.** Draw more objects to show seven. **3.** Use purple and orange to color the objects to show a way to make four. Write the numbers.

4

# 5

________ and ________

# 6

________ and ________

# 7

________ and ________

**Directions: 4.** Use green and yellow to color the objects to show a way to make five. Write the numbers. **5.** Use red and blue to color the objects to show a way to make six. Write the numbers. **6.** Look at the number. Count the objects. Circle the objects to show a way to take apart the number. Write the numbers.

Name

Lesson 6

# Make 8 and 9

**ESSENTIAL QUESTION**
How can we show a number in other ways?

## Math in My World

1 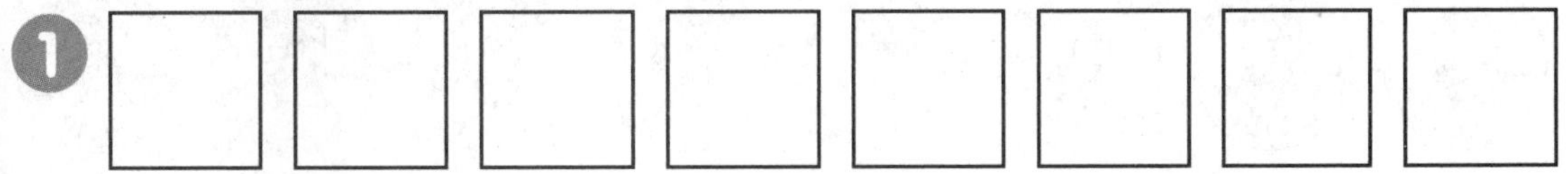

2 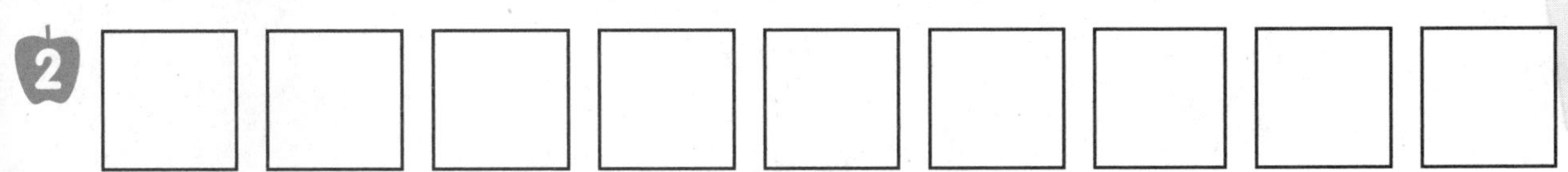

**Teacher Directions: 1.** Use [red] and [yellow] to show ways to make eight. Color the row of eight tiles red and yellow to show one way. **2.** Repeat to show ways to make nine. Color the row of tiles red and yellow to show one way.

## Guided Practice

Processes & Practices

3

8

1 and 7

4

9

6 and 3

5

9

____ and ____

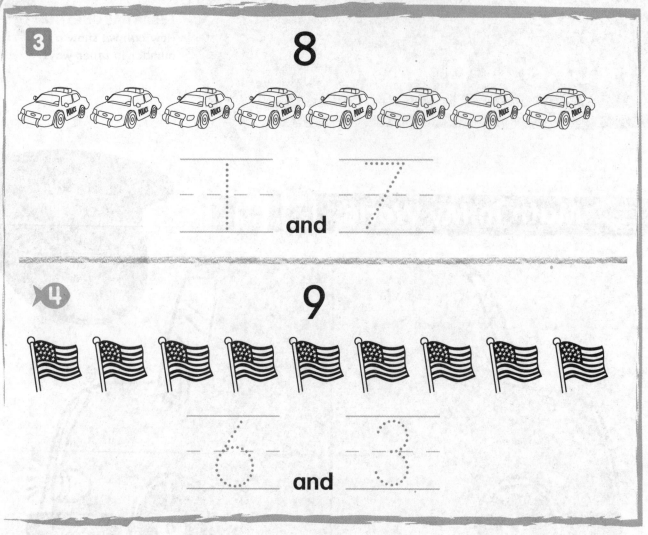

**Directions: 3–4.** Trace the numbers to show a way to make eight and nine. Use purple and green to color the objects to show this way. **5.** Use purple and green to color the objects to show a way to make nine. Write the numbers.

Name

## Independent Practice

6

8

and

7

8

and

8

9

and

**Directions: 6–8.** Use purple and green to color the objects to show a way to make eight and nine. Write the numbers.

Processes & Practices

# Brain Builders

9

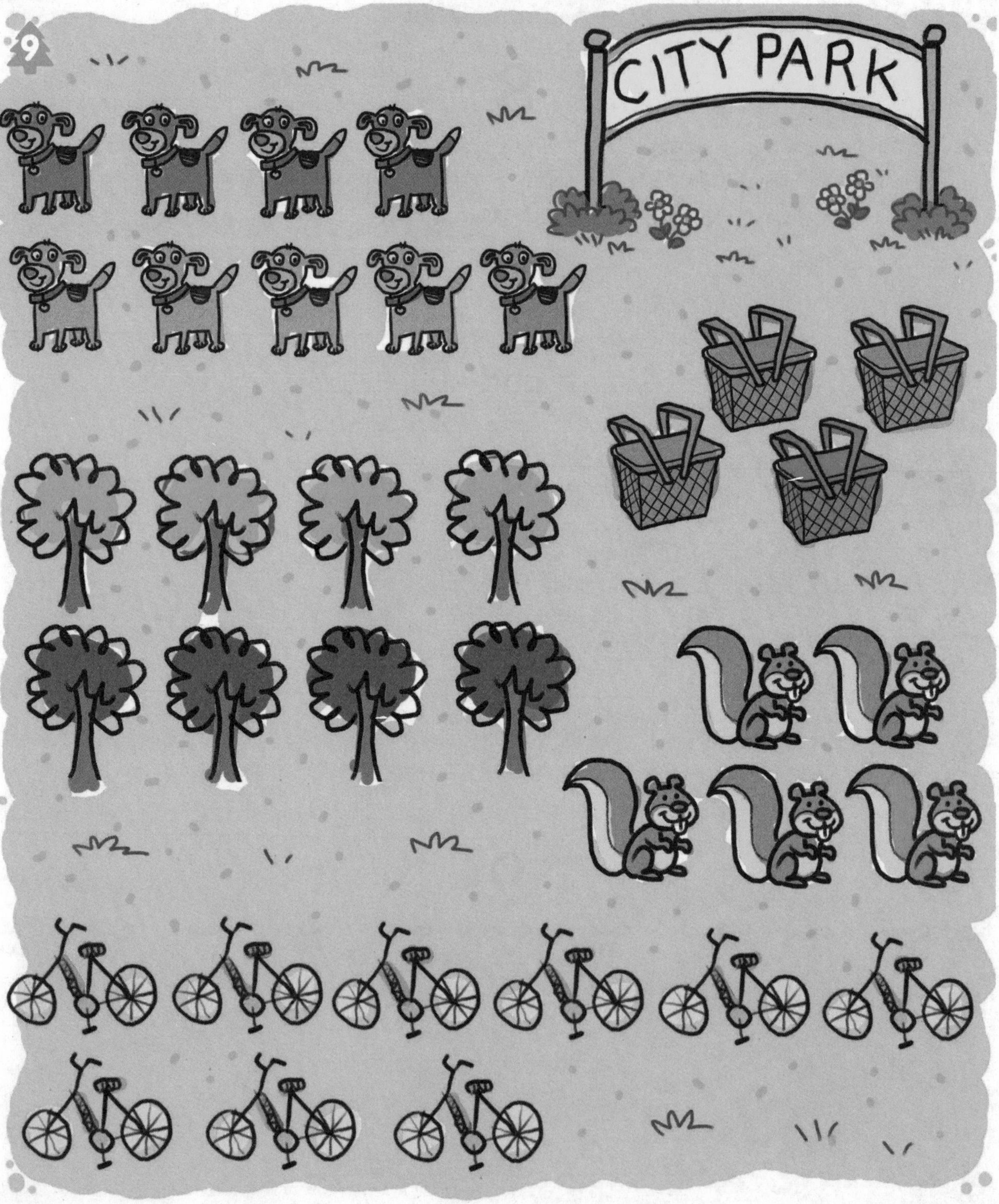

CITY PARK

**Directions: 9.** Draw a circle around the group(s) that show a way to make eight. Draw an X on the group(s) that show a way to make nine. Count with a friend the number of total objects in each group. Then take turns saying the way it shows how to make that number in each group of objects.

Name ____________________

Lesson 6

Make 8 and 9

# My Homework

## Homework Helper 

Need help? connectED.mcgraw-hill.com

**1**

8

6 and 2

**2**

8

______ and ______

**3**

9

______ and ______

**Directions: 1–3.** Use red and yellow to color the objects to show a way to make eight and nine. Write the numbers.

9

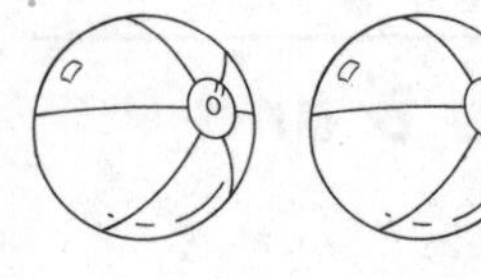
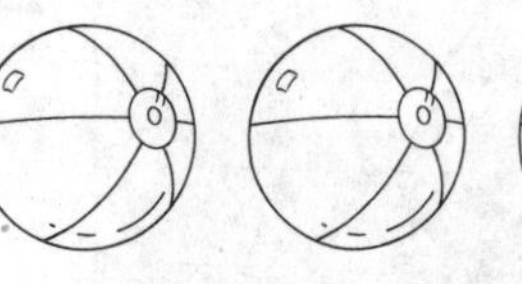
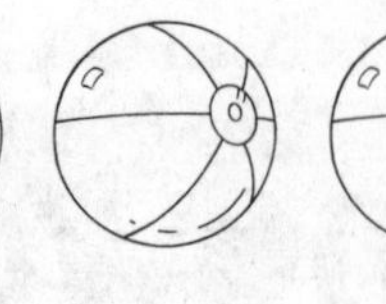
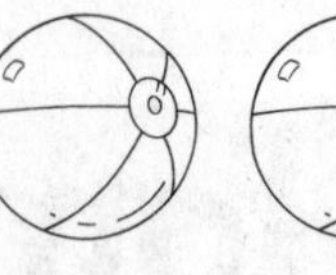
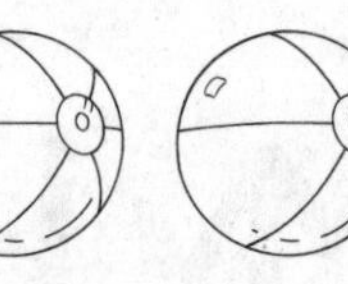

and

8

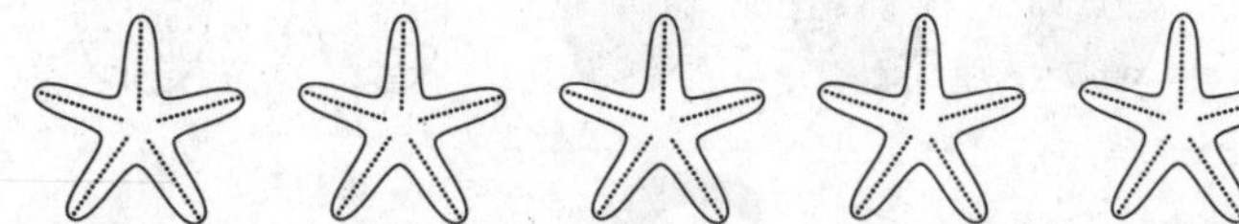
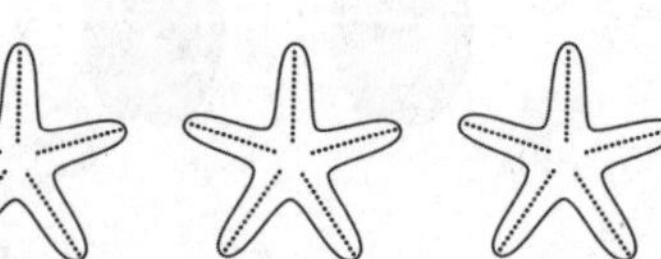

and

6

9

and

**Directions: 4–6.** Use red and yellow to color the objects to show a way to make eight and nine. Write the numbers.

**Math at Home** Draw a picture of objects such as eight dolls or eight toothbrushes. Have your child color the objects to show a way to make eight. Repeat by drawing nine objects and having your child show another way to make nine.

Name

# Lesson 7
# Take Apart 8 and 9

**ESSENTIAL QUESTION**
How can we show a number in other ways?

## Math in My World

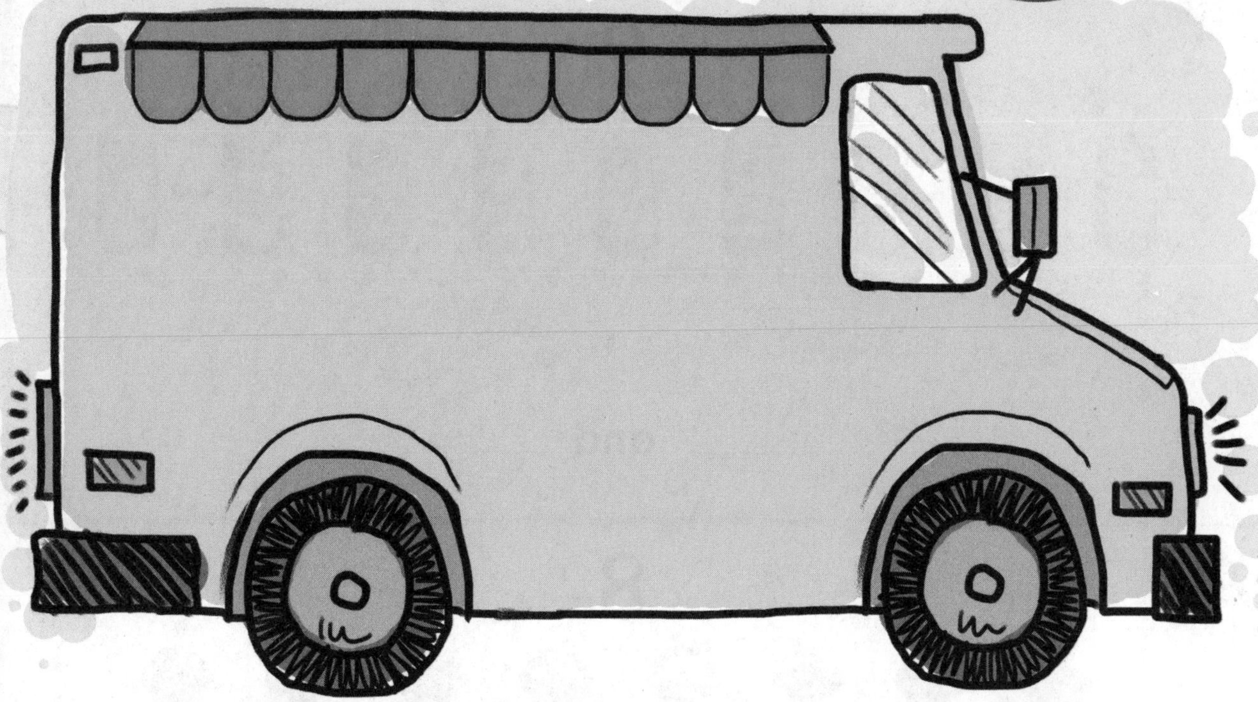

1

2

**Teacher Directions: I.** Use [tile] to show eight. Break apart the tiles into two groups to take apart eight. Circle the tiles to show a way to take apart eight. **2.** Repeat with ways to take apart nine. Circle the tiles to show a way to take apart nine.

Processes
&Practices

## Guided Practice

3

8

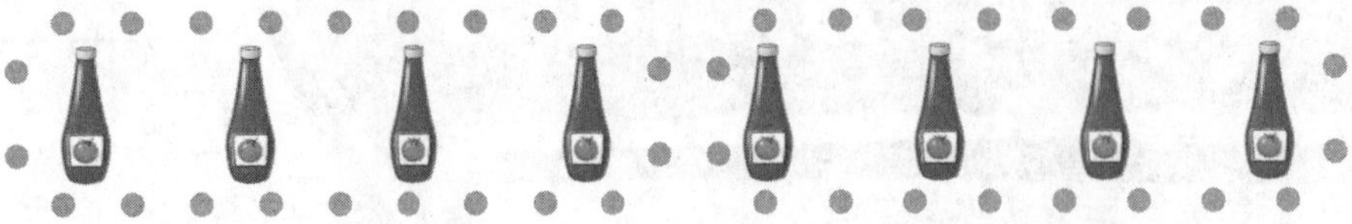

4 and 4

4

9

8 and 1

5

8

and

**Directions: 3–4.** Look at the number. Count the objects. Trace the dashed circles to show a way to take apart the number. Trace the numbers. **5.** Look at the number. Count the objects. Circle the objects to show a way to take apart the number. Write the numbers.

Name

## Independent Practice

6

9

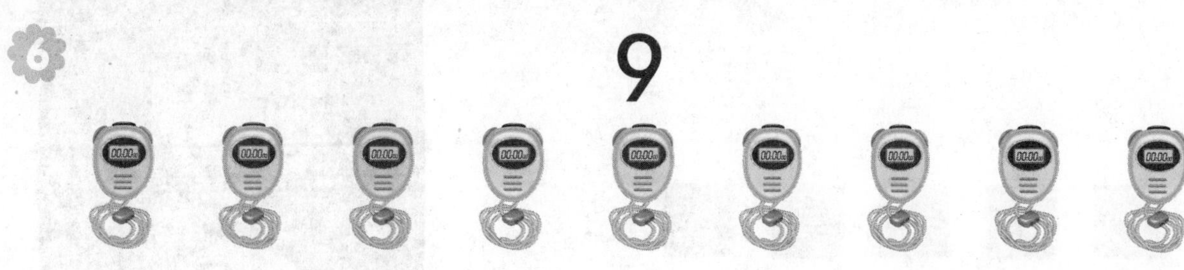

_____ _____

_____ and _____

7

8

_____ _____

_____ and _____

8

9

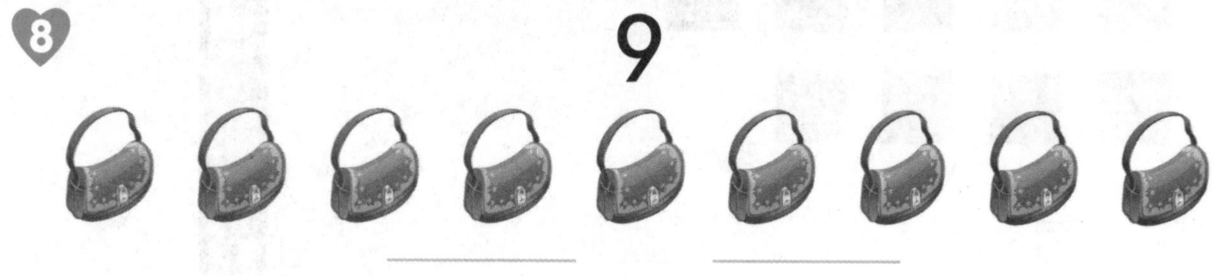

_____ _____

_____ and _____

**Directions: 6–8.** Look at the number. Count the objects. Circle the objects to show a way to take apart the number. Write the numbers.

## Brain Builders

9

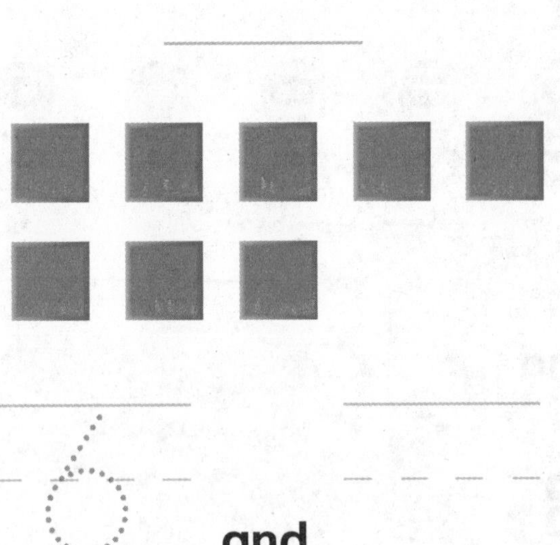

6 and

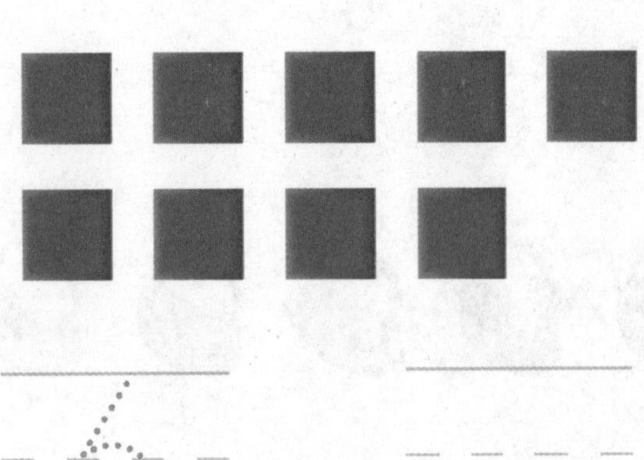

6 and

**Directions: 9–10.** Count how many tiles. Write the number above the tiles. Use tiles to show the number. Break apart a group of six tiles. Trace the number six. Count the tiles in the other group. Write the number. Circle the tiles to show each group. Show a friend eight or nine objects. Say the number of objects. Then break apart a group. Tell your friend how many there are in each of the groups.

Name ____________________

# My Homework

**Lesson 7**

**Take Apart 8 and 9**

## Homework Helper

Need help? connectED.mcgraw-hill.com

**1**

8

7 and 1

**2**

8

____ and ____

**3**

9

____ and ____

**Directions: 1–3.** Look at the number. Count the objects. Circle the objects to show a way to take apart the number. Write the numbers.

4

9

_____ _____

_____ and _____

5

8

_____ _____

_____ and _____

6

9

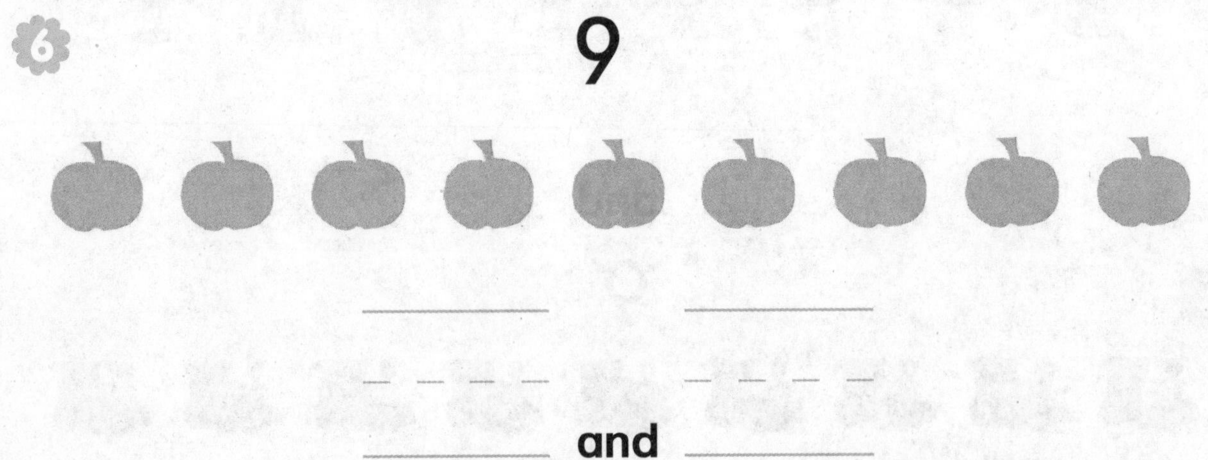

_____ _____

_____ and _____

**Directions: 4–6.** Look at the number. Count the objects. Circle the objects to show a way to take apart the number. Write the numbers.

**Math at Home** Show your child a group of eight buttons. Have your child break apart the group into two groups to show a way to take apart eight. Guide your child in writing the numbers that tell how many are in each group. Repeat using nine buttons.

Name

# Lesson 8
# Make 10

**ESSENTIAL QUESTION**
How can we show a number in other ways?

## Math in My World

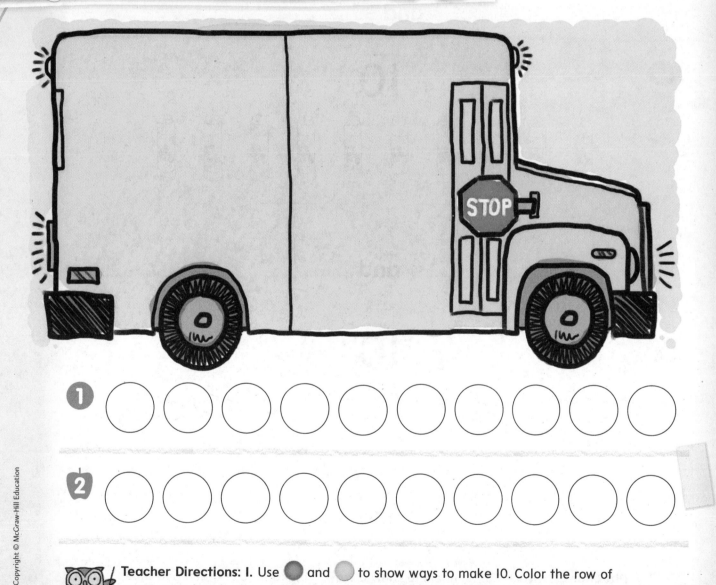

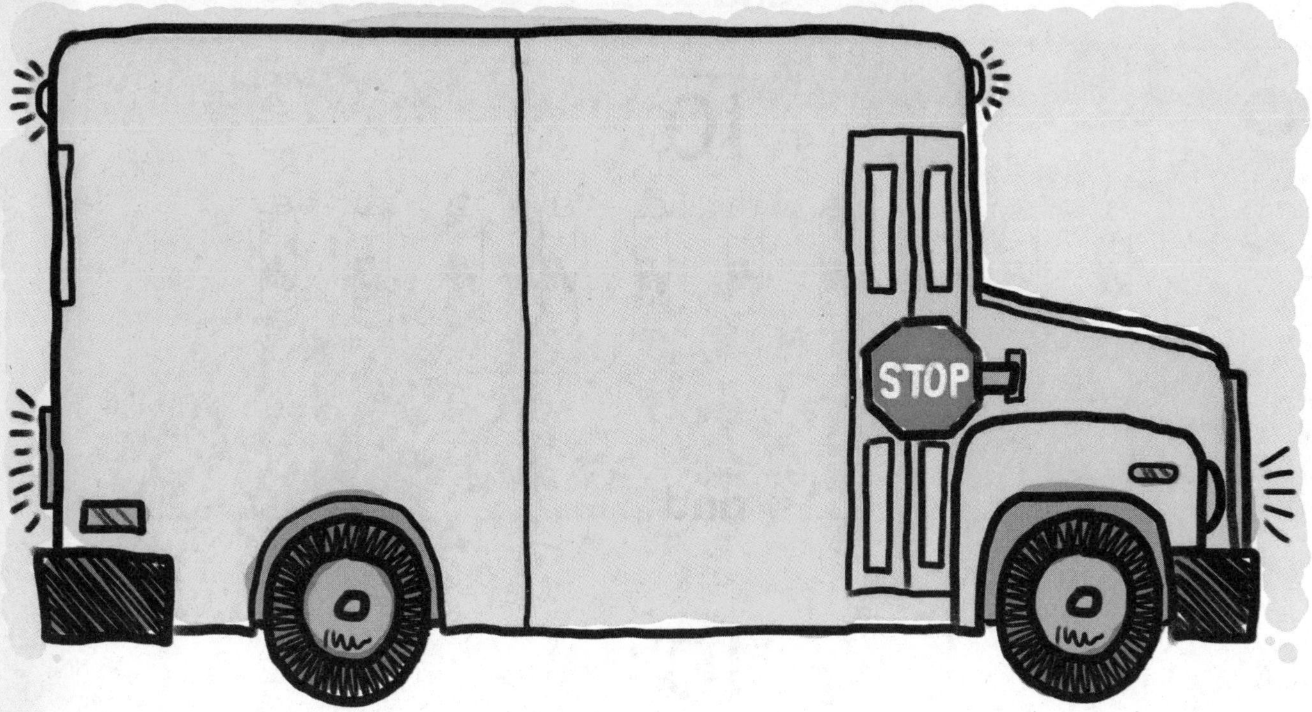

1

2

**Teacher Directions: 1.** Use red and yellow to show ways to make 10. Color the row of counters red and yellow to show one way to make 10. **2.** Repeat to show another way to make 10. Color the row of counters red and yellow to show that way to make 10.

## Guided Practice

3

10

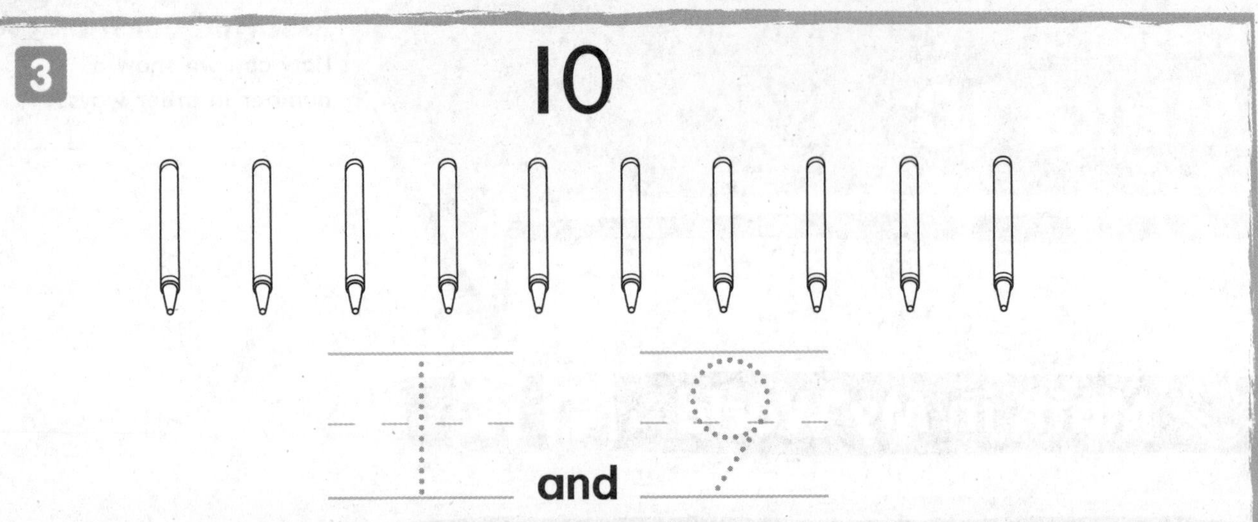

1 and 9

4

10

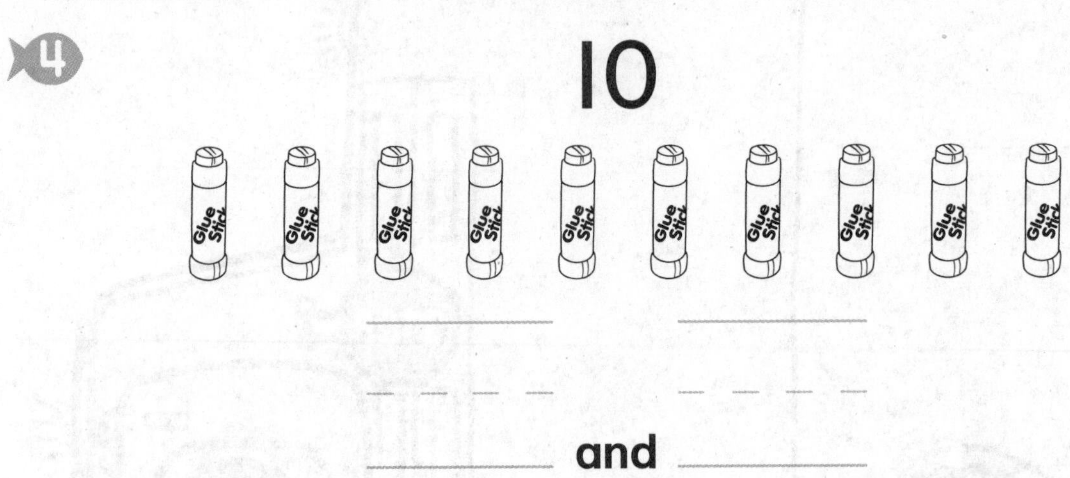

_____ and _____

5

10

_____ and _____

**Directions: 3.** Trace the numbers to show a way to make 10. Use green and blue to color the objects to show this way to make 10. **4–5.** Use green and blue to color the objects to show a way to make 10. Write the numbers.

Name

## Independent Practice

6

10

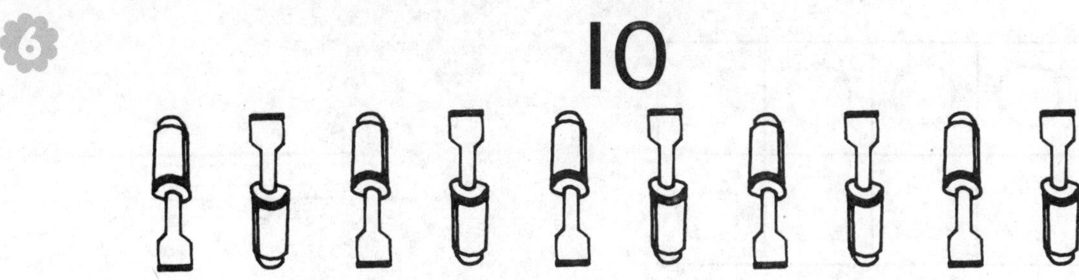

_____ _____

_____ and _____

7

10

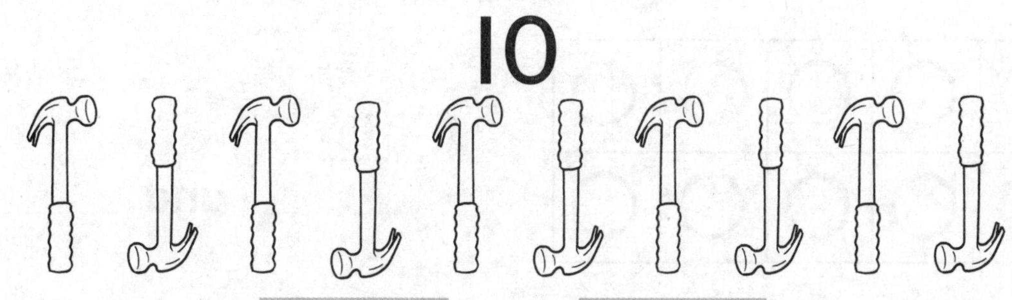

_____ _____

_____ and _____

8

10

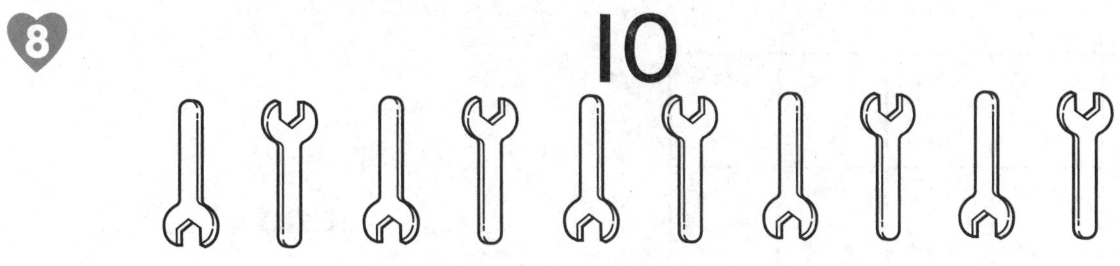

_____ _____

_____ and _____

**Directions: 6–8.** Use orange and purple to color the objects to show a way to make 10. Write the numbers.

## Brain Builders

9

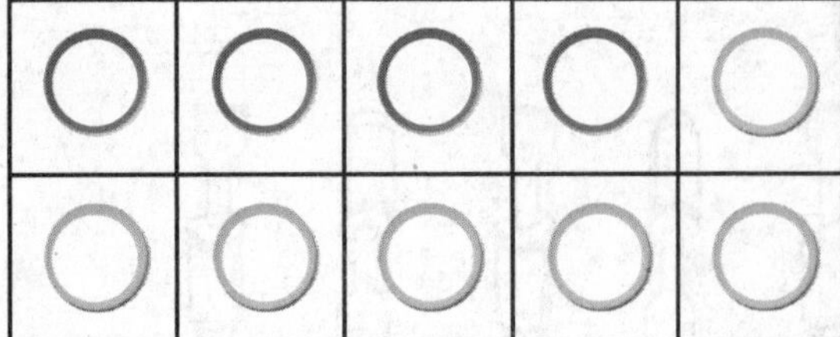

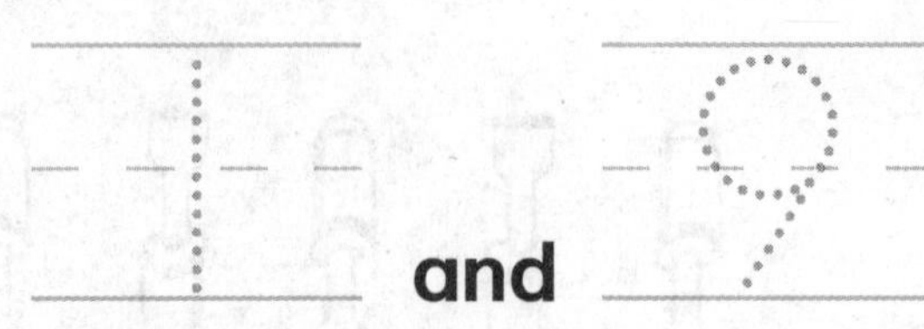

and

10

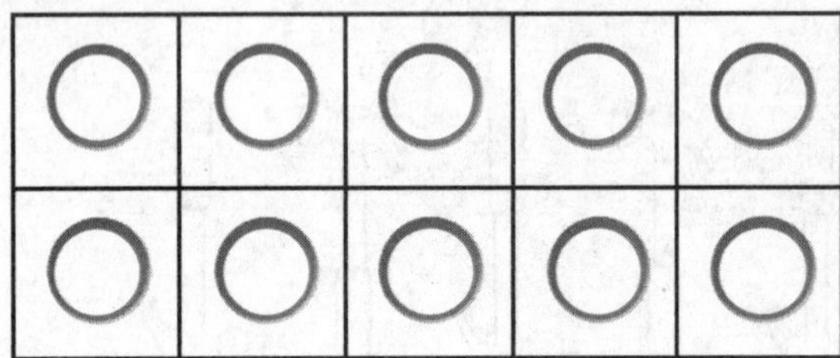

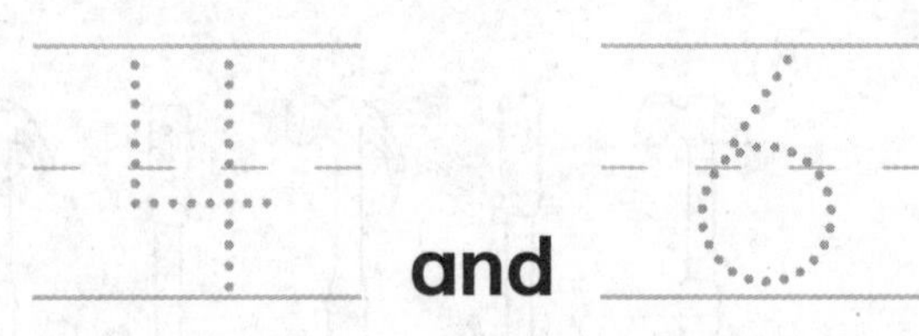

and

11

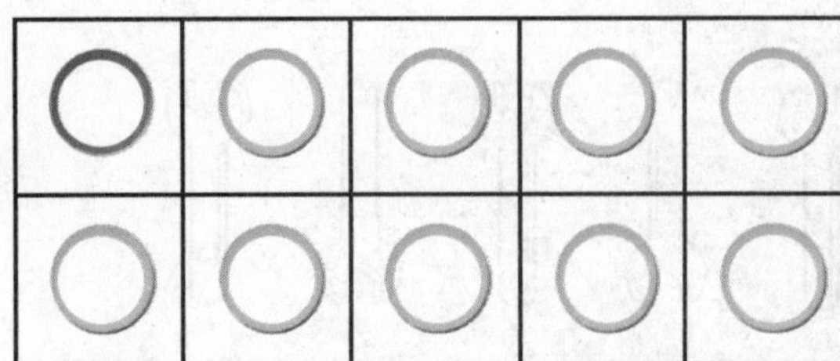

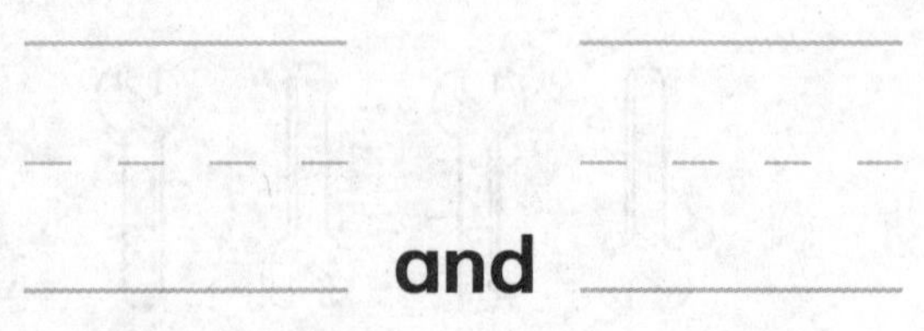

and

**Directions: 9–11.** Draw a line from the ten-frame that shows a way to make 10 to the numbers that match. Use red and yellow to finish coloring the counters to show a way to make 10. Use red and yellow counters to make 10 another way than shown on this page. Explain to a friend the numbers that make 10.

Name

# My Homework

Lesson 8

Make 10

## Homework Helper

Need help? connectED.mcgraw-hill.com

1 10

6 and 4

2 10

____ and ____

3 10

____ and ____

**Directions: 1–3.** Use orange and yellow to color the objects to show a way to make 10. Write the numbers.

4

# 10

_____ and _____

# 10

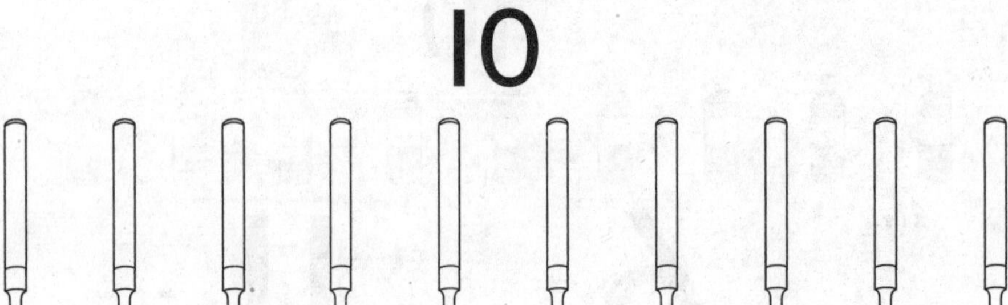

_____ and _____

# 10

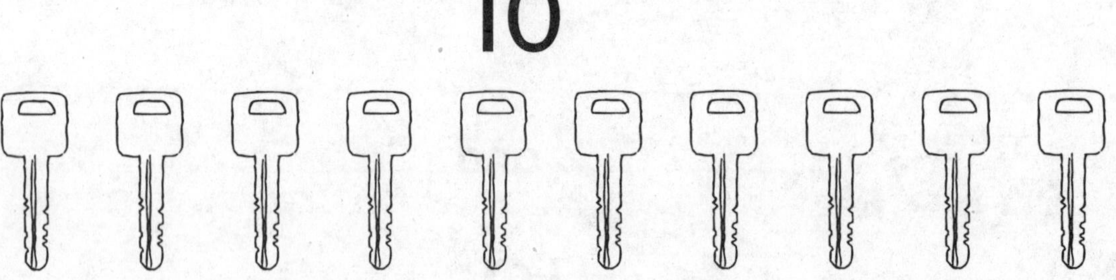

_____ and _____

**Directions: 4–6.** Use orange and yellow to color the objects to show a way to make 10. Write the numbers.

**Math at Home** Draw a picture of objects such as 10 spoons or 10 bananas. Have your child color the objects to show a way to make 10. Repeat by drawing other objects and having your child show another way to make 10.

Name

Lesson 9

# Take Apart 10

**ESSENTIAL QUESTION**
How can we show a number in other ways?

## Math in My World

CITY HALL

1

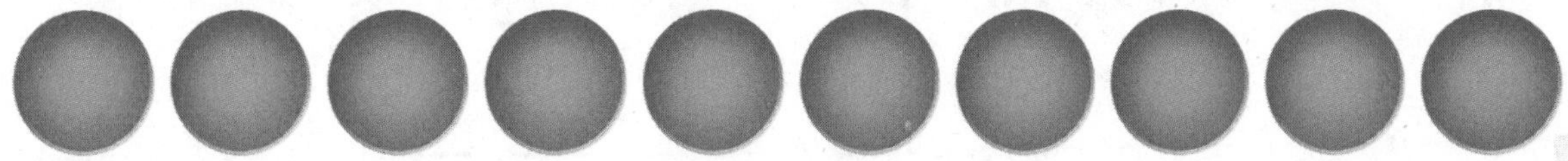

2

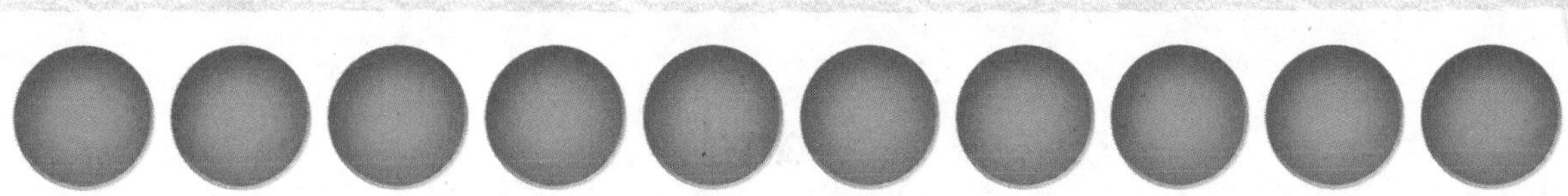

**Teacher Directions: 1.** Use ● to show 10. Break apart the counters into two groups to take apart 10. Circle the counters to show a way to take apart 10. **2.** Repeat. Show another way to take apart 10. Circle the counters to show that way to take apart 10.

## Guided Practice

Processes & Practices

3

10

1 and 9

4

10

and

5

10

and

**Directions: 3.** Look at the number. Count the objects. Trace the dashed circles to show a way to take apart the number. Trace the numbers. **4–5.** Look at the numbers. Count the objects. Circle the objects to show a way to take apart 10. Write the numbers.

Name

## Independent Practice

6

10

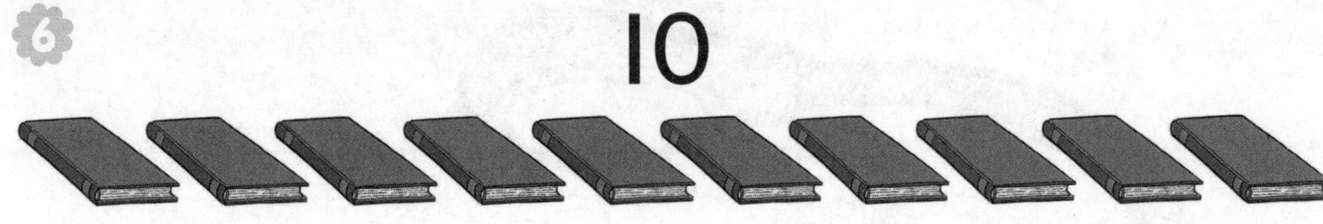

and

7

10

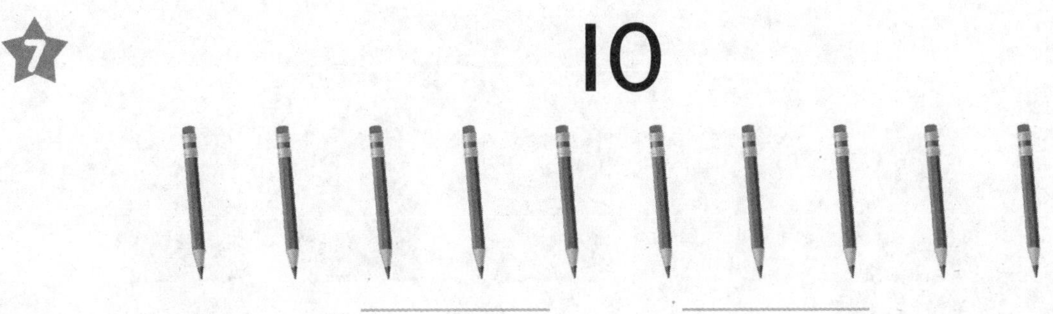

and

8

10

and

**Directions: 6–8.** Look at the number. Count the objects. Circle the objects to show a way to take apart 10. Write the numbers.

## Brain Builders

9

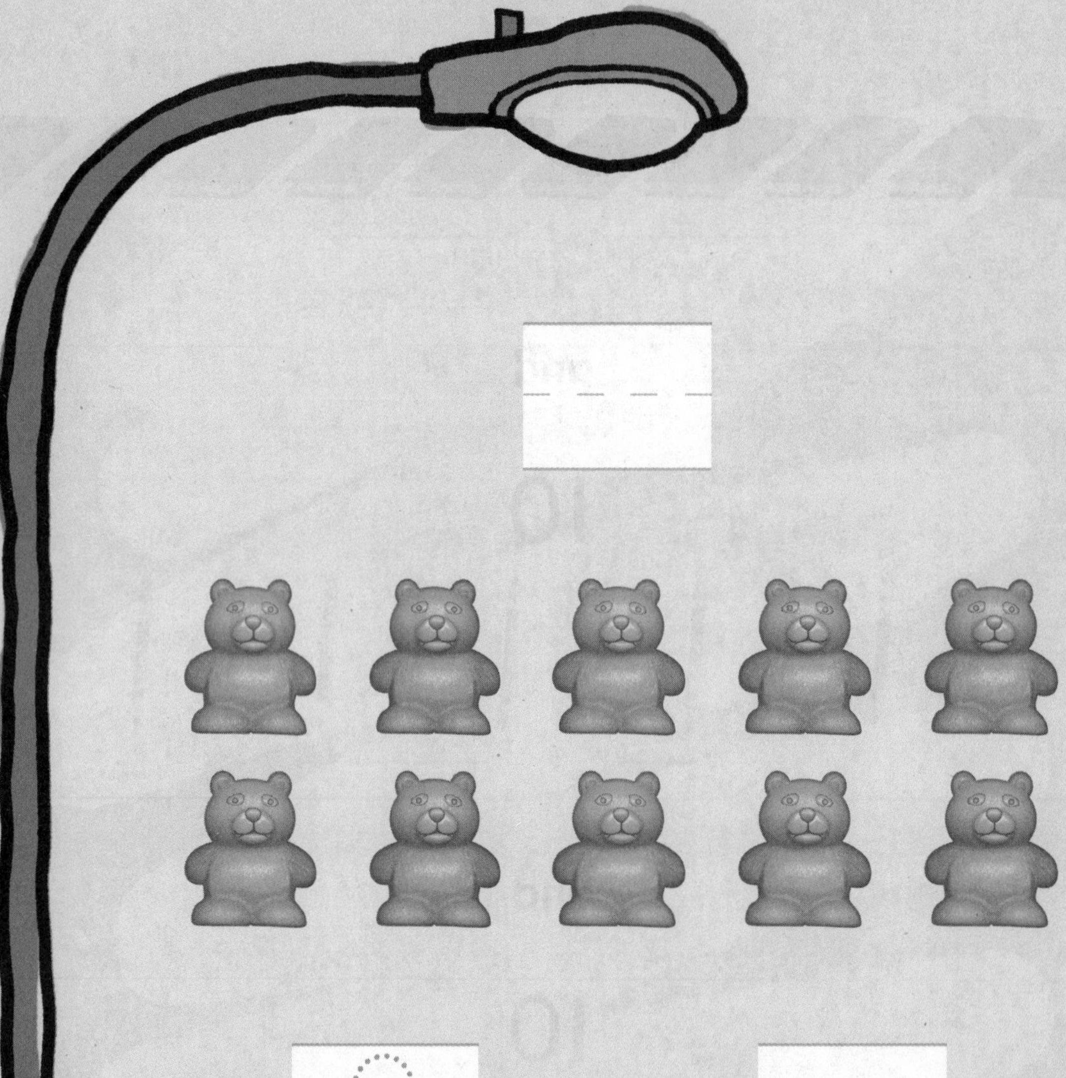

3

and

**Directions: 9.** Count how many bears. Write the number above the bears. Use counting bears to show 10. Take apart a group of three bears. Trace the three. Count the bears in the other group. Write the number. Circle the bears to show each group. Discuss with a friend different ways to take apart 10. Work together to show each of the ways using bears.

Name ..........................................

Lesson 9

Take Apart 10

# My Homework

## Homework Helper

eHelp

Need help? connectED.mcgraw-hill.com

1

2 and 8

2

10

_____ and _____

3

10

_____ and _____

**Directions: 1–3.** Look at the number. Count the objects. Circle the objects to show a way to take apart 10. Write the numbers.

4

10

and

5

10

and

6

10

and

**Directions: 4–6.** Look at the number. Count the objects. Circle the objects to show a way to take apart 10. Write the numbers.

**Math at Home** Show your child a group of 10 spoons. Have your child break apart the group into two groups to show a way to take apart 10. Guide your child in writing the numbers that tell how many are in each group.

Name

# My Review



## Vocabulary Check 

eight

four

ten

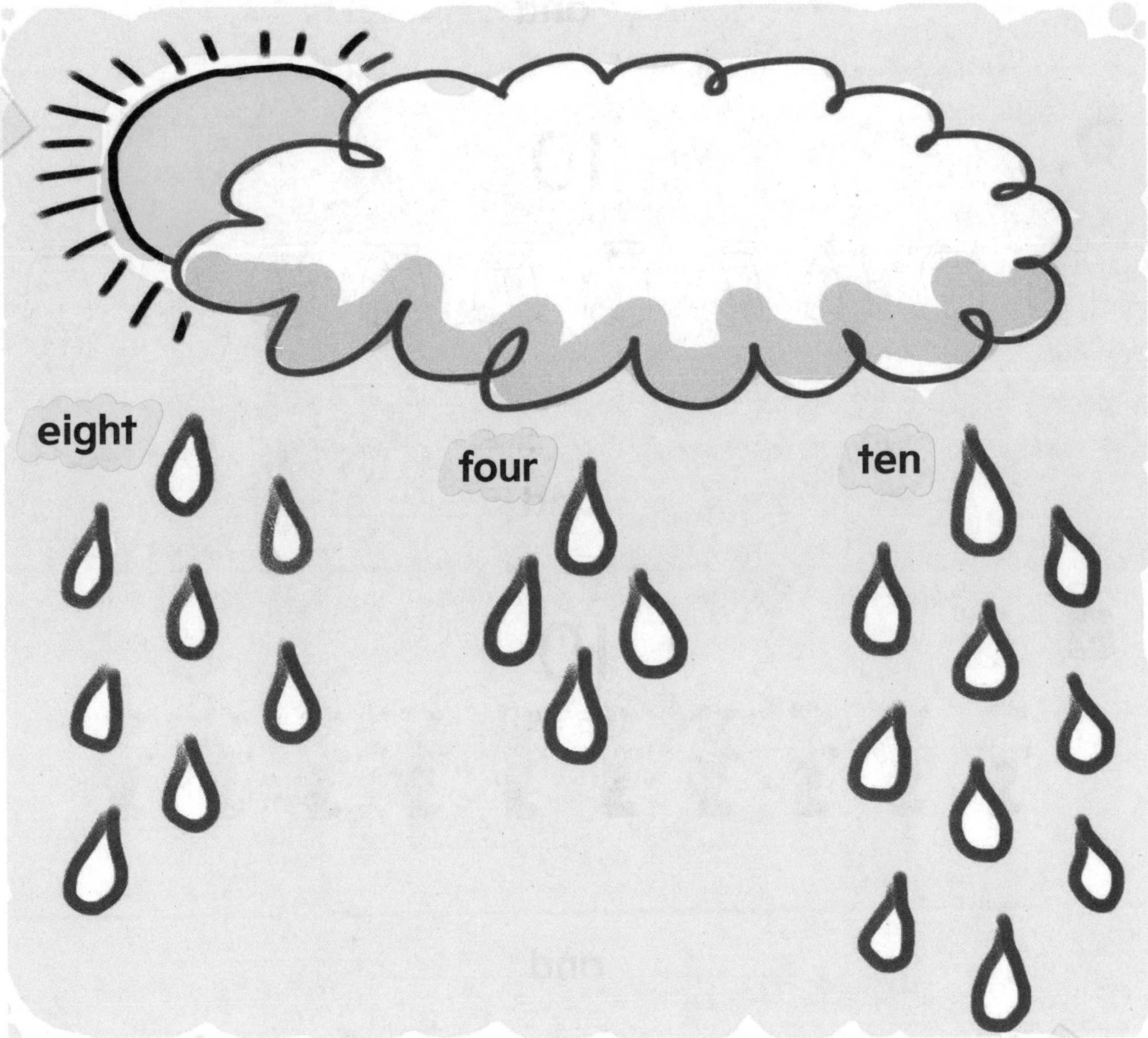

**Directions:** Color the group of four raindrops yellow. Color the group of eight raindrops orange. Color the group of 10 raindrops purple.

# Concept Check

1 9

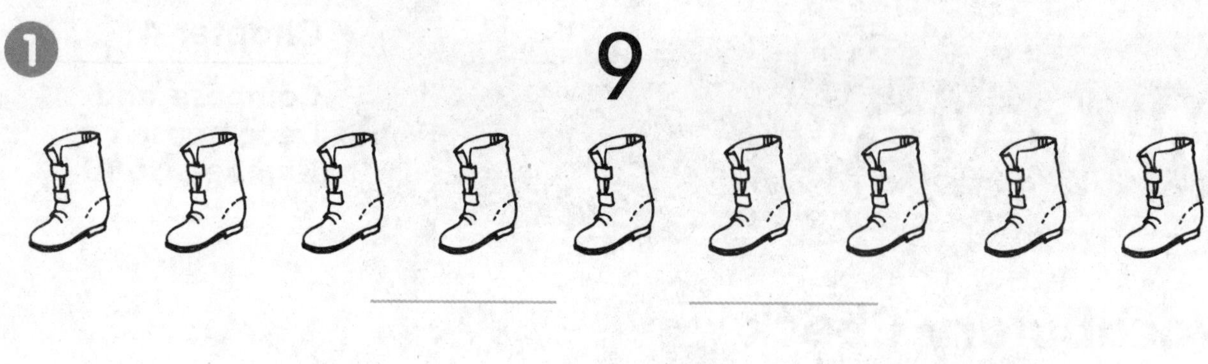

______ ______

______ and ______

2 10

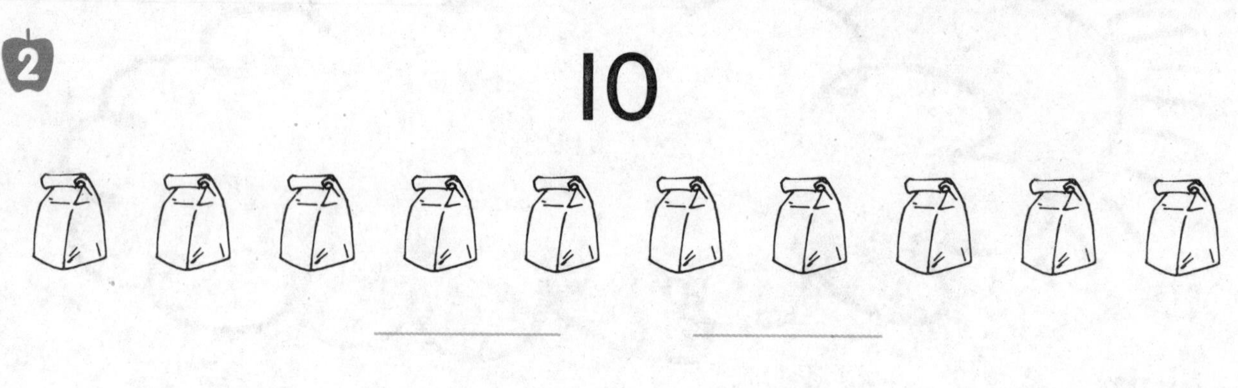

______ ______

______ and ______

3 10

______ ______

______ and ______

**Directions: 1–2.** Use red and yellow to color the objects to show a way to make nine and 10. Write the numbers. **3.** Look at the number. Count the objects. Circle the objects to show a way to take apart 10. Write the numbers.

Name

## Brain Builders

4

______ and ______

5

______ and ______



**Directions: 4.** Use red and yellow counters to show a way to make seven. Color the counters to show that way. Write the numbers. **5.** Use red and yellow counters to show another way to make seven. Color the counters to show that way. Write the numbers.

# Reflect

**Chapter 4**

**ESSENTIAL QUESTION**

How can we show a number in other ways?

**1**

4

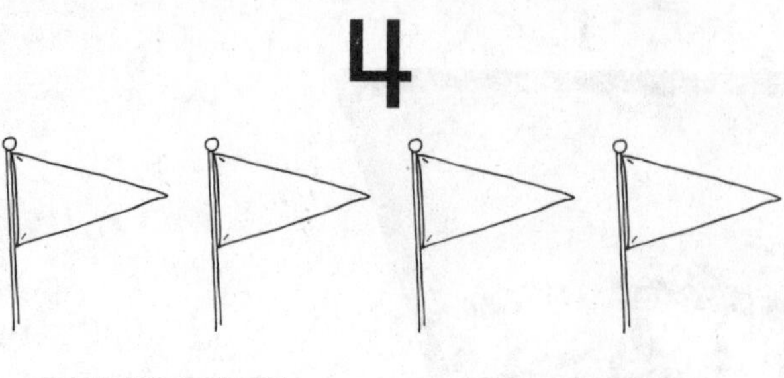

_______ _______

_______ and _______

**2**

6

_______ _______

_______ and _______

**3**

9

_______ _______

_______ and _______

**Directions: 1–2.** Color the objects red and yellow to show a way to make that number. Write the numbers. **3.** Look at the number. Count the objects. Circle the objects to show a way to take apart nine. Write the numbers.

Name

Date

Score

# Performance Task

## At The Park

You are visiting the park.

**Show all of your work to receive full credit.**

### Part A

There are 5 birds. Circle the birds to show one way to make 5.

Write the numbers.

________ and ________

Circle the birds to show another way to make 5.
Write the numbers.

________ and ________

## Part B

Show one way to make 9. Color some of the counters red and some of the counters blue to show one way. Write the numbers.

_____ and _____

## Part C

There are 8 leaves. Circle the leaves to show a way to take apart the number. Write the numbers.

_____ and _____

## Part D

There are 10 marbles. Circle the marbles to show a way to take apart 10. Write the numbers.

_____ and _____

Chapter

# Addition

**ESSENTIAL QUESTION**

How can I use objects to add?

We Have Lots to Celebrate!

Watch a video!

Name

# Chapter 5 Project

## Collections Poster

1. Decide which collection of objects your group would like to draw in each of the rows.
2. Draw two groups of objects that when joined equals 10.
3. Write the numbers. Trace the symbols.
4. When your group is finished with your poster, present one collection to the class.

1

___ + ___ = 10

2

___ + ___ = 10

3

___ + ___ = 10

Name

# Am I Ready?

1

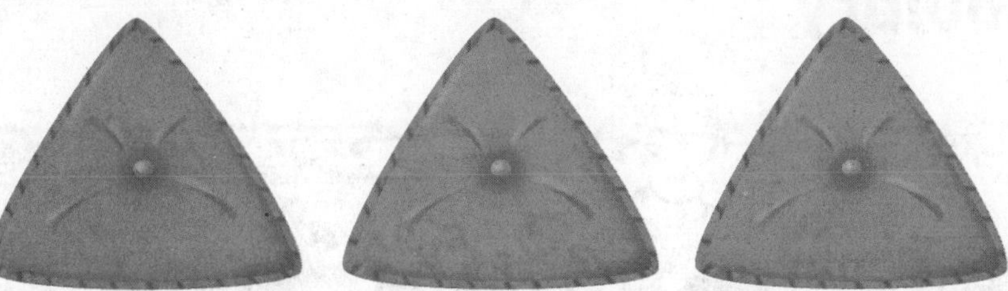

2

3

4

**Directions: 1–2.** Count the objects. Write the number. **3–4.** Say the number. Trace the number. Color the objects to show a way to make the number.

Name

# My Math Words

## Review Vocabulary

10 ten

**Directions:** Count the ants. Trace the number and the word. Draw more apples to show ten.

# My Vocabulary Cards

## add

3 2

5 in all

## equals sign =

3 + 2 = 5

## in all

2 1

3 **in all**

## join

## plus sign +

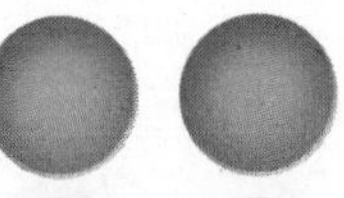

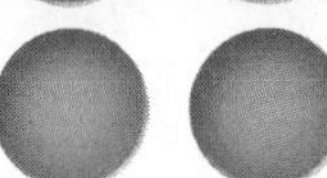

4 + 2 is 6.

**Teacher Directions:**
**Ideas for Use**

- Have students think of words that rhyme with some of the words.
- Have students name the letters in each word.
- Have students create an addition story and draw the pictures on their blank card. Have students choose a classmate to model their story.

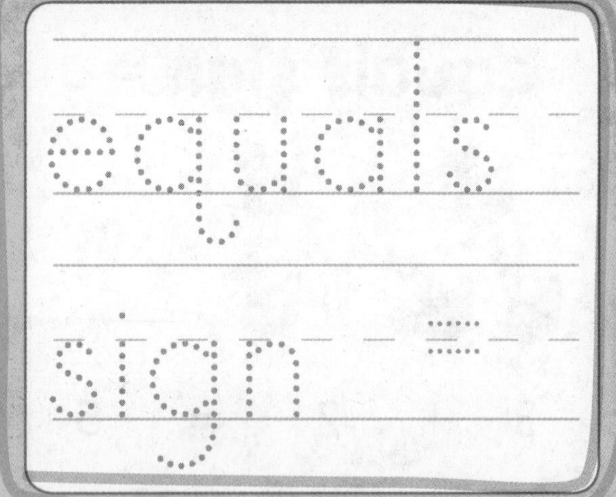

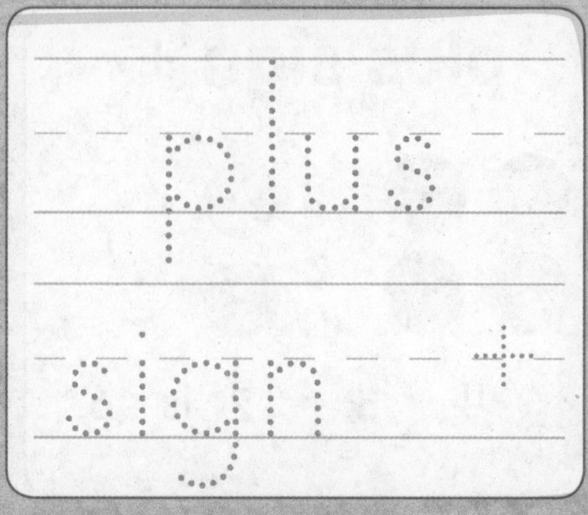

# My Foldable

FOLDABLES® Follow the steps on the back to make your Foldable.

☐ + ☐ = ☐

☐ + ☐ = ☐

☐ + ☐ = ☐

☐ + ☐ = ☐

☐ + ☐ = ☐

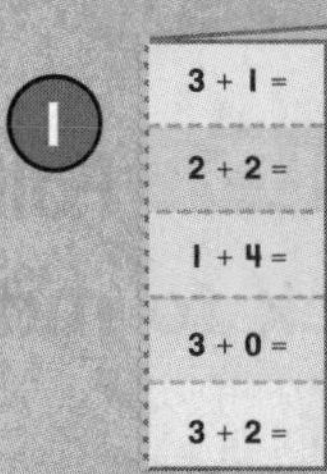


2

3 + 1 =
2 + 2 =
1 + 4 =
3 + 0 =
3 + 2 =

3

2 + 2 =
1 + 4 =
3 + 0 =
3 + 2 =

$3 + 1 = \_\_\_\_$

$2 + 2 = \_\_\_\_$

$1 + 4 = \_\_\_\_$

$3 + 0 = \_\_\_\_$

$3 + 2 = \_\_\_\_$

Name

Lesson 1
# Addition Stories

**ESSENTIAL QUESTION**
**How can I use objects to add?**

## Math in My World

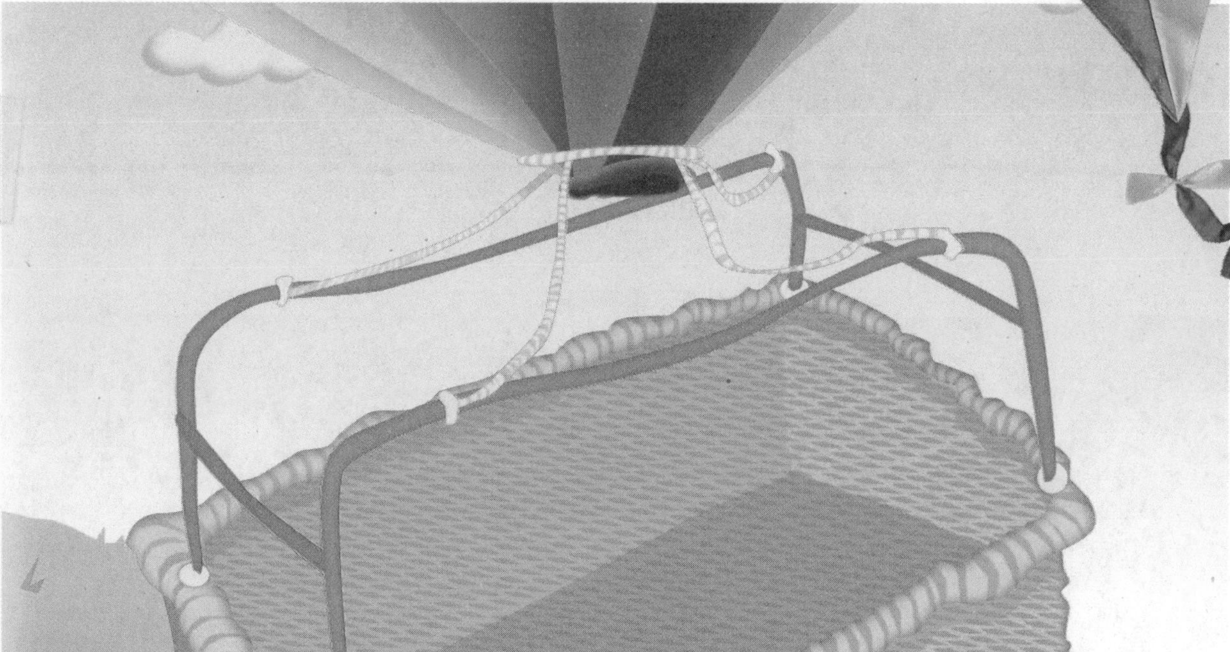

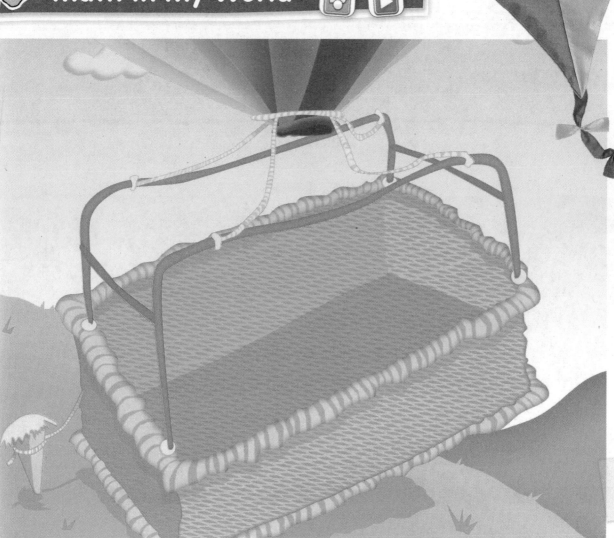

**Teacher Directions:** Use ◯ to model the addition story. Two people are inside the basket of the hot air balloon. Three more people join them. Trace the counters to model the story. How many people are there in all?

## Guided Practice

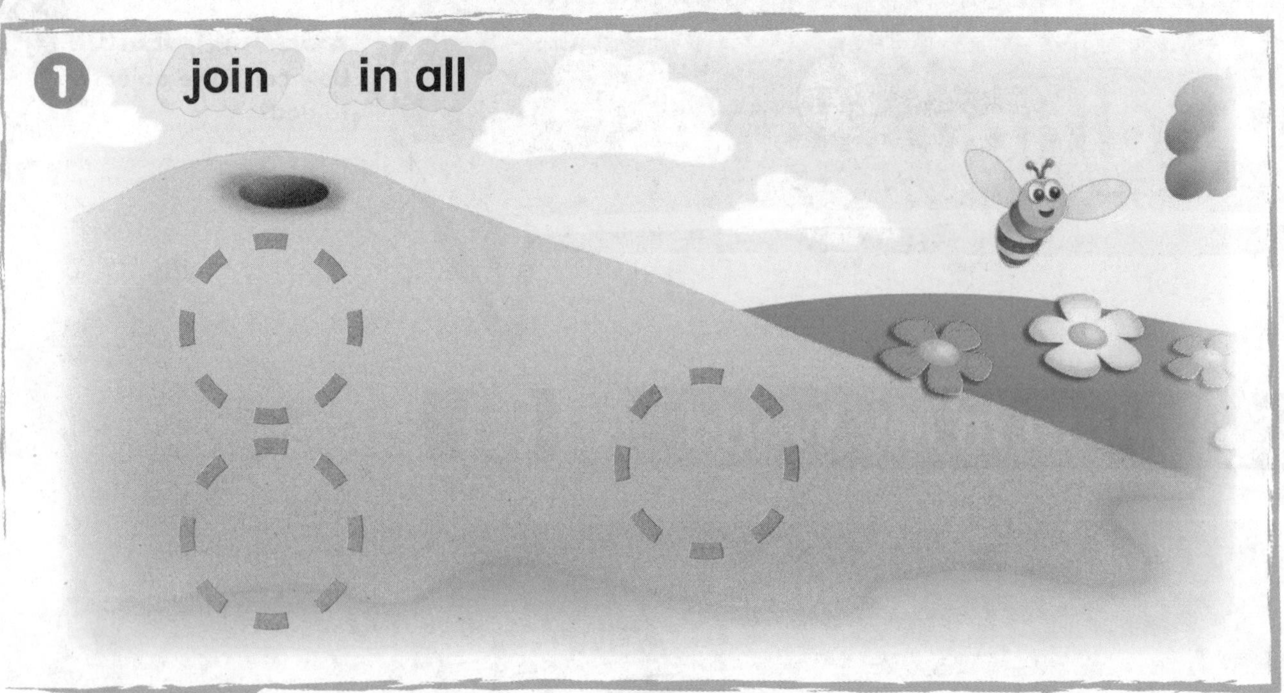

**Directions:** Use counters to model each addition story. Trace the counters. **1.** Two red ants are climbing up the hill. One yellow ant joins them. How many ants are there in all?
**2.** Three red plates are on the table. Three yellow plates are also on the table. How many plates are there in all?

Name

## Independent Practice

3

4

**Directions:** Use counters to model each addition story. Trace the counters to show your work. **3.** Four children are near the slide. Two more children join them. How many children are there in all? **4.** Three children are on the swings. Two more children join them. How many children are there in all?

# Brain Builders

Processes &Practices

5

6

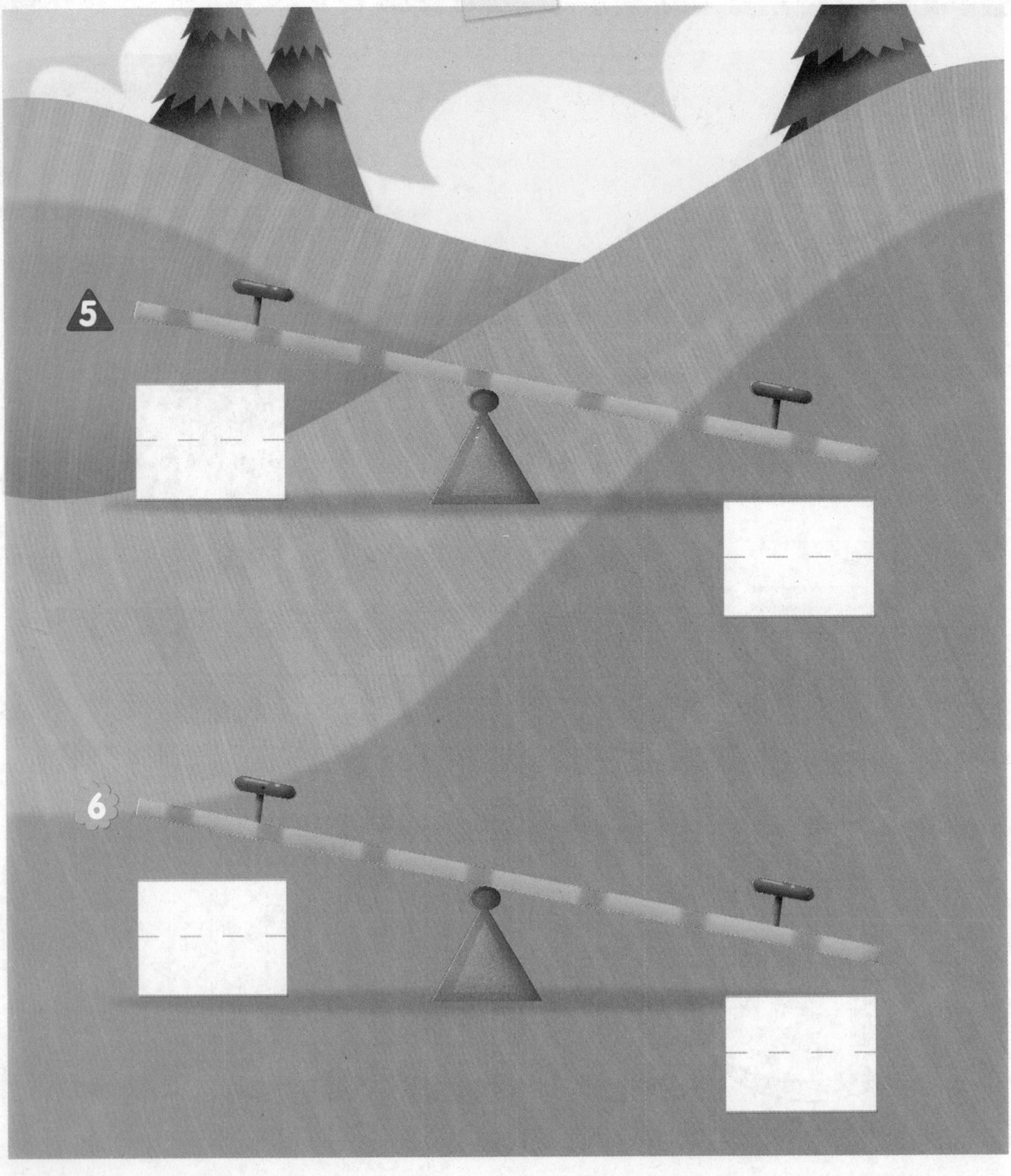

**Directions:** Use counters to model each addition story. Trace the counters to show your work. **5.** Two children are playing on the teeter-totters. Two more children join them. Write the numbers. How many are there now? **6.** Three children are playing on the teeter-totters. One more child joins them. Write the numbers. How many are there now? Tell a friend how many children are on both teeter-totters now. Explain how you found the answer.

Name



# My Homework

## Homework Helper

Need help? connectED.mcgraw-hill.com

1

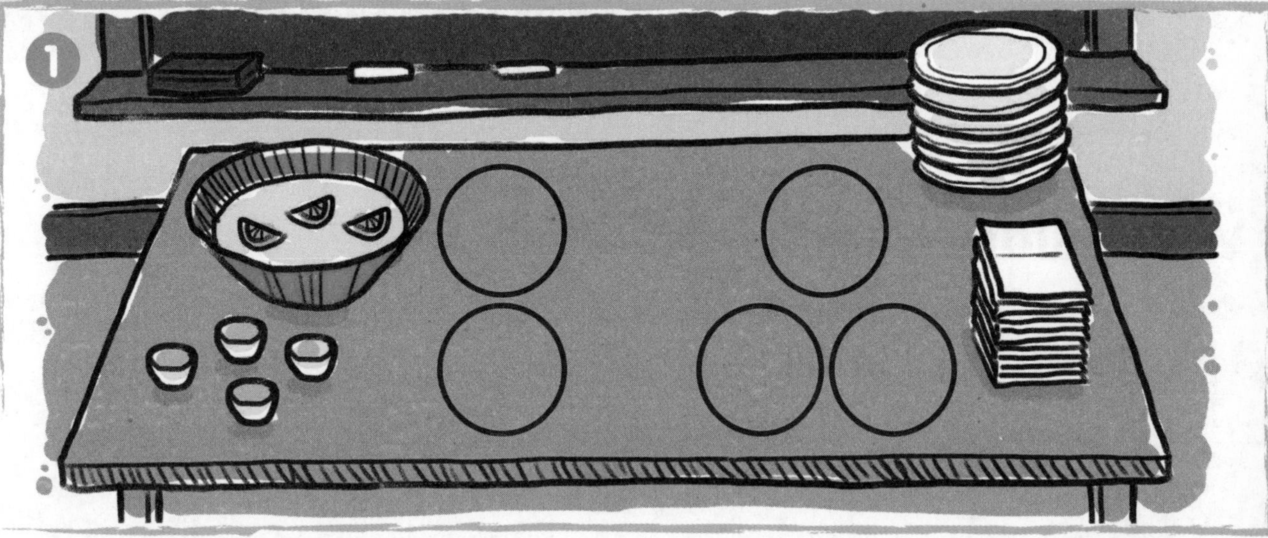

2

**Directions:** Use pennies to model each addition story. Trace the pennies. **1.** Two pizzas are on the table. Three more pizzas are put on the table. How many pizzas are there in all? **2.** Four children jump in the pool. Three more children jump in the pool. How many children are there in all?

3

## Vocabulary Check

4 join in all

_____

_____ in all

**Directions: 3.** Use pennies to model the addition story. Trace the pennies. Five children are playing volleyball. Two more children join them. How many children are there in all? **4.** Draw a group of three blue balloons. Draw a group of five red balloons. Join the two groups. How many balloons are there in all? Write the number.

**Math at Home** Tell your child an addition story. Have your child use pennies to model the story. Have him or her tell you how many pennies there are in all.

Name

Lesson 2

# Use Objects to Add

**ESSENTIAL QUESTION**
How can I use objects to add?

## Math in My World

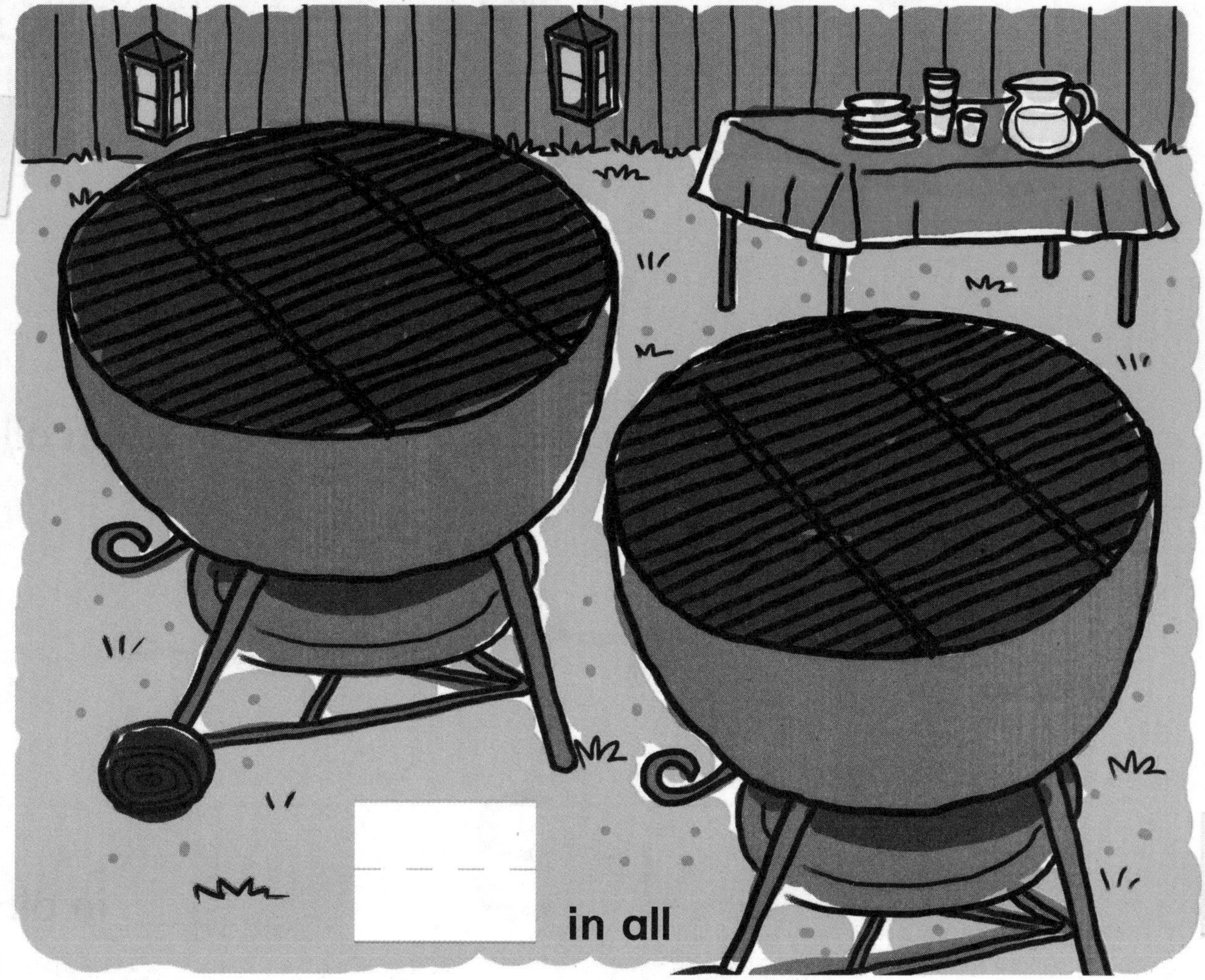

in all

**Teacher Directions:** Use ● to model the addition story. Four hamburgers are on one grill. Five hot dogs are on another grill. Write the number that tells how many hamburgers and hot dogs there are in all.

## Guided Practice

1 add

3 2 5 in all

2

2 1 ____ in all

3

1 3 ____ in all

**Directions: 1.** Use counters to model. Three birds are in a tree. Two more birds join them. How many birds are there in all? Trace the number. **2–3.** Use counters to model joining groups of birds. Write the number that tells how many there are in all.

Name

## Independent Practice

4

4 2 ______ in all

5

3 5 ______ in all

6

4 3 ______ in all

**Directions: 4–6.** Use counters to model joining groups of birds. Write the number that tells how many there are in all.

Processes & Practices

# Brain Builders

7

2 3

8

5 4

9

1 3

**Directions: 7–9.** Write the number that tells how many there are in all. For each problem, tell a friend another pair of numbers that makes the same number in all.

Name

# My Homework

**Lesson 2**

Use Objects to Add

## Homework Helper

Need help? connectED.mcgraw-hill.com

1

3 2 5 in all

2

1 3 ___ in all

3

4 3 ___ in all

**Directions: 1.** Use pennies to model. Trace the pennies. Three bugs are on a branch. Two more bugs join them. How many bugs are there in all? Write the number.
**2–3.** Use pennies to model joining groups of bugs. Write the number that tells how many.

4

5 4

______

_ _ _ _

______ in all

5

6 2

______

_ _ _ _

______ in all

## Vocabulary Check

6 **add**

______

_ _ _ _

______ in all

**Directions: 4–5.** Use pennies to model joining groups of bugs. Write the number that tells how many there are in all. **6.** Count the row boats. Draw three more. Write the number that tells how many there are in all.

**Math at Home** Tell your child an addition story. Using buttons or pennies, ask your child to model the story and tell how many there are in all.

Name

## Vocabulary Check

1 join in all

## Concept Check

2

**Directions: 1.** Draw a group of two balls. Draw a group of four balls. Join the two groups. How many balls are there in all? Write the number. **2.** Use counters to model the addition story. There are two fish swimming in the lake. Three more come to join them. How many fish are there in all? Write the number.

3

4 2 ____ in all

4

1 6 ____ in all

5

3 6 ____ in all

**Directions:** Use counters to model. **3.** There are four sailboats. Two more sailboats join them. How many sailboats are there in all? Write the number. **4–5.** Use counters to model joining groups. Write the number that tells how many there are in all.

Name

Lesson 3

# Use the + Symbol

**ESSENTIAL QUESTION**
How can I use objects to add?

##  Math in My World 

DOGS ARE US

☐ + ☐ is ☐.

**Teacher Directions:** Count the dogs in the first group. Write the number. Trace the plus sign. Count the dogs in the second group. Write the number. Use ● to model each group. Join the two groups. Write how many dogs there are in all.

## Guided Practice

1

plus sign +

1 + 2 is 3.

2

___ + ___ is ___.

3

___ + ___ is ___.

**Directions: 1.** Count the bears in each group. Trace the numbers and trace the plus sign. Circle the groups to join them. Trace how many bears there are in all. **2–3.** Count the bears in each group. Write the numbers and trace the plus sign. Circle the groups to join them. Write how many bears there are in all.

Name

## Independent Practice

4

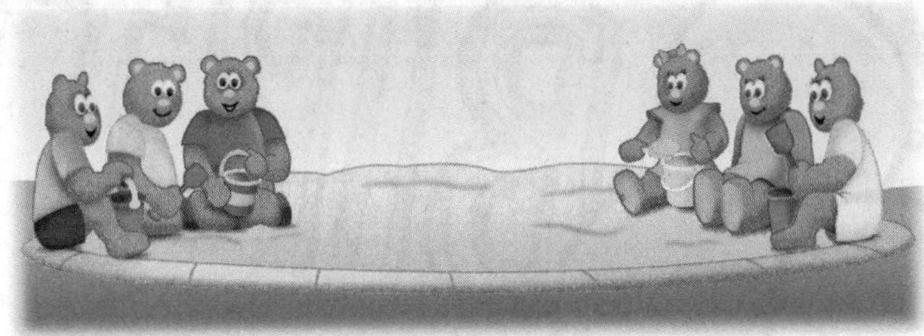

+

is .

5

+

is .

6

+

is .

**Directions: 4–6.** Count each group of bears. Write the numbers and trace the plus sign. Circle the groups to join them. Write how many bears there are in all.

Processes & Practices

# Brain Builders

7

______ + ______

______ ______ is ______.

**Directions: 7.** Count the cupcakes. Write the number on the line below the cupcakes. Trace the plus sign. Draw a group of cupcakes that is less than six cupcakes. Write the number. Write how many cupcakes there are in all. Draw a group of presents that is less than seven presents. Ask a friend to tell you how many presents there are in all. Talk with your friend about what a plus sign is used for.

Name

Lesson 3

Use the + Symbol

# My Homework

## Homework Helper

Need help? connectED.mcgraw-hill.com

1

2  2 is 4.

2

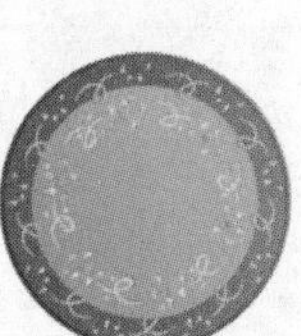

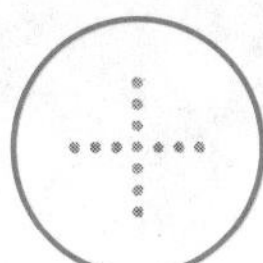 is ____.

3

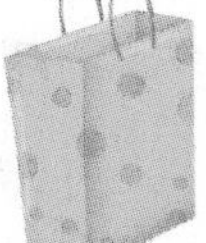

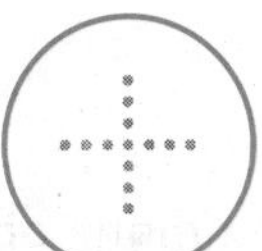 is ____.

**Directions:** I–3. Count the objects in each group. Write the numbers and trace the plus sign. Circle the groups to join them. Write how many objects there are in all.

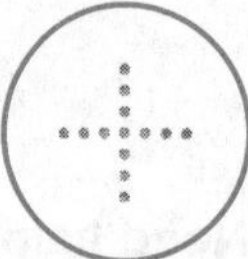

____ + ____ is ____.

____ + ____ is ____.

## Vocabulary Check

**plus sign (+)**

6 + 3 is 9.

**Directions: 4–5.** Count the objects in each group. Write the numbers and trace the plus sign. Circle the groups to join them. Write how many objects there are in all. **6.** Put a circle around the plus sign. Trace the plus sign.

**Math at Home** Write the words *Teddy Bear*. Have your child count and write the number of letters in each word. Have your child write a plus sign between the numbers. Tell your child to write how many letters there are in all.

Name

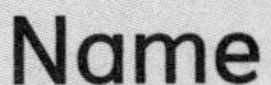

# Lesson 4
# Use the = Symbol

**ESSENTIAL QUESTION**
How can I use objects to add?

## Math in My World

____ + ____ = ____

**Teacher Directions:** Use ● to model each group. There are five cups of punch. There are two cups of milk. Write the numbers on the lines. Trace the equals sign. Write how many cups there are in all.

# Guided Practice

Processes & Practices

1

equals sign =

4 + 2 = 6

2

\+ =

3

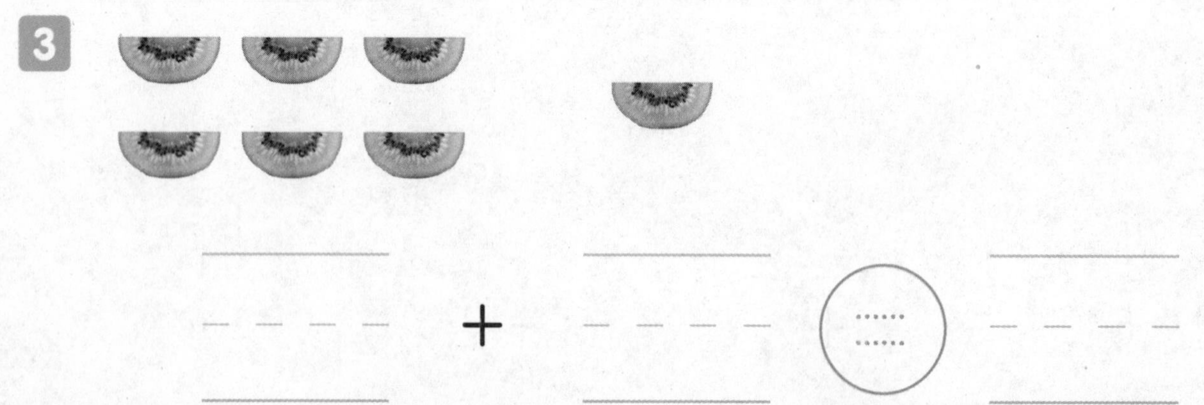

\+ =

**Directions: 1.** Count the fruit slices in each group. Trace the numbers and trace the equals sign. Circle the groups to join them. Trace how many fruit slices there are in all. **2–3.** Count the fruit slices in each group. Write the numbers and trace the equals sign. Circle the groups to join them. Write how many fruit slices there are in all.

Name

## Independent Practice

\+ 

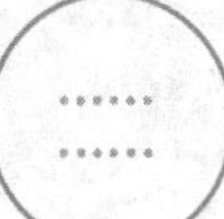

5

\+

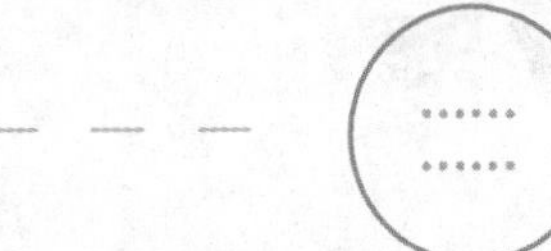

6

\+

   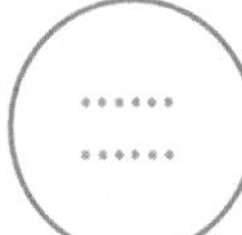

**Directions: 4–6.** Count each group of toys. Write the numbers and trace the equals sign. Circle the groups to join them. Write how many toys there are in all.

## Brain Builders

7

___ + ___ = ___

**Directions: 7.** Count the hats on the table. Write the number on the line below the hats. Draw a group of hats that is less than nine hats. Write the number. Trace the equals sign. Write how many hats there are in all. Tell a friend what the equals sign means.

Name

Lesson 4

Use the = Symbol

# My Homework

## Homework Helper

Need help? connectED.mcgraw-hill.com

1

6 + 2 = 8

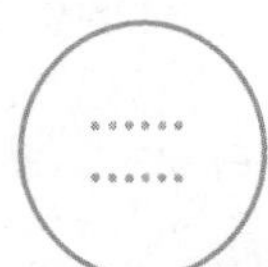

2

+ =

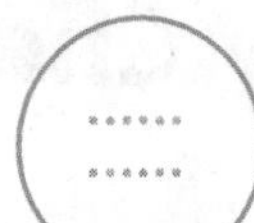

3

+ =

**Directions: 1–3.** Count the objects in each group. Write the numbers on the line below each picture. Trace the equals sign. Circle the groups to join them. Write how many objects there are in all.

4

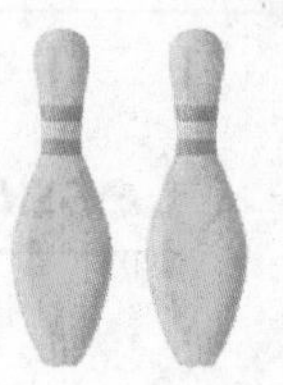

____ + ____ ◯ ____

5

____ + ____ ◯ ____

## Vocabulary Check

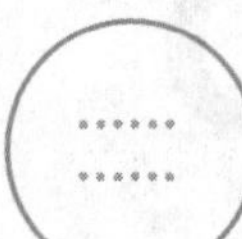

6 **equals sign**

4 + 1 = 5

**Directions: 4-5.** Count the objects in each group. Write the numbers on the line below each picture. Trace the equals sign. Circle the groups to join them. Write how many objects there are in all. **6.** Put a box around the equals sign.

Name

Lesson 5

# How Many in All?

**ESSENTIAL QUESTION**
How can I use objects to add?

## Math in My World

___ + ___ = ___

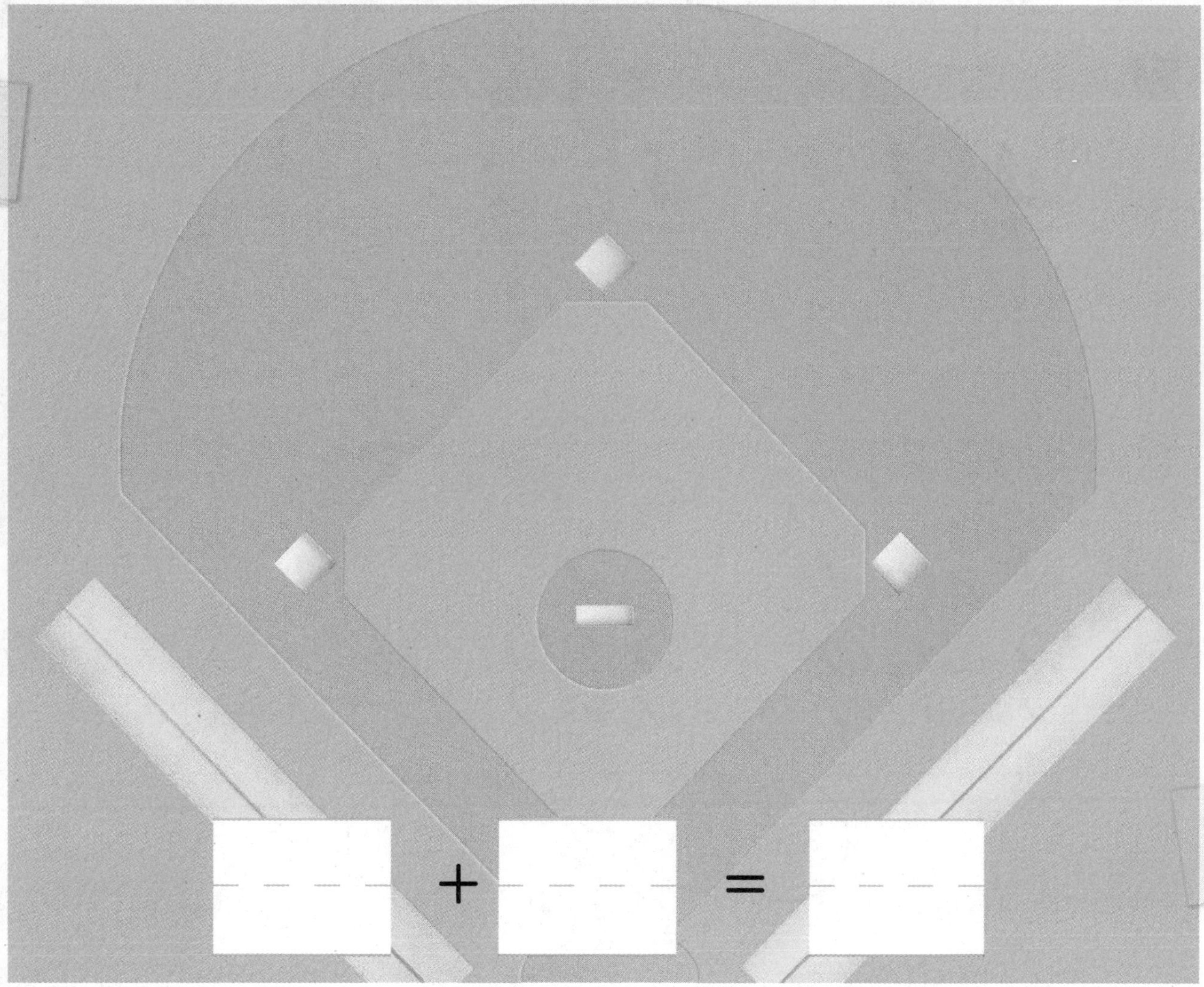

**Teacher Directions:** Use ● to model each group. There are four players on the field. Four more players come to play. Write the numbers. Write how many players there are in all.

# Guided Practice

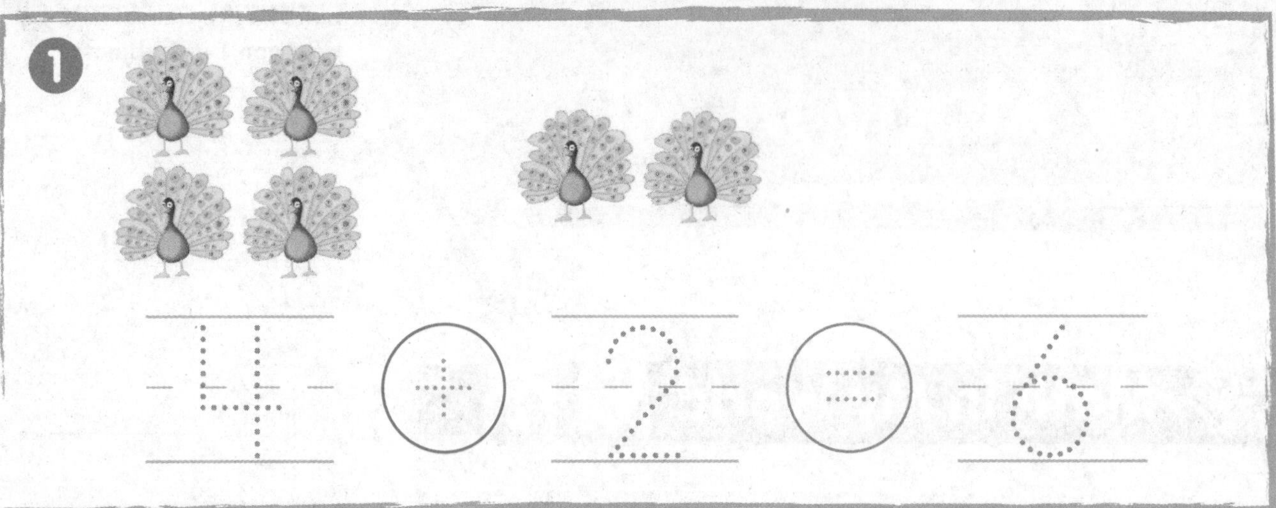


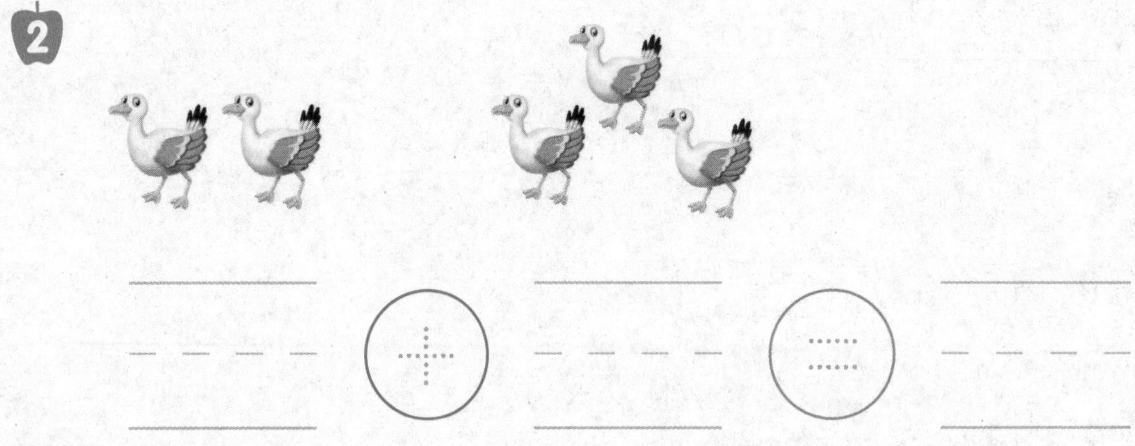


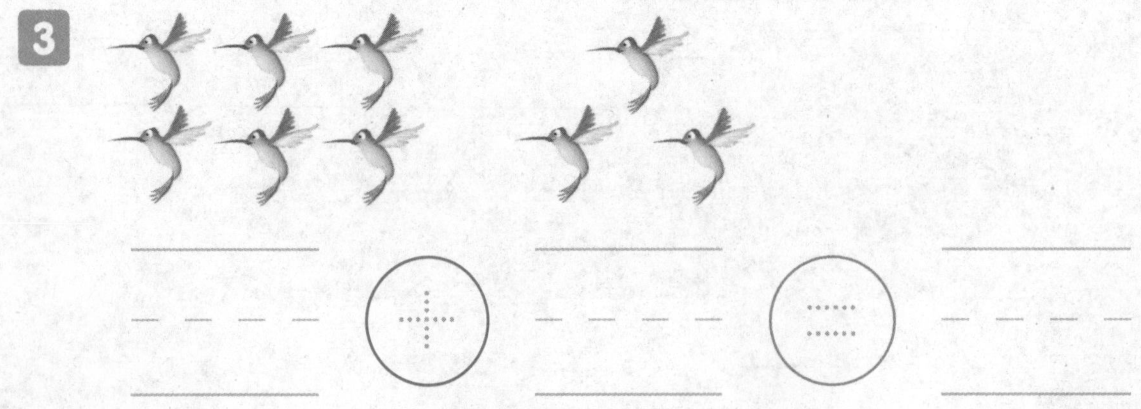

**Directions: I.** Use counters to model the addition sentence. Trace the numbers and the symbols. Trace how many there are in all. **2-3.** Use counters to model the addition sentence. Write the numbers and trace the symbols. Write how many there are in all.

Name

## Independent Practice

4

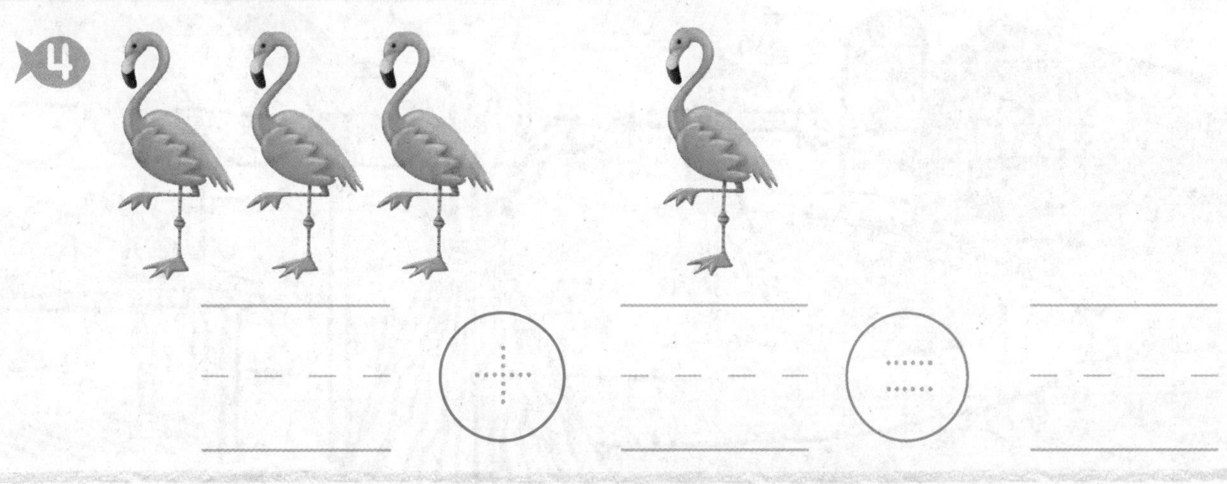

5

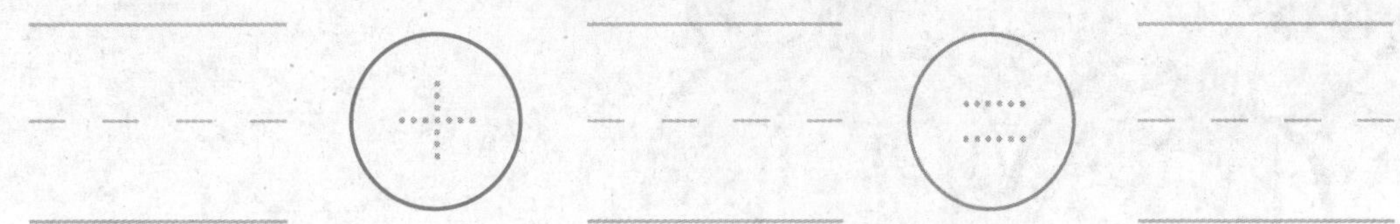

6

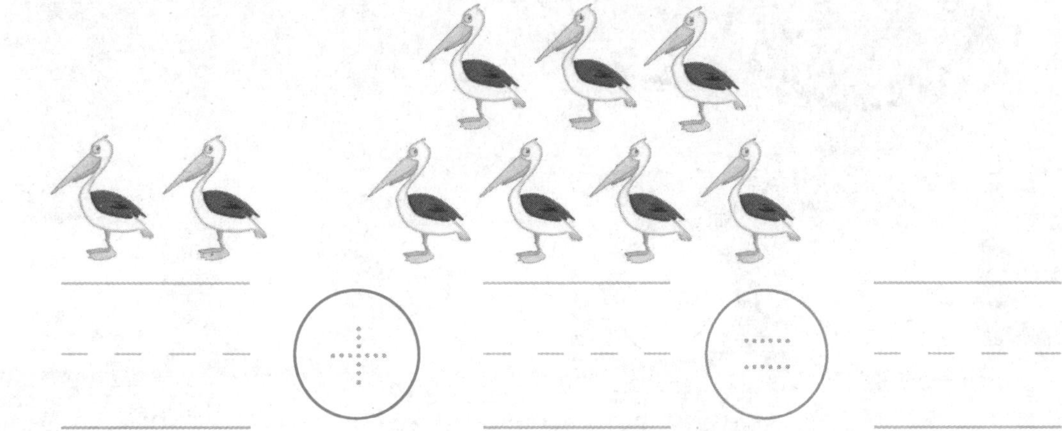



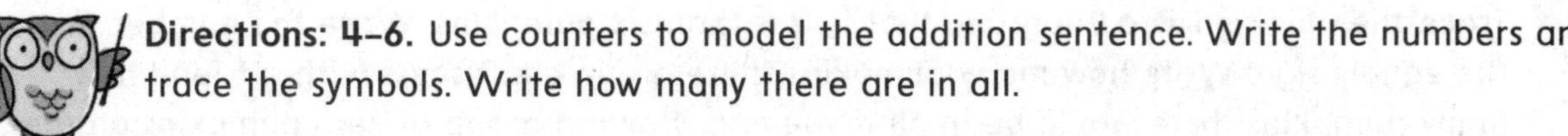

**Directions: 4–6.** Use counters to model the addition sentence. Write the numbers and trace the symbols. Write how many there are in all.

# Brain Builders

7

**Directions: 7.** Count the pumpkins. Write the number on the line below the pumpkins. Trace the plus sign. Draw a group that is less than six pumpkins. Write the number. Trace the equals sign. Write how many pumpkins there are in all. Discuss with a friend how many pumpkins there would be in all if you had drawn a group of zero pumpkins on the page.

Name

# My Homework

Lesson 5

How Many in All?

## Homework Helper

Need help? connectED.mcgraw-hill.com

1

2 + 3 = 5

2

 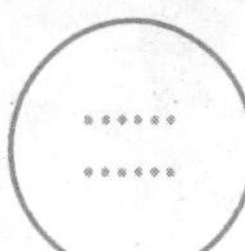

3

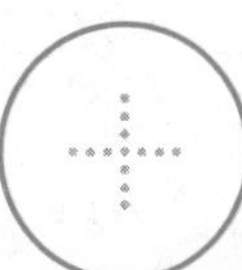 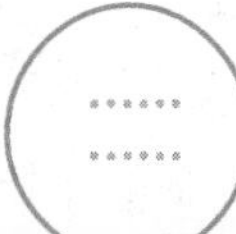

**Directions: 1–3.** Count the objects in each group. Write the numbers and trace the symbols. Write how many there are in all.

4

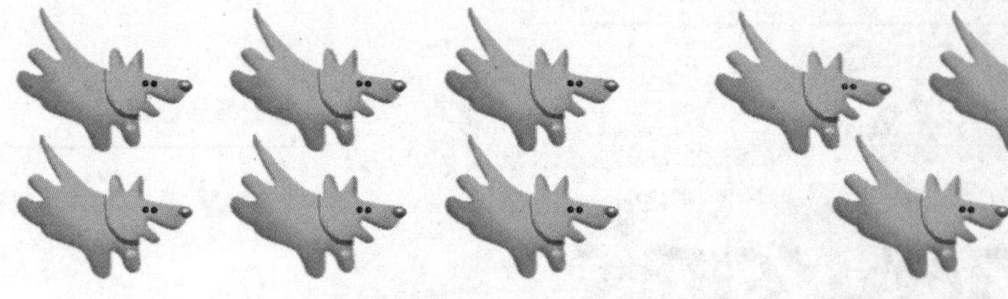

$\_\_\_ + \_\_\_ = \_\_\_$

5

$\_\_\_ + \_\_\_ = \_\_\_$

6

$\_\_\_ + \_\_\_ = \_\_\_$

**Directions: 4–6.** Count the objects in each group. Write the numbers and trace the symbols. Write how many there are in all.

**Math at Home** Have your child make a group of three items and a group of four items. Tell your child to join the groups and tell how many there are in all.

Name

Lesson 6

# Problem Solving

## STRATEGY: Write a Number Sentence

**ESSENTIAL QUESTION**
How can I use objects to add?

## How many are there in all?

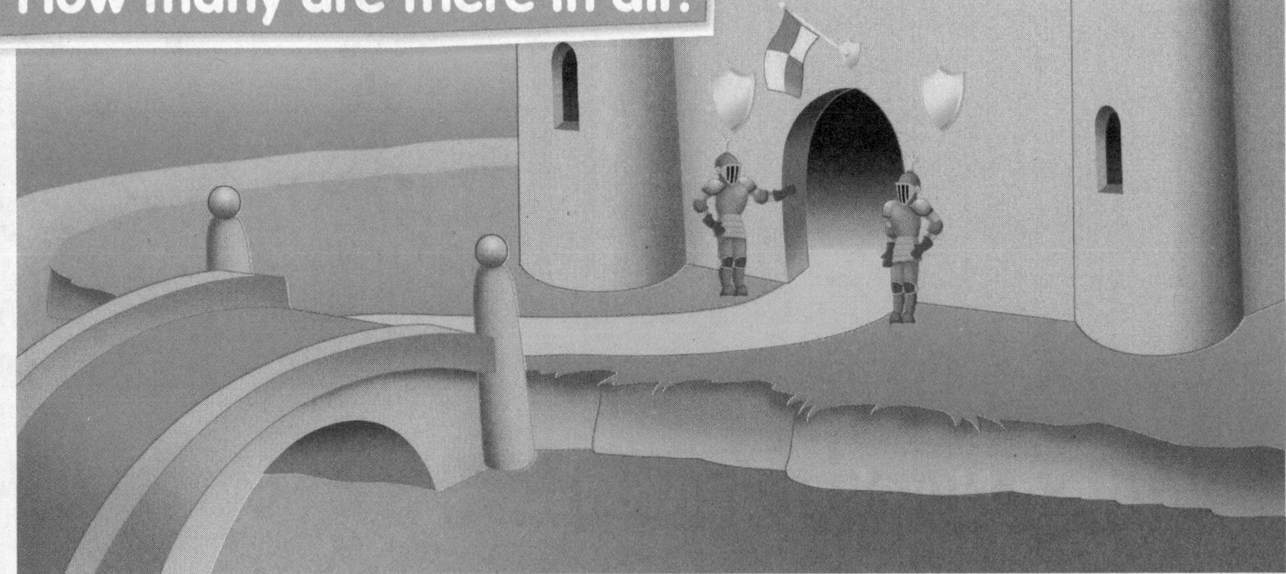

## Write a Number Sentence

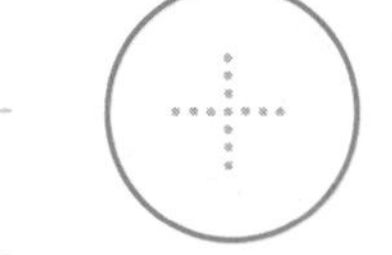
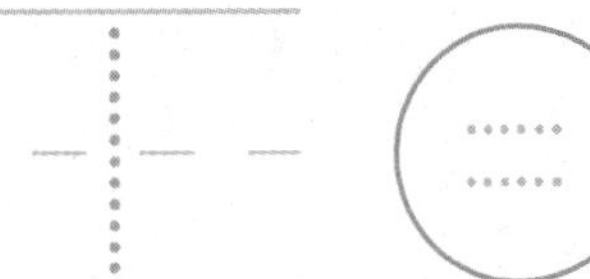
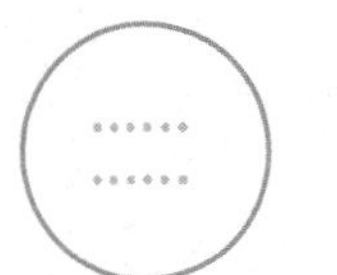
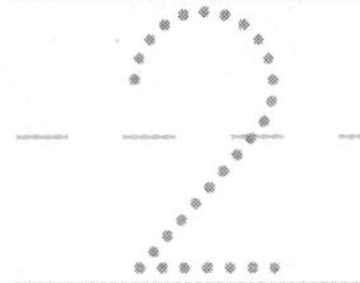

1 + 1 = 2

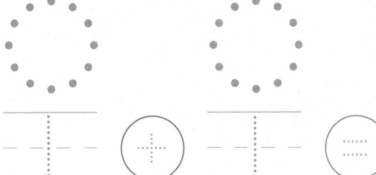

**Teacher Directions:** Use ● to model joining groups. There is one guard at the gate. One more guard joins him. Trace the counters. Trace the number sentence to tell how many there are in all.

## How many are there in all?

## Write a Number Sentence

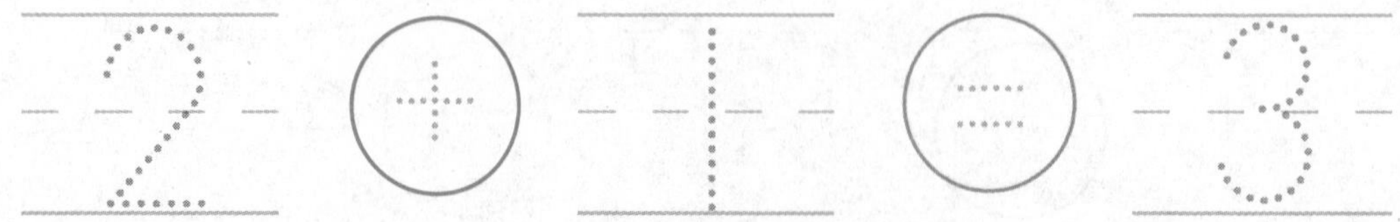

**Directions:** Use counters to model joining groups. Two astronauts are on the moon. One more joins them. Trace the counters. Trace the number sentence to tell how many astronauts there are in all.

Name

## How many are there in all?

## Write a Number Sentence

**Directions:** Use counters to model joining groups. Five snowboarders were coming down the mountain. Four skiers joined them. Trace the counters. Write a number sentence to tell how many there are in all. If you drew another skier how many would be there in all? Tell a friend your answer.

## How many are there in all?

## Write a Number Sentence

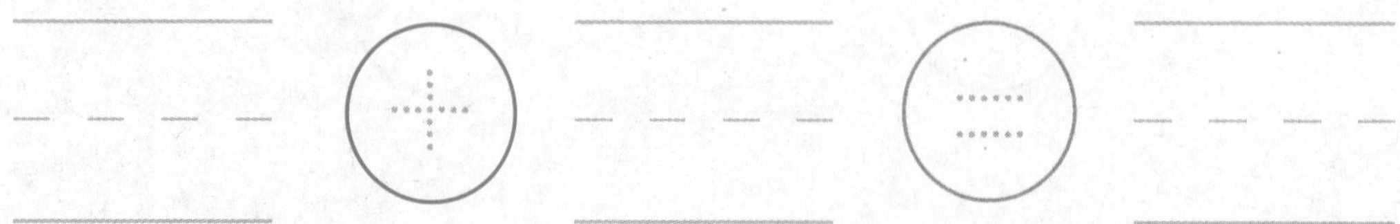

**Directions:** Use counters to model joining groups. Six scuba divers jumped into the water. Two more scuba divers joined them. Trace the counters. Write a number sentence to tell how many there are in all.

Name

Lesson 6

Problem Solving: Write a Number Sentence

# My Homework

## How many are there in all?

## Write a Number Sentence

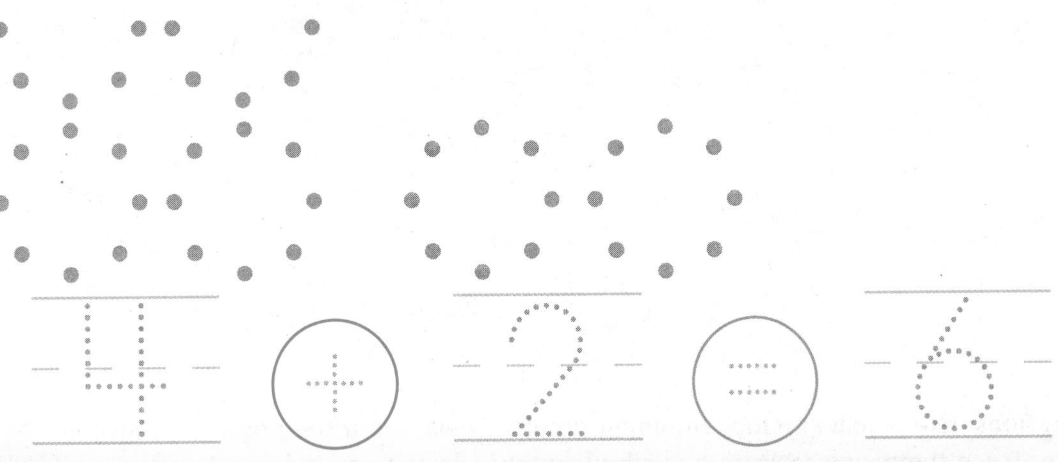

4 + 2 = 6

**Directions:** Use pennies to model joining groups. There are four firefighters. Two firefighters join them. Trace the pennies. Trace the number sentence to tell how many there are in all.

## How many are there in all?

## Write a Number Sentence

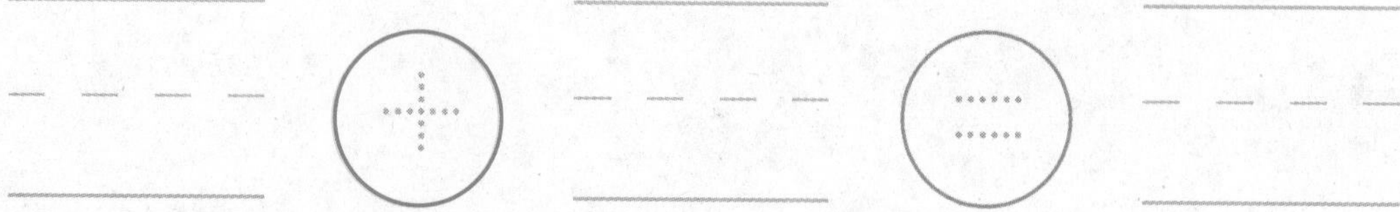

**Directions:** Use pennies to model joining groups. There are three dogs. Six dogs join them. Trace the pennies. Write a number sentence to tell how many there are in all.

**Math at Home** Take advantage of problem-solving opportunities during daily routines such as riding in the car, bedtime, doing laundry, putting away groceries, and so on.

Name

Lesson 7

# Add to Make 10

**ESSENTIAL QUESTION**
How can I use objects to add?

## Math in My World

____ + ____ = 10

**Teacher Directions:** Use ● to model making 10. There are two children with maracas. There are eight children with drums. Write the numbers to show how you made 10.

# Guided Practice

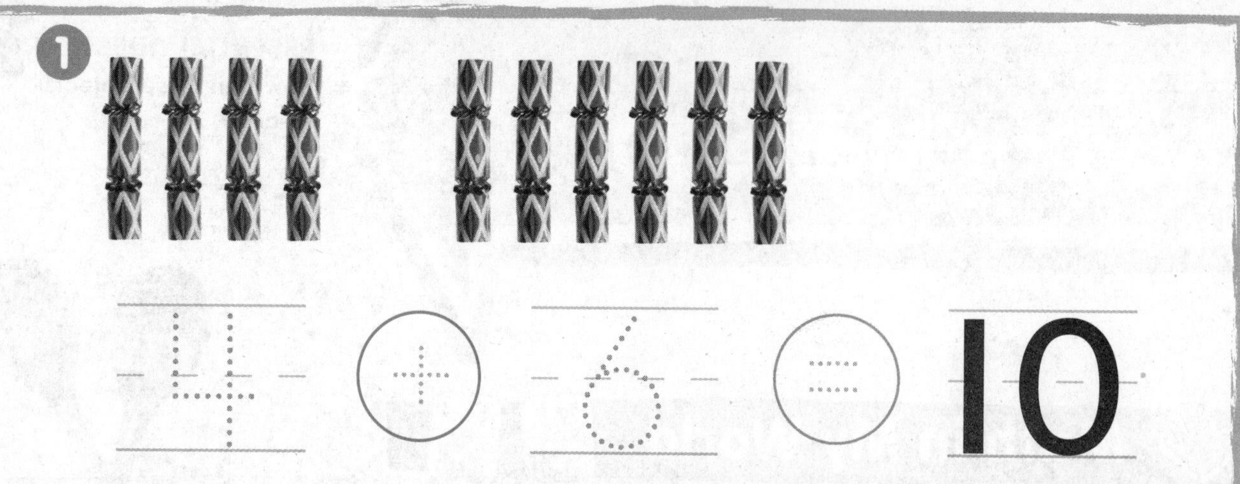


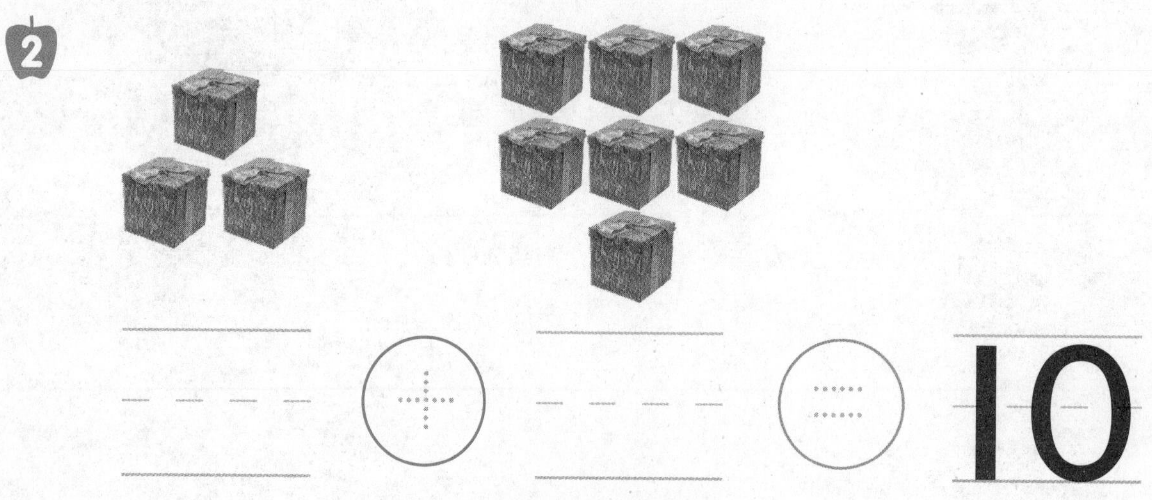


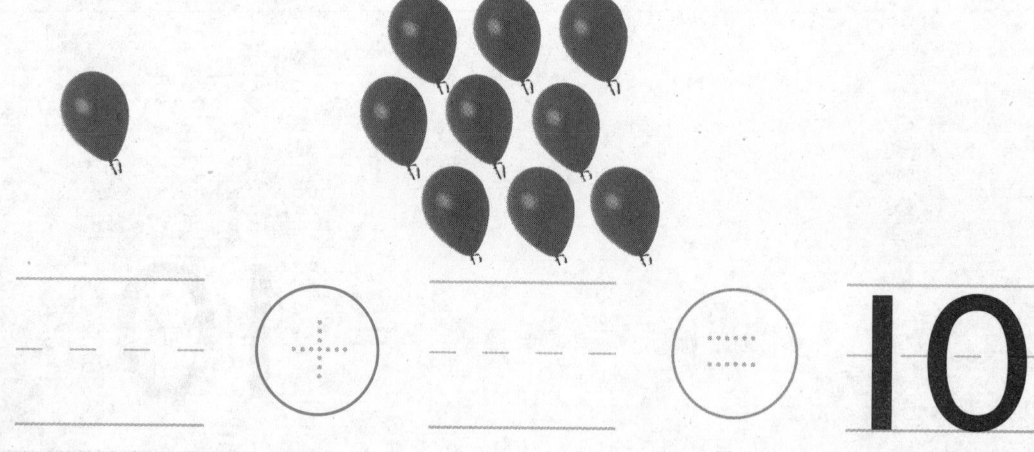

**Directions: I.** Count the objects and trace the numbers and symbols. **2–3.** Count the objects and write the numbers. Trace the symbols.

Name

## Independent Practice

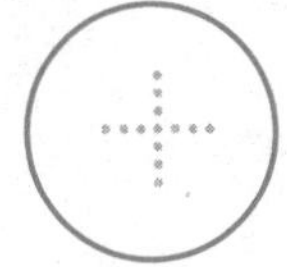

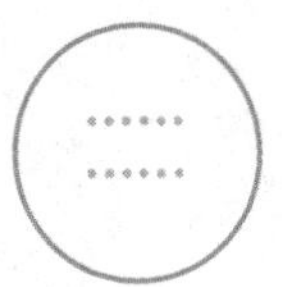

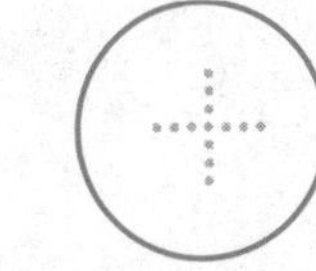

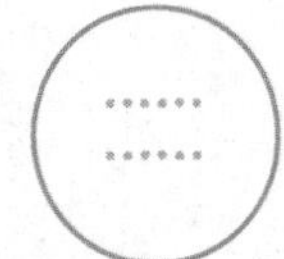

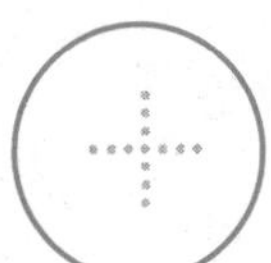

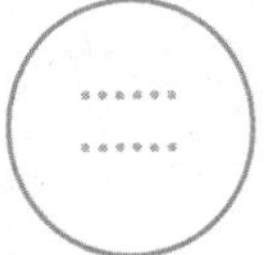

**Directions: 4–6.** Count the objects and write the numbers. Trace the symbols.

Processes &Practices

## Brain Builders

7

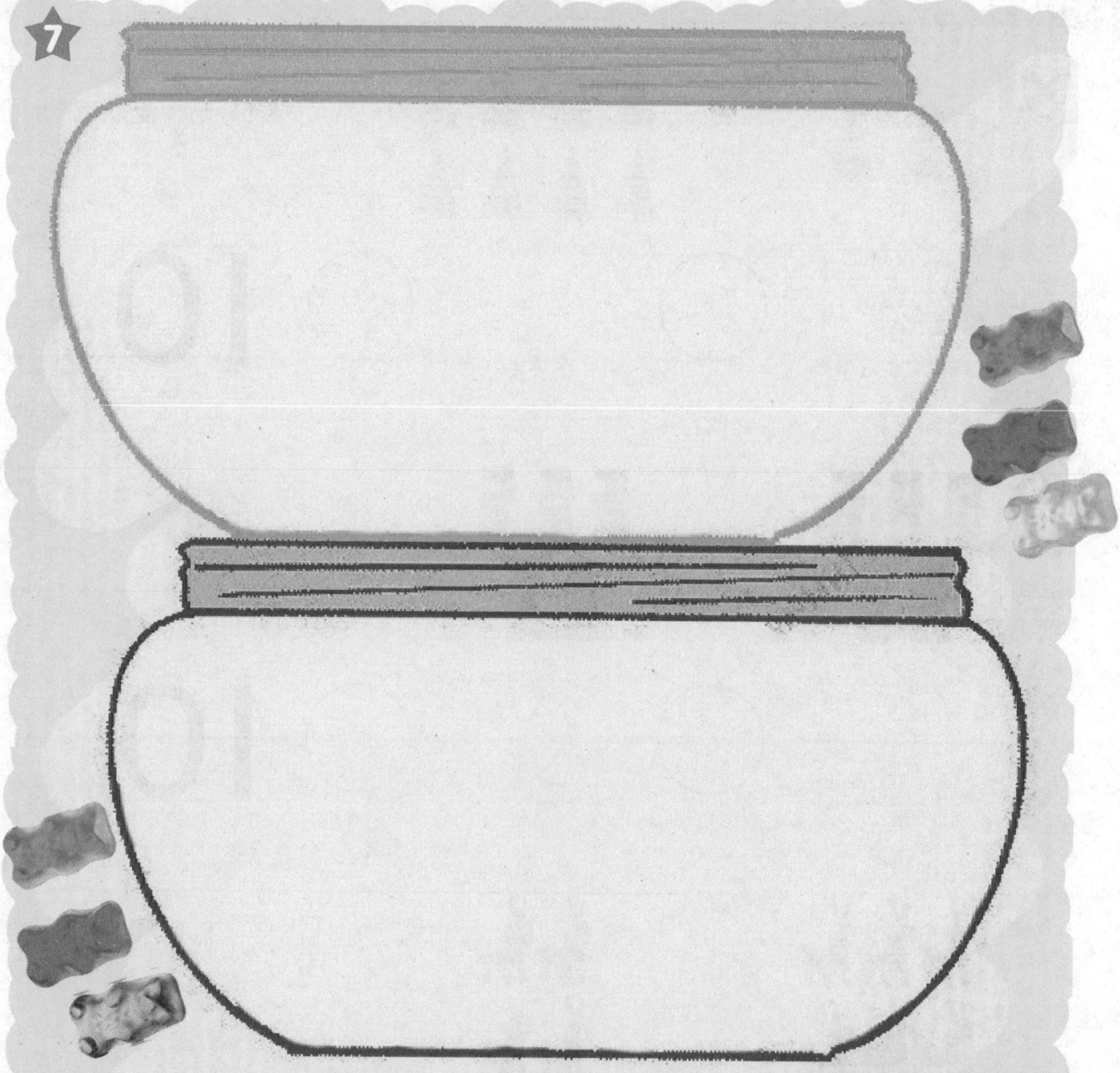

+ = 10

**Directions: 7.** Use counters to model making 10. Draw the first group in the top jar. Write the number. Draw the second group in the bottom jar. Write the number. Compare your answer to a friend's. Discuss whether they are the same. Explain to your friend why there is more than one possible answer.

Name

# My Homework

Lesson 7

Add to Make 10

## Homework Helper

Need help? connectED.mcgraw-hill.com

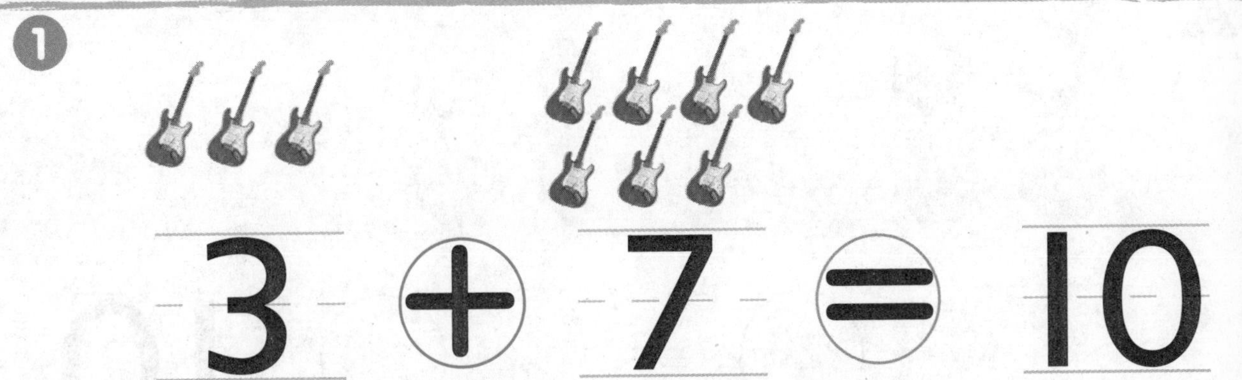

1

3 + 7 = 10

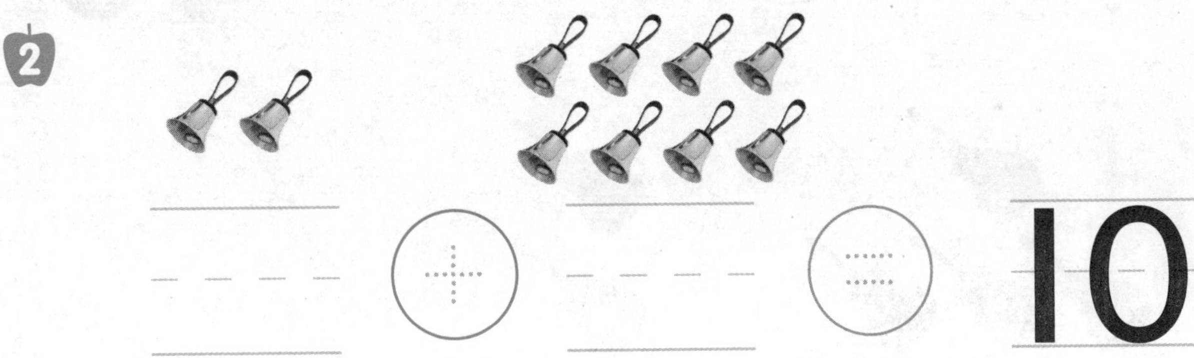

2

+ = 10

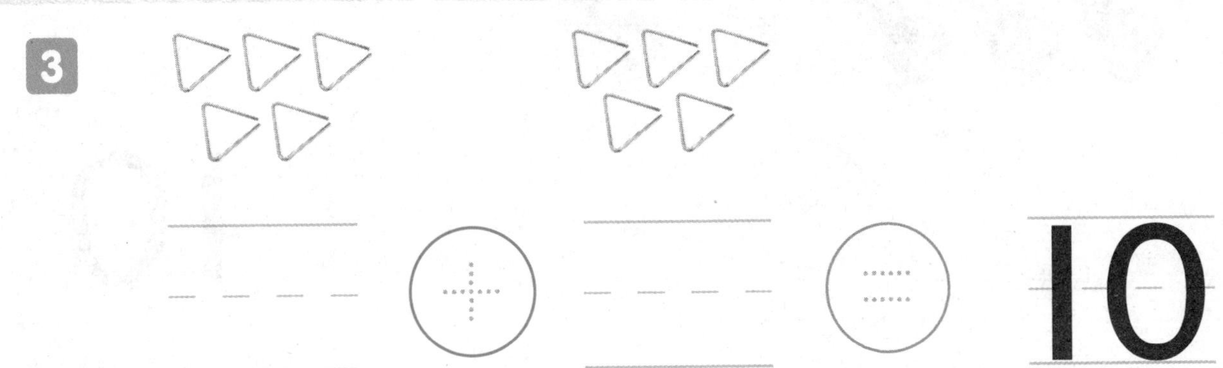

3

+ = 10

**Directions: 1–3.** Count the objects and write the numbers. Trace the symbols.

4

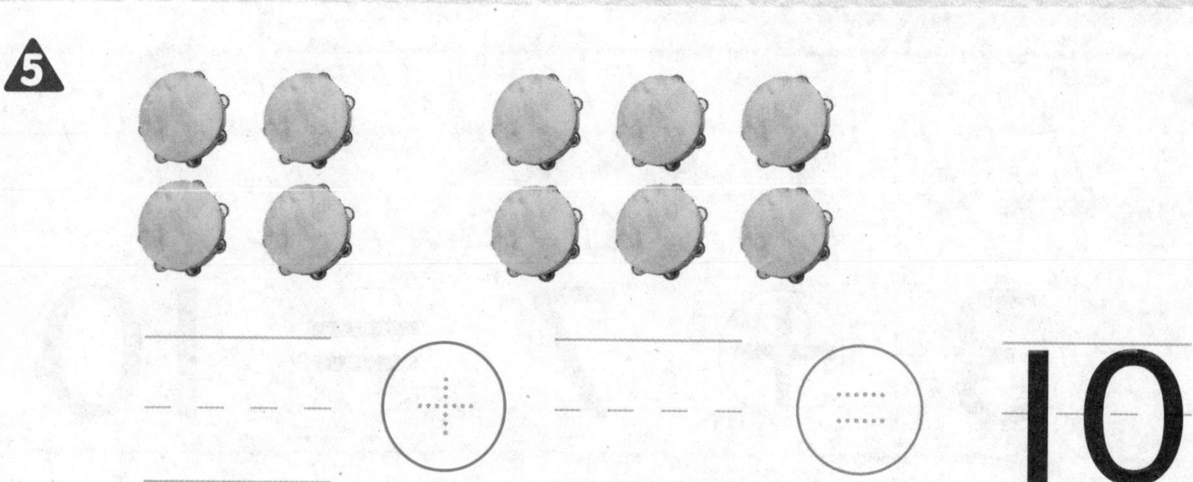

____ + ____ = 10

5

____ + ____ = 10

6

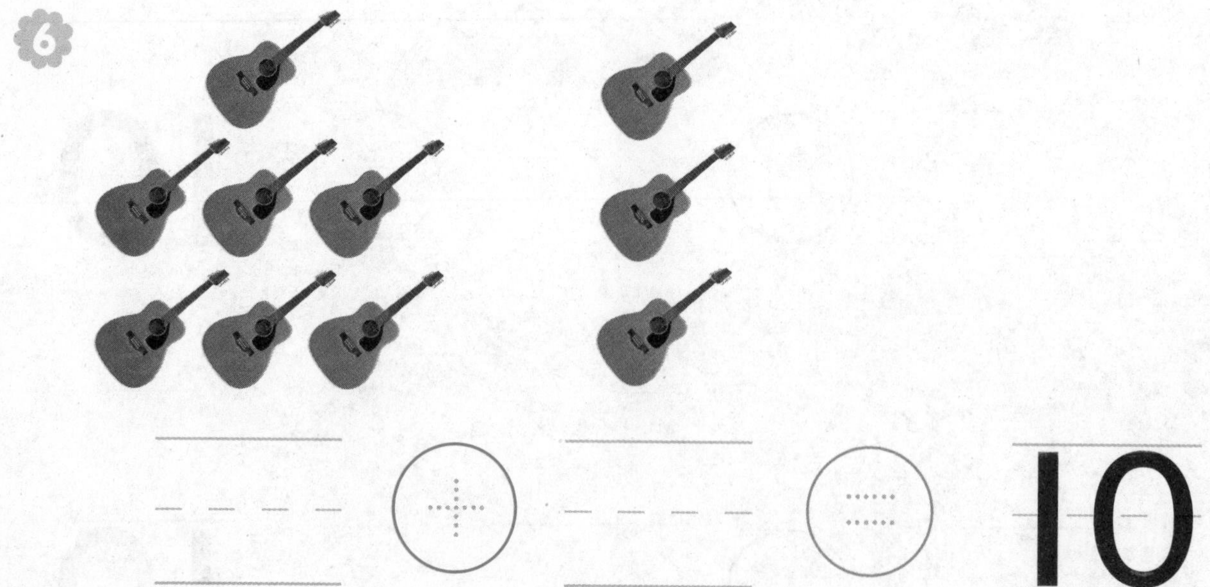

____ + ____ = 10

**Directions: 4–6.** Count the objects and write the numbers. Trace the symbols.

**Math at Home** Give your child a group of six items and a group of four items. Have your child add the groups together to tell how many in all.

Name

# Fluency Practice

1

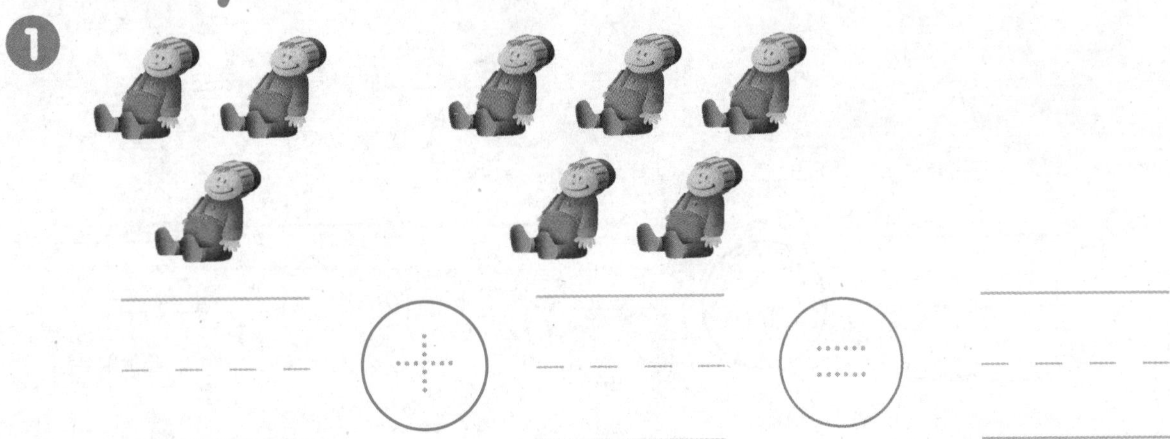

2

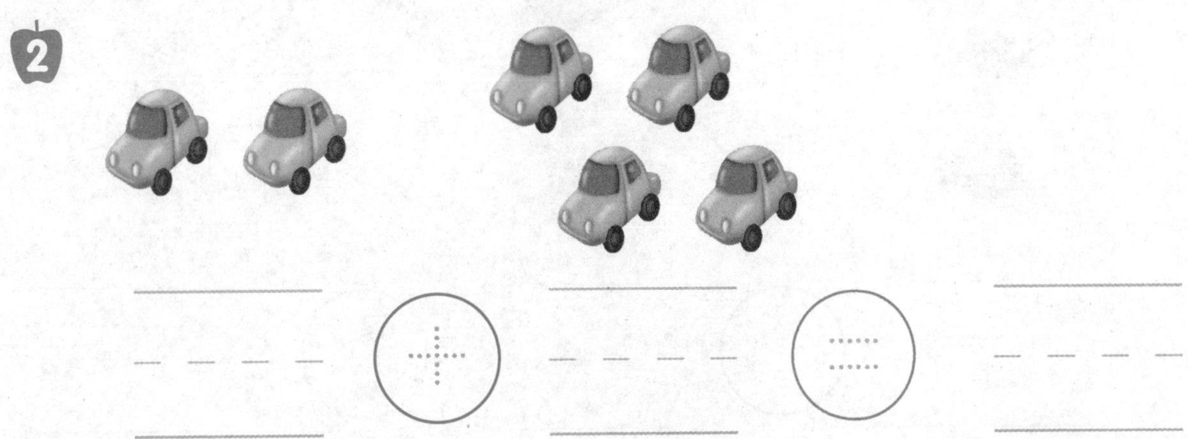

3

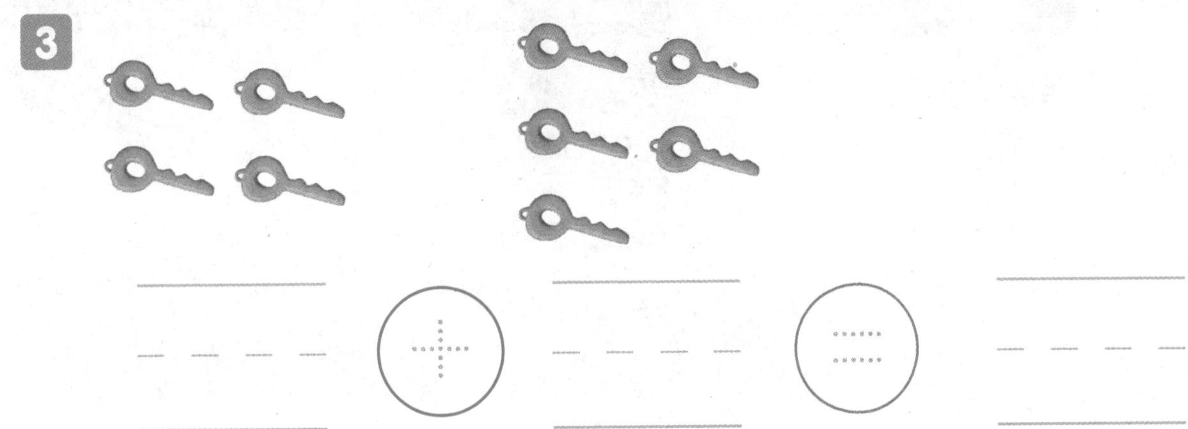

**Directions: 1–3.** Count the objects in each group. Write the numbers and trace the symbols. Write how many there are in all.

# Fluency Practice

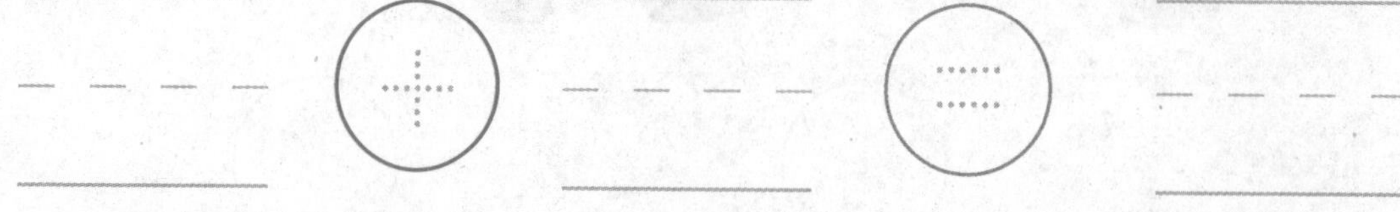

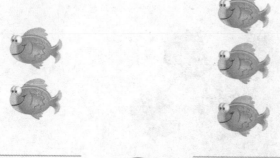

**Directions: 4–6.** Count the objects in each group. Write the numbers and trace the symbols. Write how many there are in all.

Name

# My Review

Chapter 5

Addition

## Vocabulary Check 

1 plus sign

+ =

2 equals sign

+ =

3 

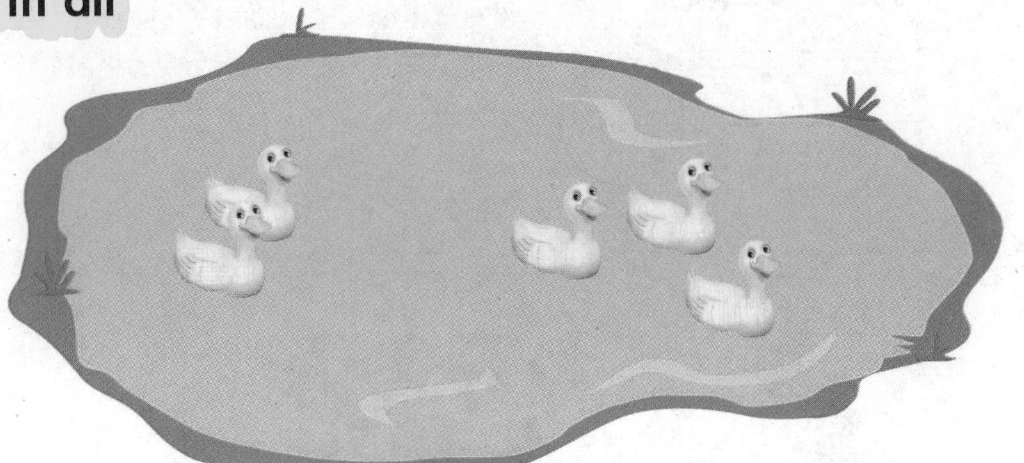

+ 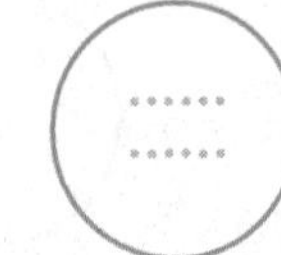

**Directions:** **1.** Circle the plus sign. **2.** Circle the equals sign. **3.** Count the ducks in each group. Write the numbers and trace the symbols. Write how many ducks there are in all.

# Concept Check

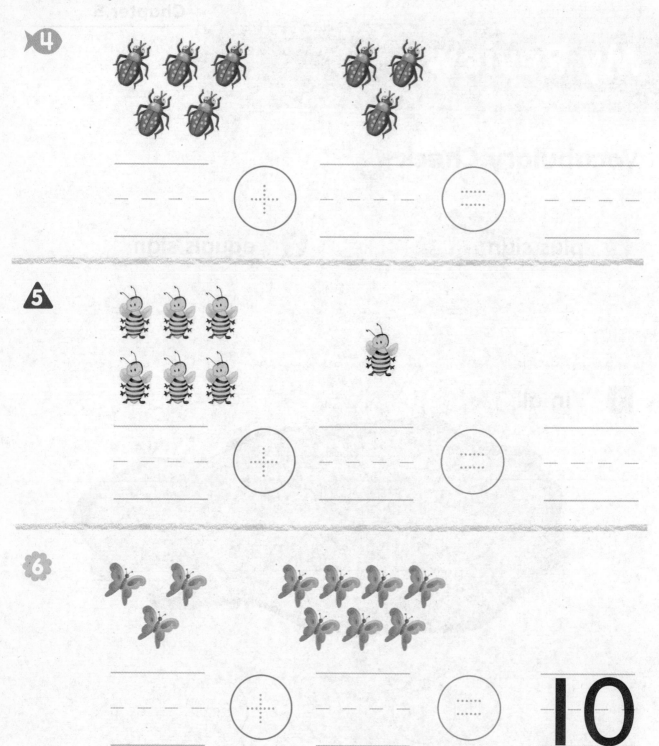

4

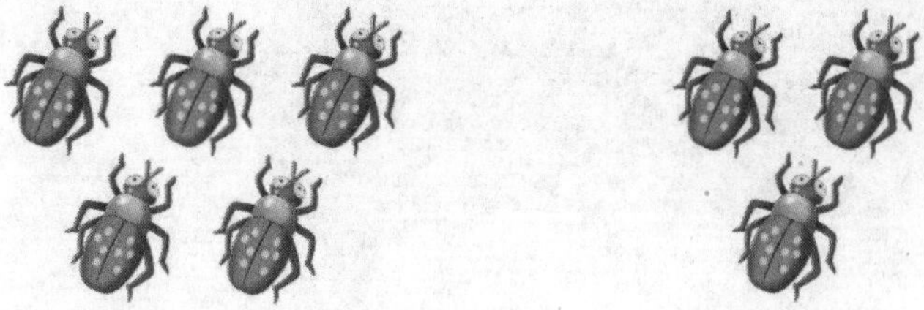

+ =

5

+ =

6

+ = 10

**Directions: 4–5.** Count the insects in each group. Write the numbers and trace the symbols. Circle the groups to join them. Write how many there are in all. **6.** Count the objects and write the numbers. Trace the symbols.

Name

# Brain Builders

7

**Directions: 7.** There are two horses to ride. Draw four more horses to show another group. Write the numbers and trace the symbols.

# Reflect

**Chapter 5**

**ESSENTIAL QUESTION**

**How can I use objects to add?**

HAPPY BIRTHDAY

____ + ____ = 10

**Directions:** Count the cupcakes and write the number. Draw more to make 10. Write the number. Trace the symbols.

Name

Date

Score

# Performance Task

## At the Party

The class is planning a party. They make a list of party items they will need to buy.

**Show all of your work to receive full credit.**

### Part A

Count the number of objects in each group. Write the numbers. Trace the symbols. Circle the groups to join them. Write how many objects there are in all.

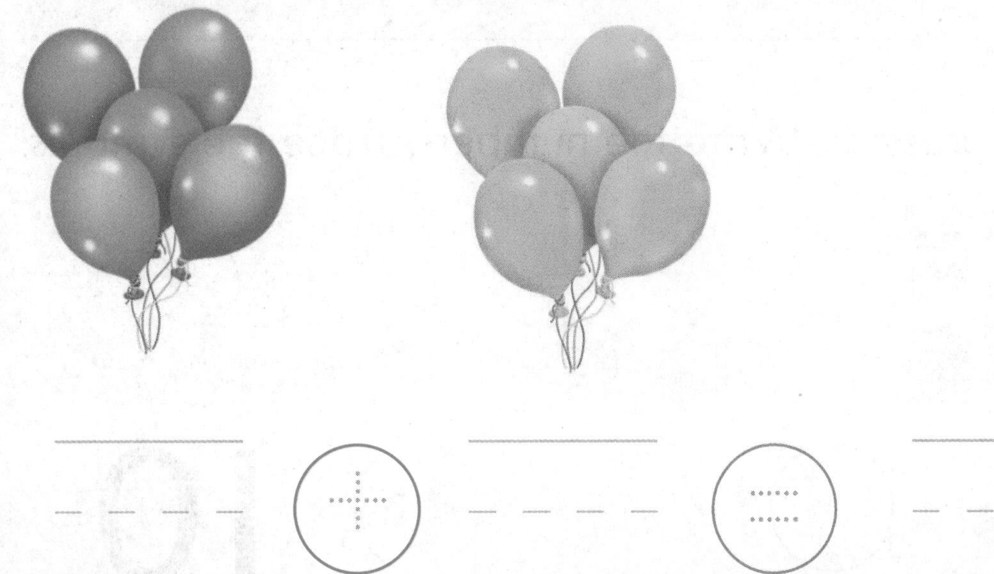

## Part B

Count each group of cups. Write the numbers and trace the symbols. Write how many cups there are in all.

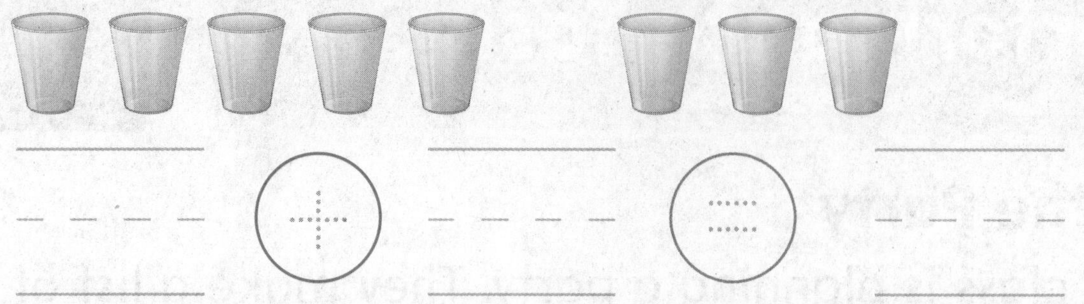

## Part C

Count the hats in each group. Write the numbers and trace the symbol. Write how many hats there are in all.

is

## Part D

Count the presents. Write the numbers. Trace the symbols.

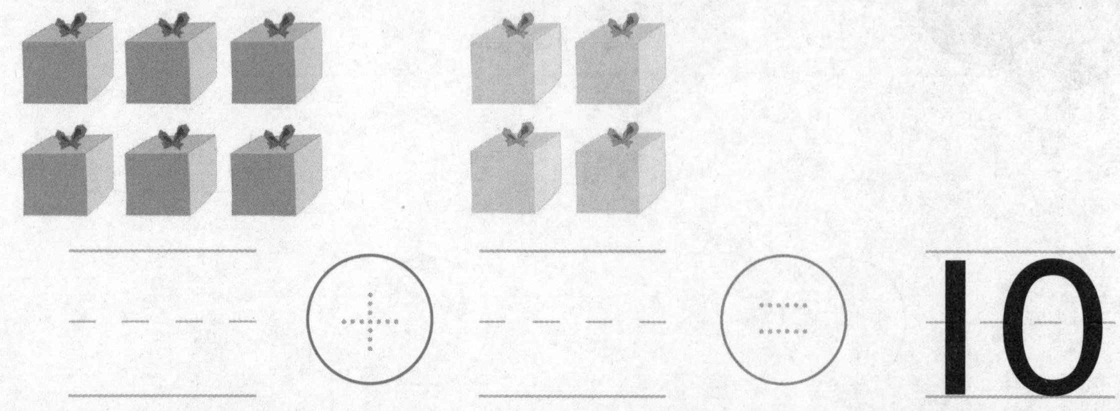

10

# Glossary/Glosario

| English | Spanish/Español |
|---|---|

## above

above

## encima

encima

## add

3 ducks 2 more join 5 ducks in all

## sumar

3 patos se unen 2 más 5 patos en total

## afternoon

## tarde

## Aa

### alike (same)

### igual

### are left

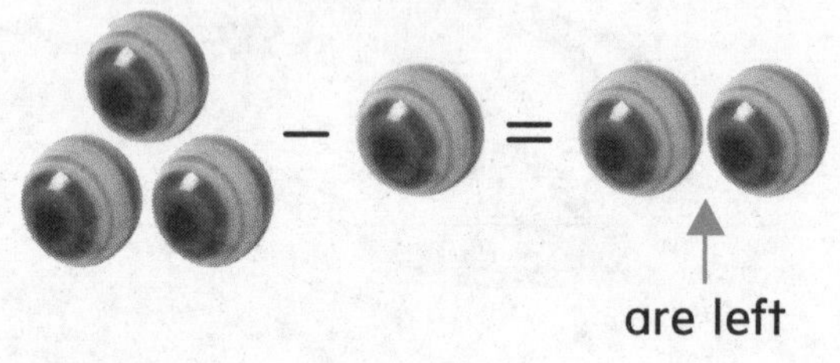


### quedan

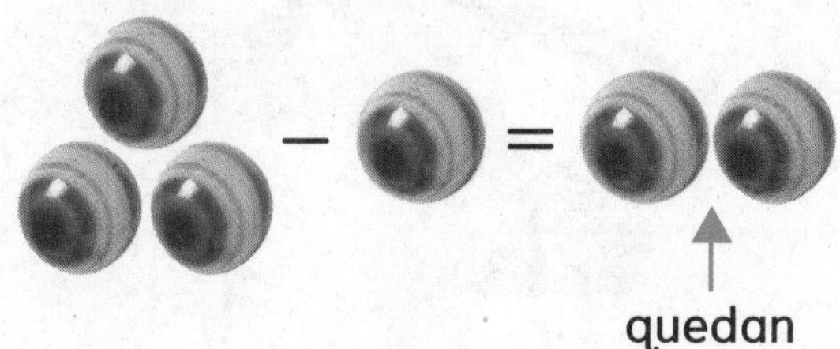


## Bb

### behind

### detrás

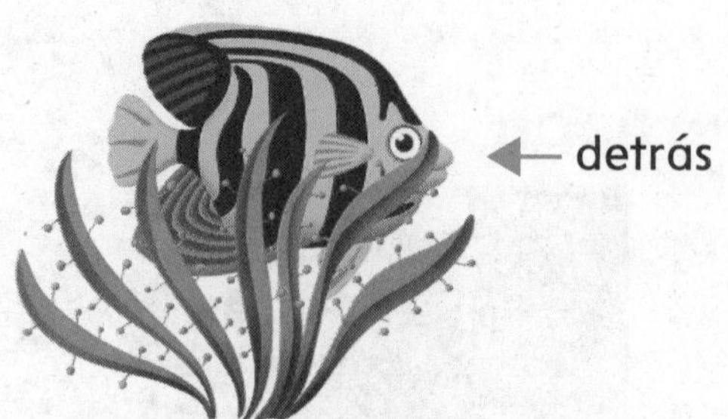

## below

below

## debajo

debajo

## beside

The cat is beside the dog.

## al lado

El gato está al lado del perro.

# Cc

## calendar

April

| Sunday | Monday | Tuesday | Wednesday | Thursday | Friday | Saturday |
|---|---|---|---|---|---|---|
| | | 1 | 2 | 3 | 4 | 5 |
| 6 | 7 | 8 | 9 | 10 | 11 | 12 |
| 13 | 14 | 15 | 16 | 17 | 18 | 19 |
| 20 | 21 | 22 | 23 | 24 | 25 | 26 |
| 27 | 28 | 29 | 30 | | | |

## calendario

abril

| domingo | lunes | martes | miércoles | jueves | viernes | sábado |
|---|---|---|---|---|---|---|
| | | 1 | 2 | 3 | 4 | 5 |
| 6 | 7 | 8 | 9 | 10 | 11 | 12 |
| 13 | 14 | 15 | 16 | 17 | 18 | 19 |
| 20 | 21 | 22 | 23 | 24 | 25 | 26 |
| 27 | 28 | 29 | 30 | | | |

## Cc

| | |
|---|---|
| **capacity** | **capacidad** |
|   holds more holds less |   contiene más contiene menos |
| **circle** | **círculo** |
|  |  |
| **compare** | **comparar** |
| 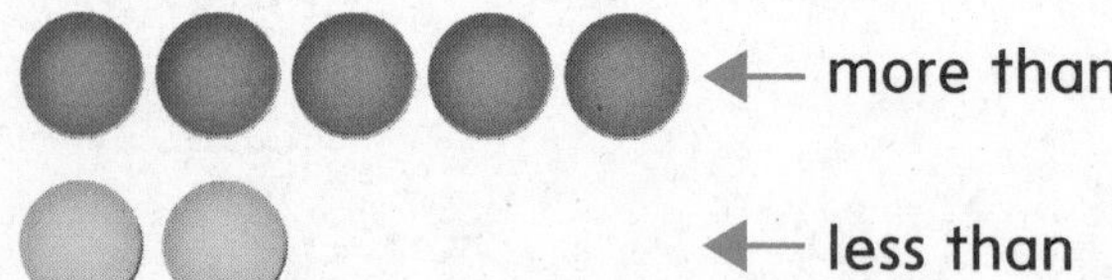  | 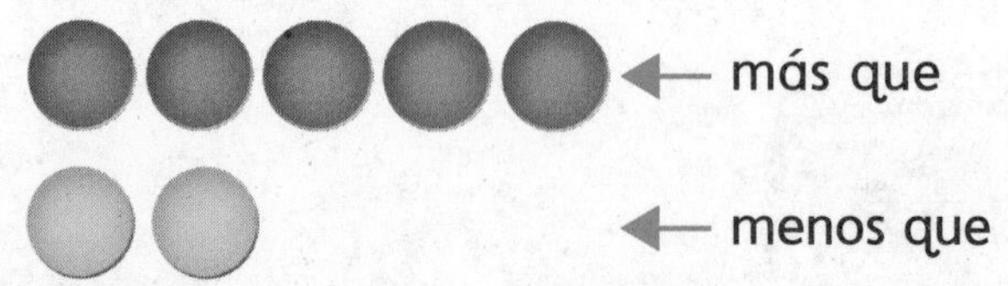  |
| **cone** | **cono** |
| 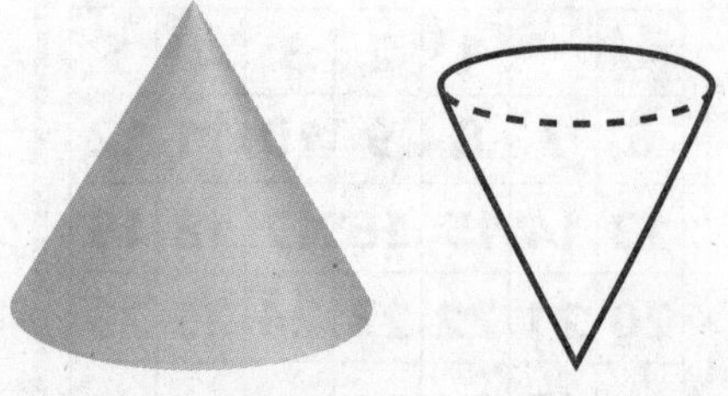 | 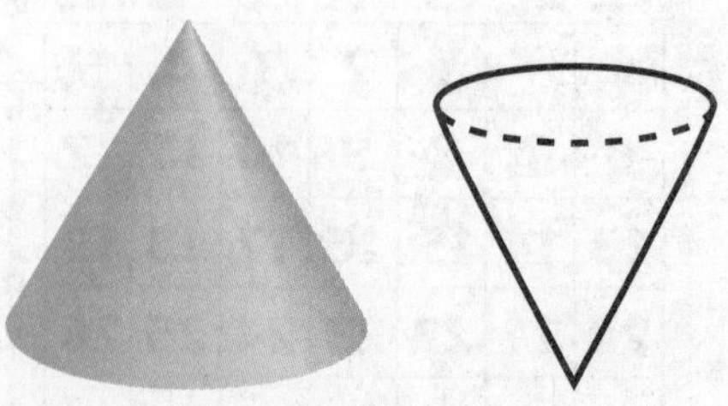 |

## corner

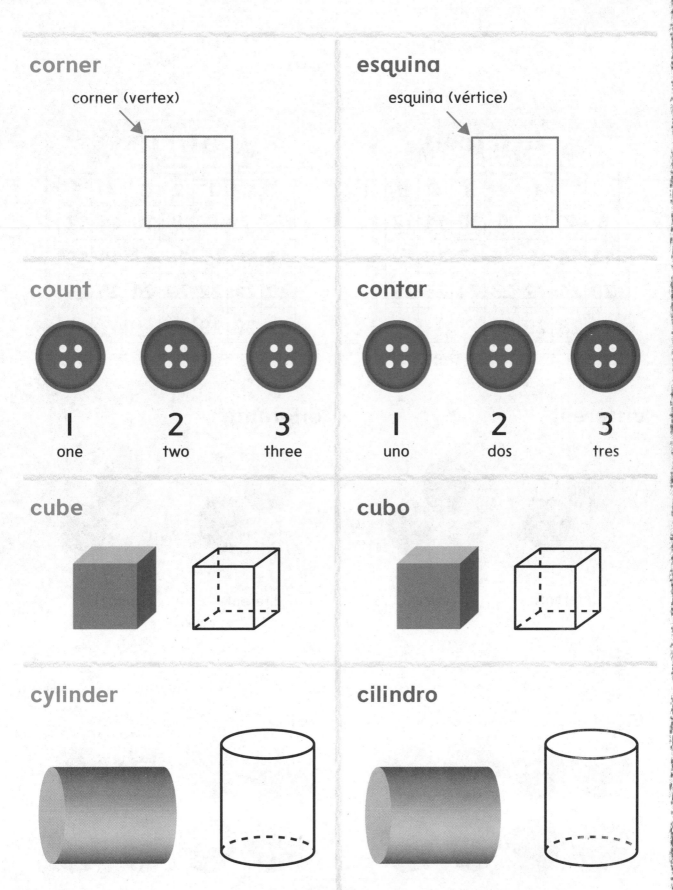

corner (vertex)

## esquina

esquina (vértice)

## count

1 one 2 two 3 three

## contar

1 uno 2 dos 3 tres

## cube

## cubo

## cylinder

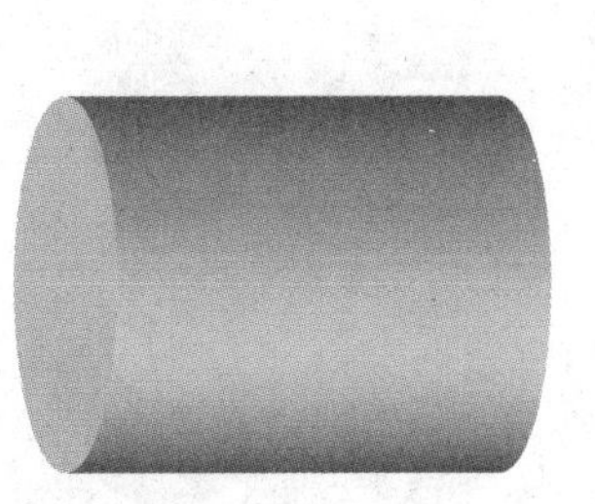
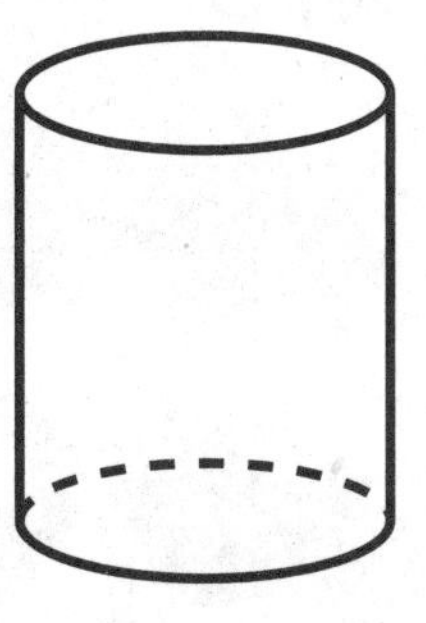

## cilindro

## Dd

### day

### día

### different

different alike

### diferente

diferentes iguales

## Ee

### eight

### ocho

**eighteen**

**dieciocho**

**eleven**

**once**

**equals sign (=)**

4 + 1 = 5

equals

**signo igual (=)**

4 + 1 = 5

es igual a

**equal to**

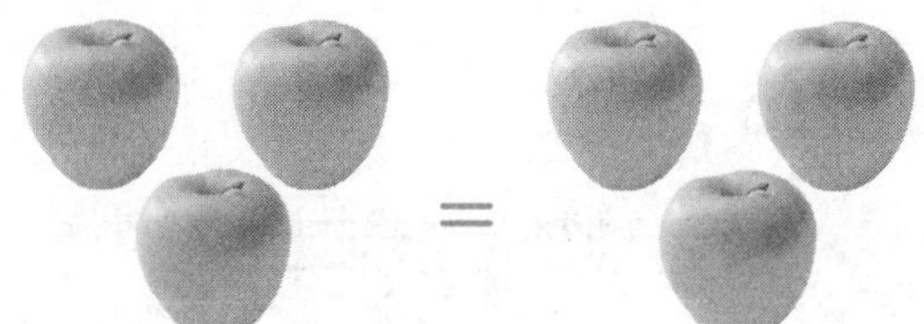

**igual a**

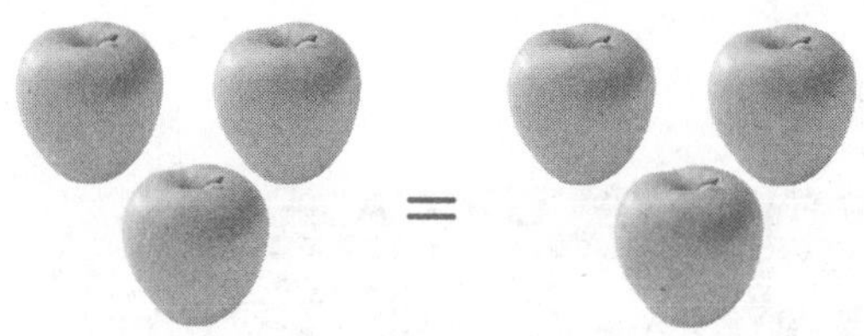

## Ee

**evening** | **noche**

**fifteen** | **quince**

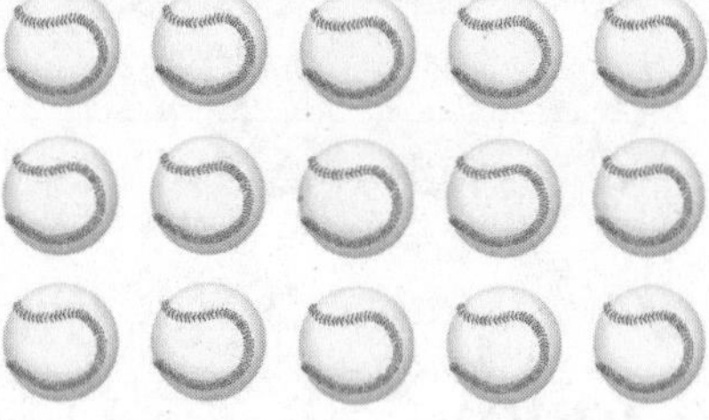

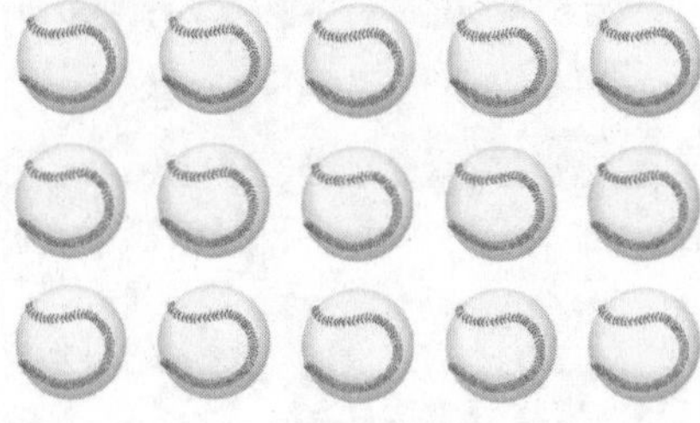

**five** | **cinco**

**four** | **cuatro**

**fourteen**

**catorce**

## Gg

**greater than**

**mayor que**

## Hh

**heavy (heavier)**

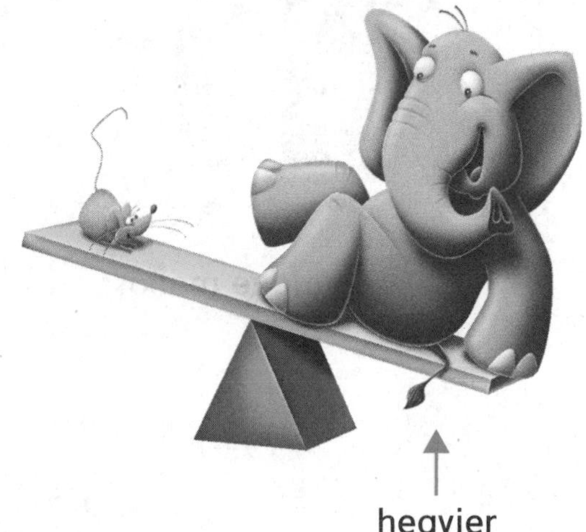


**pesado (más pesado)**

## Hh

### height

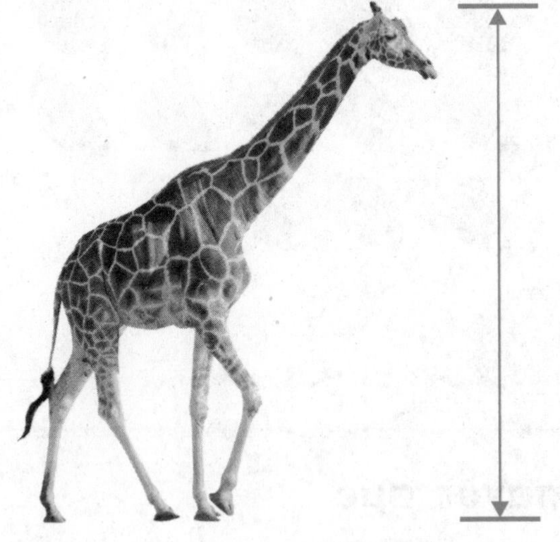

### altura

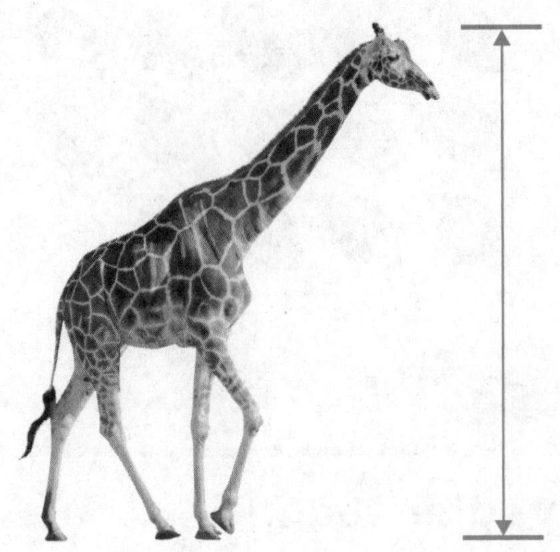

### hexagon

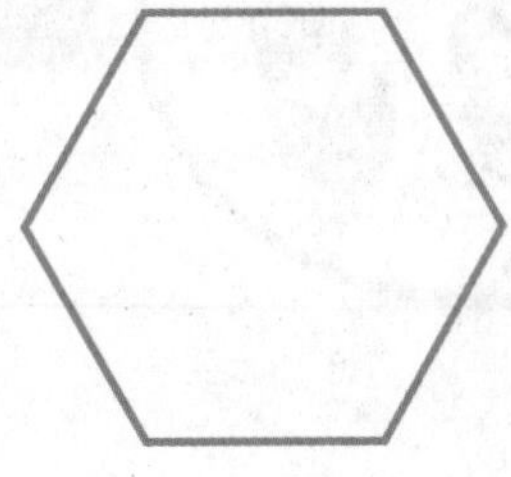

### hexágono

### holds less

### contiene menos

## holds more

holds more

## contiene más

contiene más

## holds the same

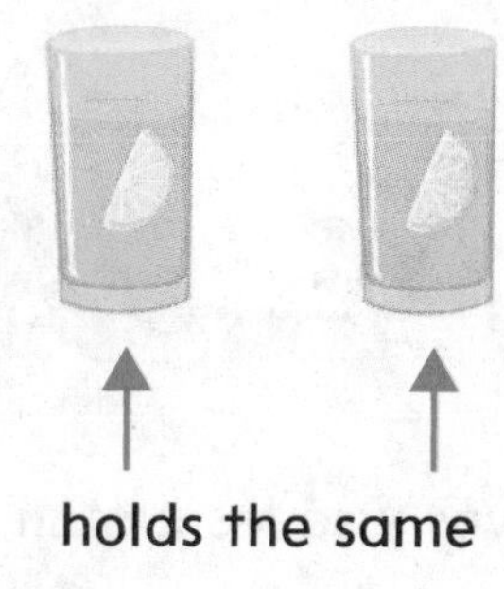

holds the same

## contiene la misma cantidad

contiene la misma cantidad

## in all

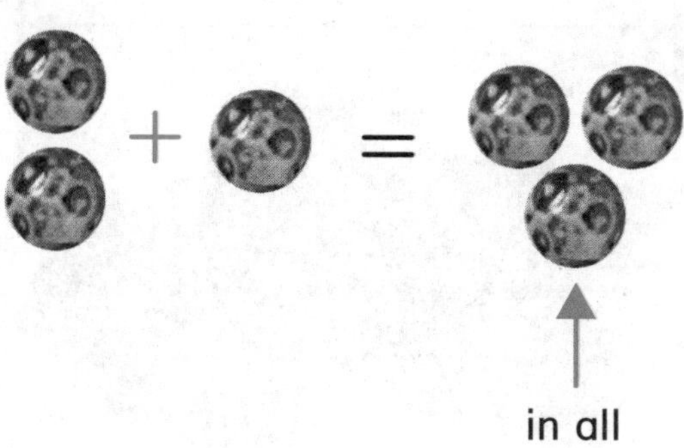


in all

## en total

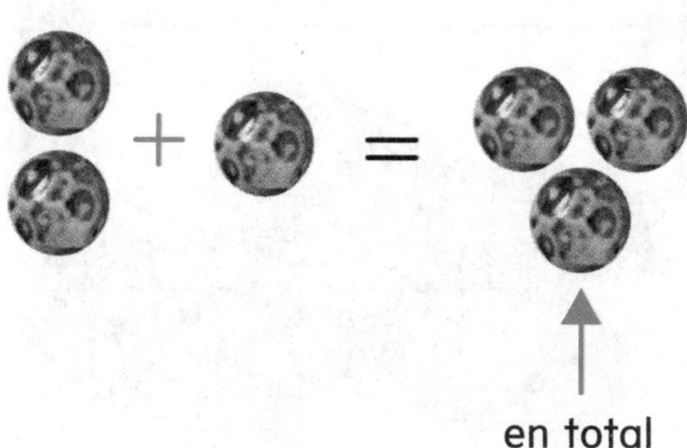


en total

## in front of

## en frente de

← en frente de

Jj

## join

3 birds and 2 birds join.

## juntar

Hay 3 aves y se les juntan 2 más.

Ll

## length

length

## longitud

longitud

## less than

## menor que

## light (lighter)

## liviano (más liviano)

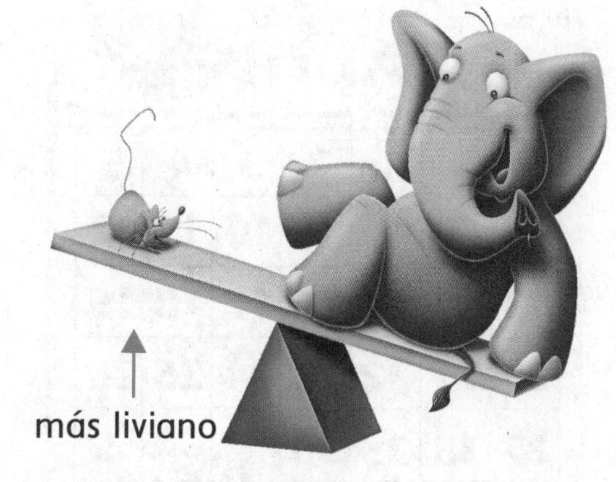


## long (longer)

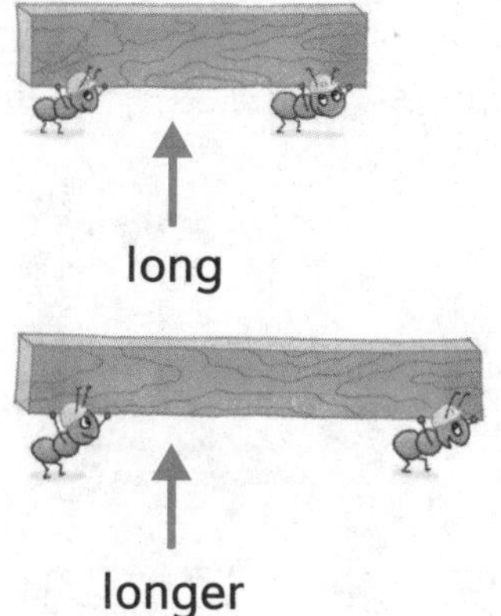


## largo (más largo)

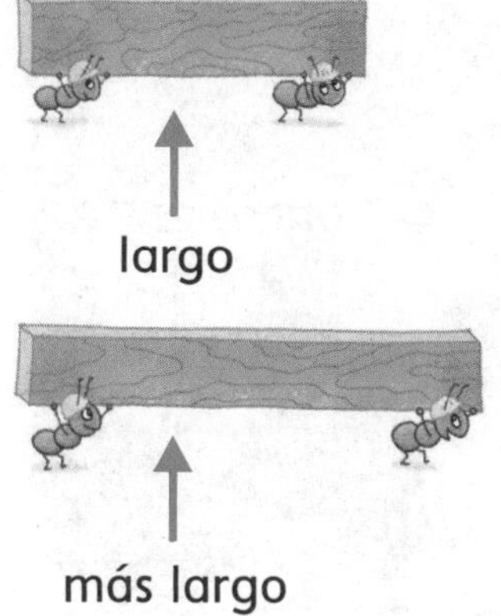

## minus sign (−)

$5 - 2 = 3$

minus

## signo menos (−)

$5 - 2 = 3$

menos

## month

month

## mes

mes

## morning

## mañana

## Nn

### next to

The cat is next to the dog.

### junto a (al lado)

El gato está junto al perro.

### nine

### nueve

### nineteen

### diecinueve

### number

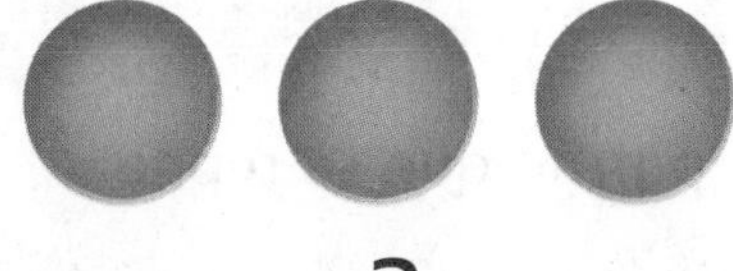

3

tells how many

### número

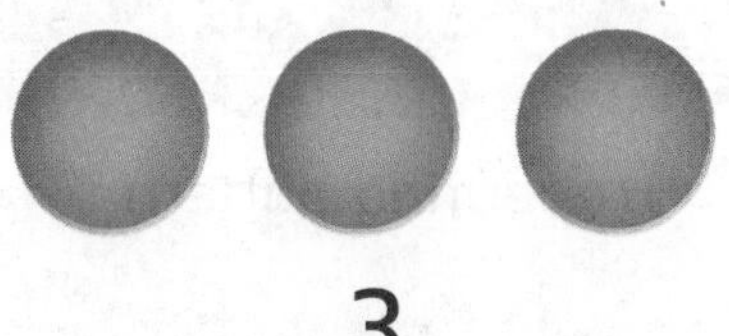

3

dice cuántos hay

## Oo

### one

### uno

### ordinal numbers

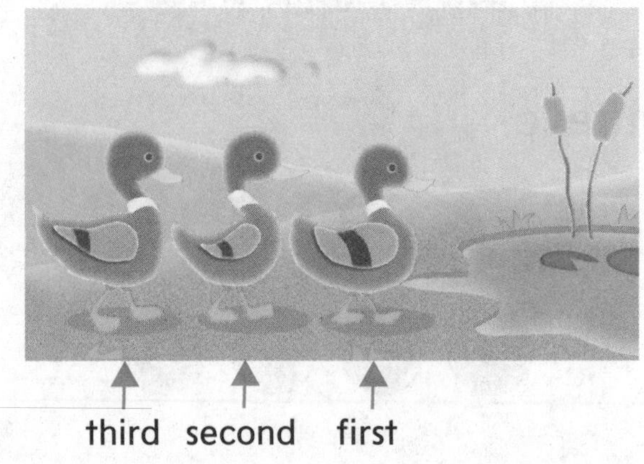

third second first

### números ordinales

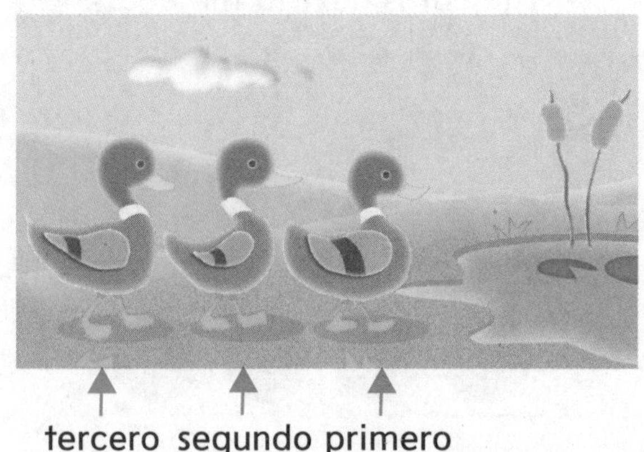

tercero segundo primero

## Pp

### pattern

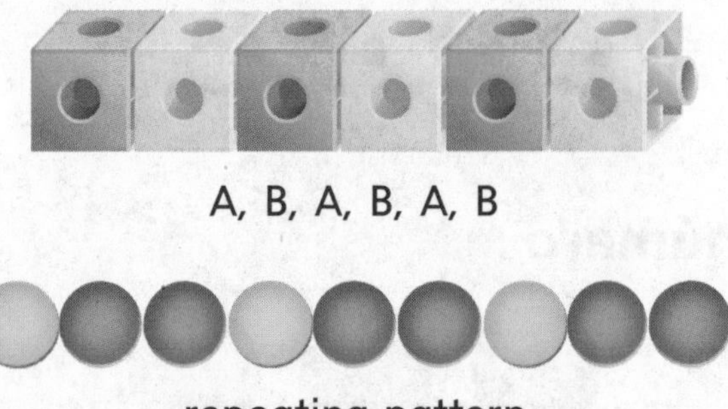

A, B, A, B, A, B

repeating pattern

### patrón

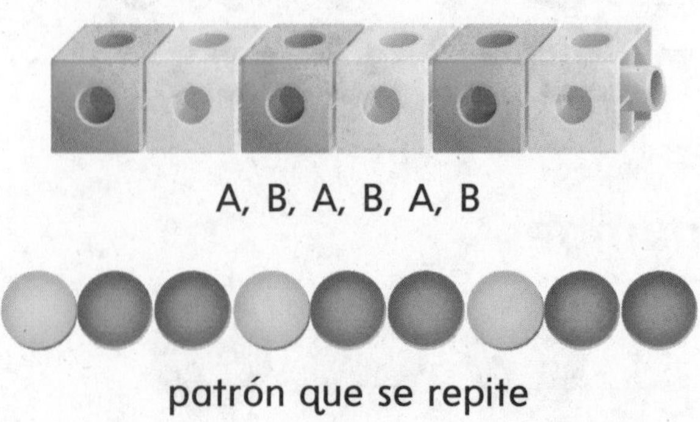

A, B, A, B, A, B

patrón que se repite

## plus sign (+)

$5 + 2 = 7$

plus

## signo más (+)

$5 + 2 = 7$

más

## position

## posición

## rectangle

## rectángulo

**repeating pattern**

repeating pattern

**patrón que se repite**

patrón que se repite

**roll**

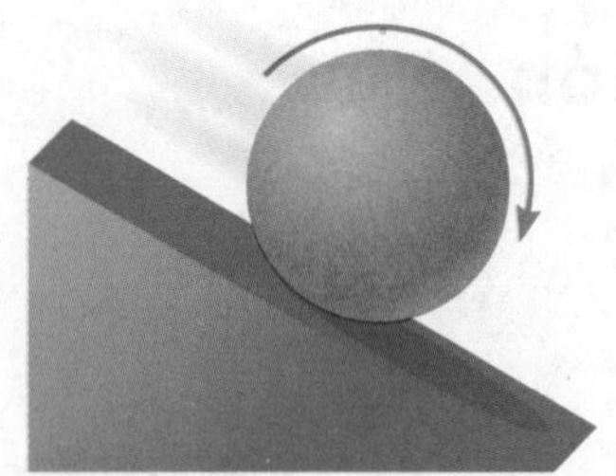

**rodar**

**round**

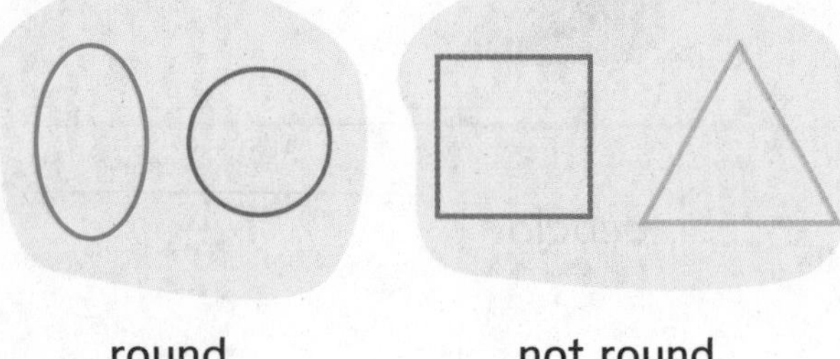

round not round

**redondo**

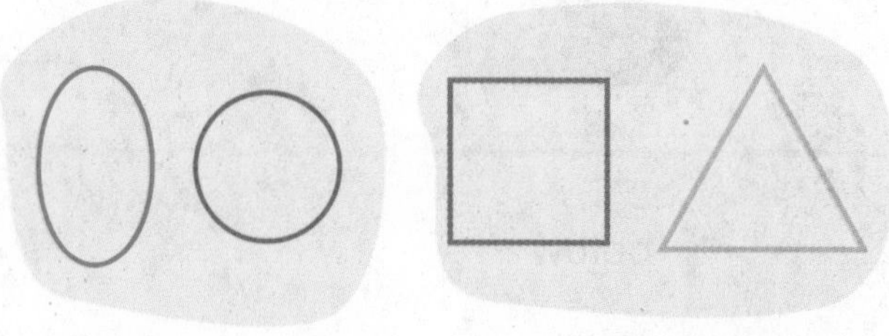

redondo no redondo

## Ss

**same number**

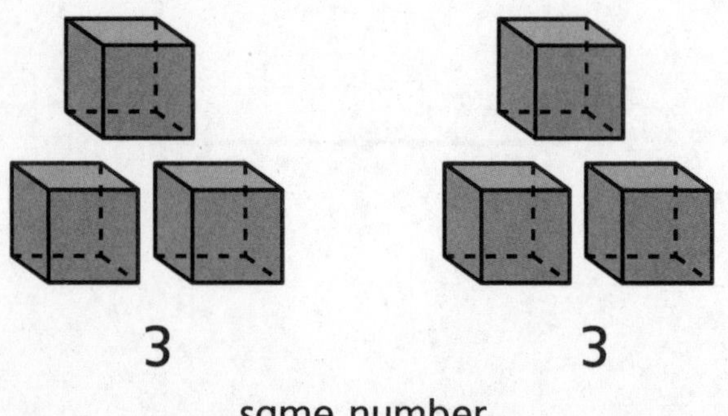

3 3

same number

**el mismo número**

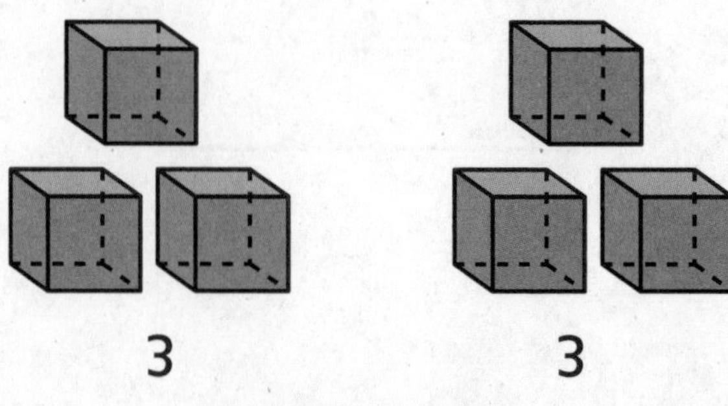

3 3

el mismo número

separate

separar

seven

siete

seventeen

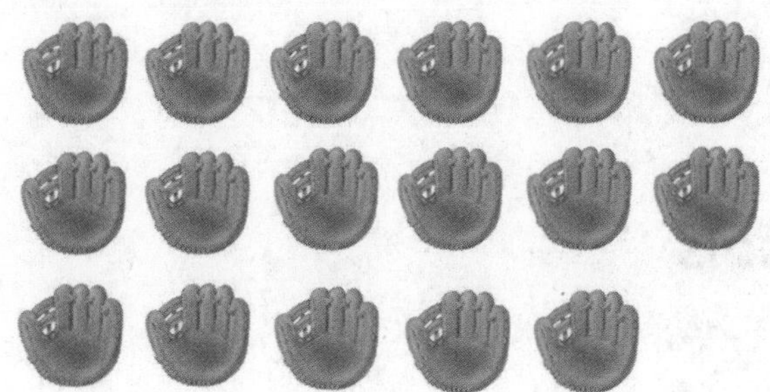

diecisiete

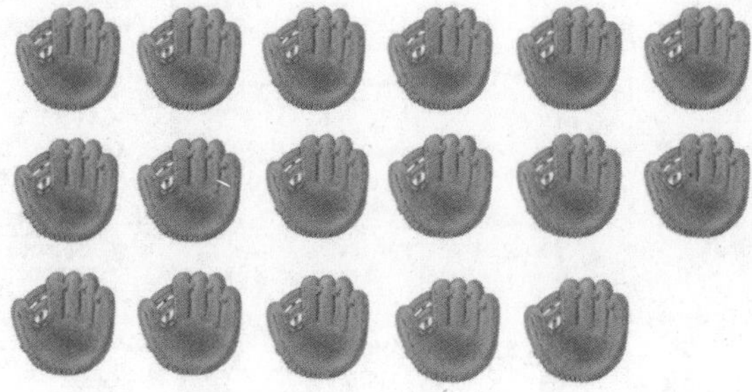

shape

figura

## short (shorter)

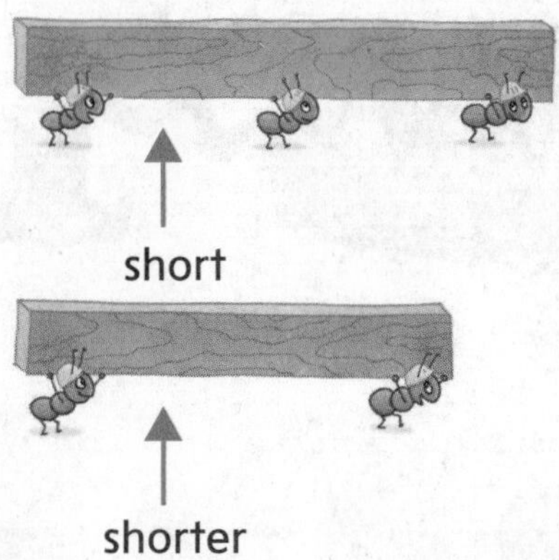


## corto (más corto)

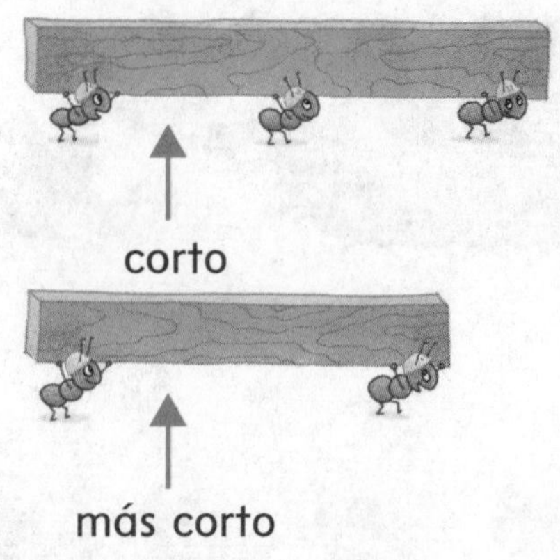


## side

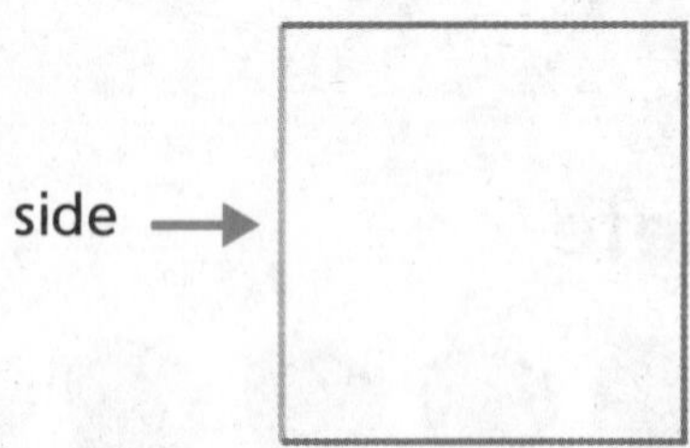


## lado

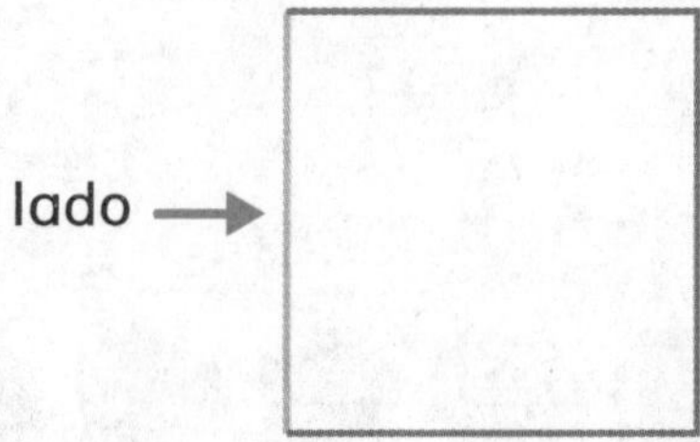


## six

## seis

## sixteen

## dieciséis

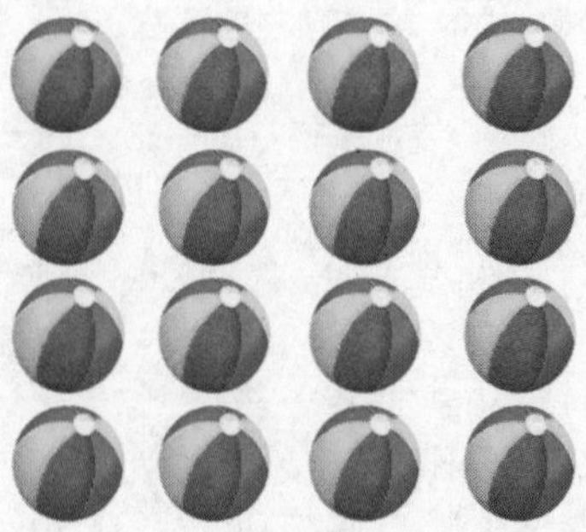

**size**

**tamaño**

**slide**

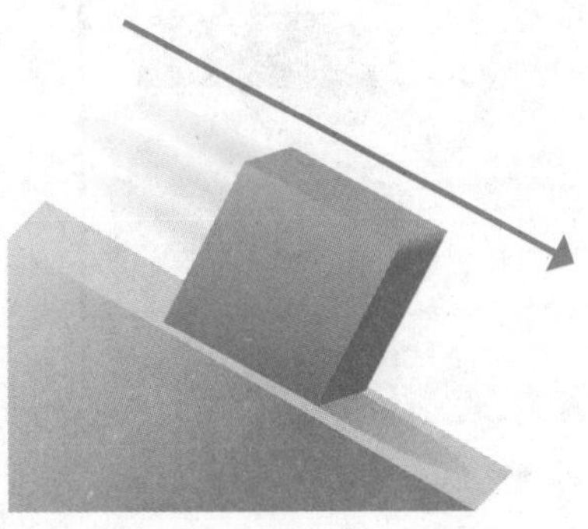

**deslizar**

**sort**

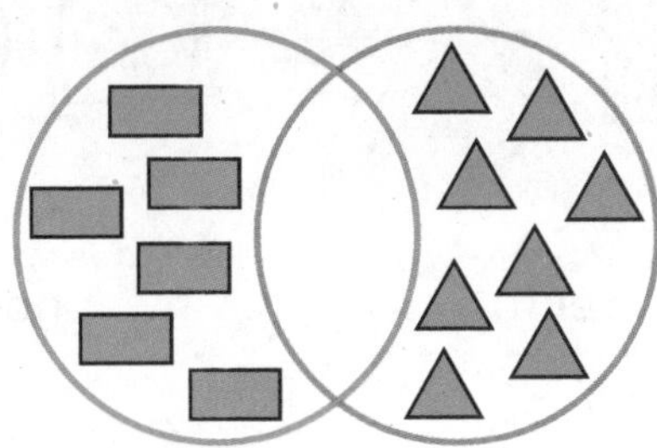

sorted or grouped by shape

**ordenar**

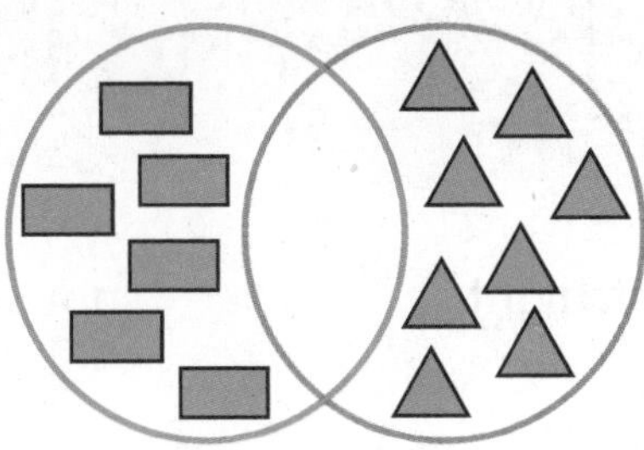

ordenados o agrupados por su forma

**sphere**

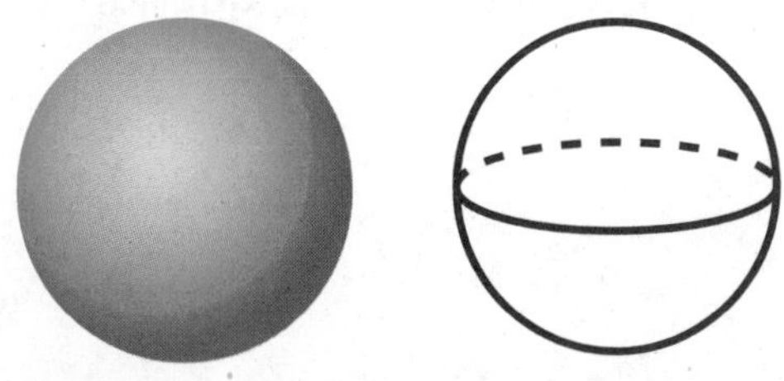

**esfera**

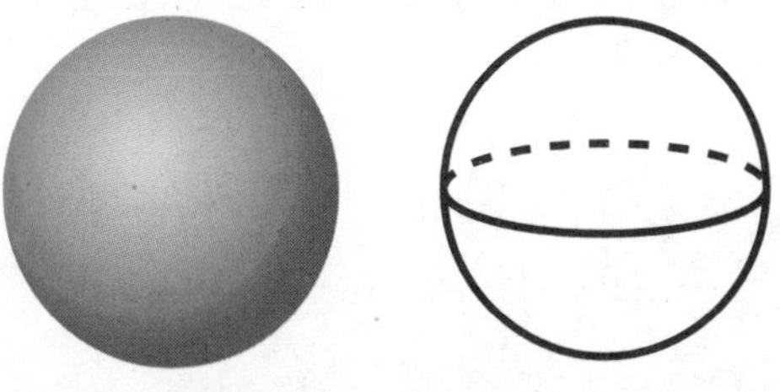

**square**

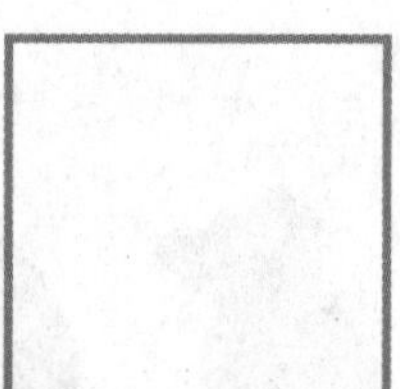

**cuadrado**

**stack**

**pila**

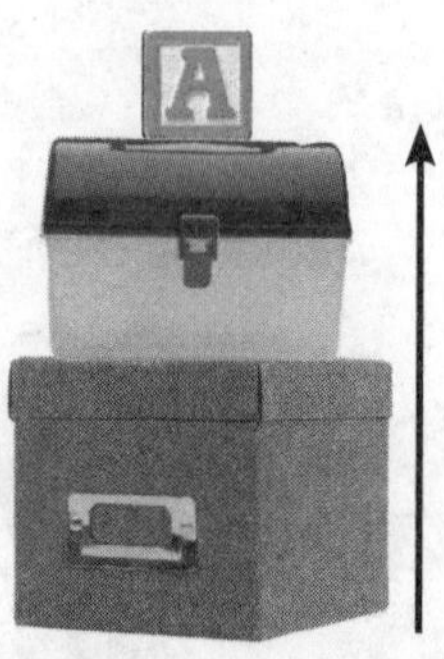


**straight**

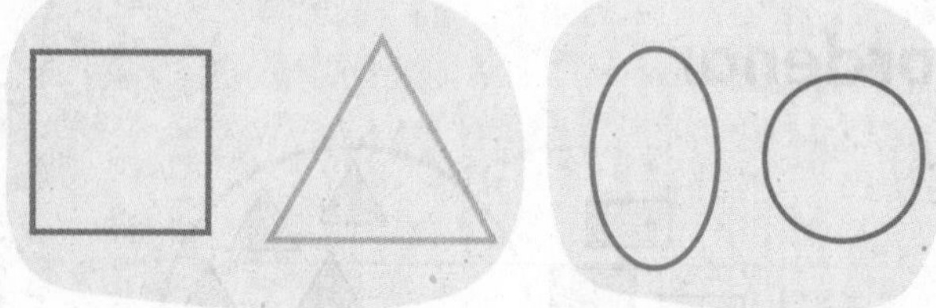

straight	not straight

**recto**

recto	no recto

**subtract (subtraction)**

5 take away 3 is 2. 2 are left.

**restar (resta)**

Si a 5 le quitamos 3, quedan 2.

# Tt

**tall (taller)**

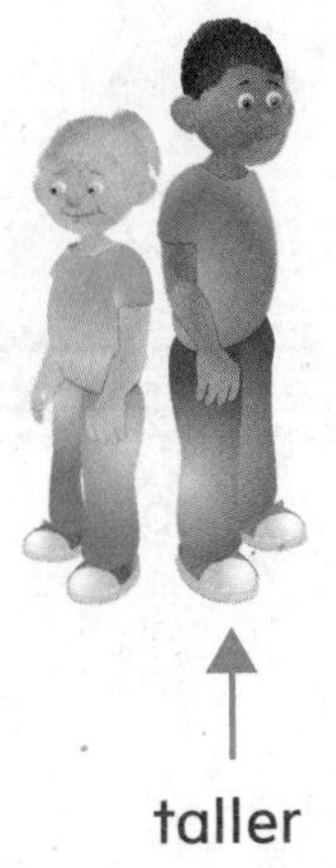


**alto (más alto)**

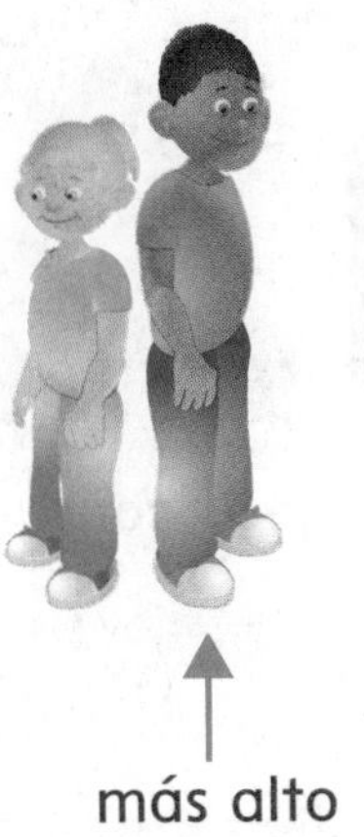


**ten**

**diez**

**thirteen**

**trece**

**three**

**tres**

## three-dimensional shape

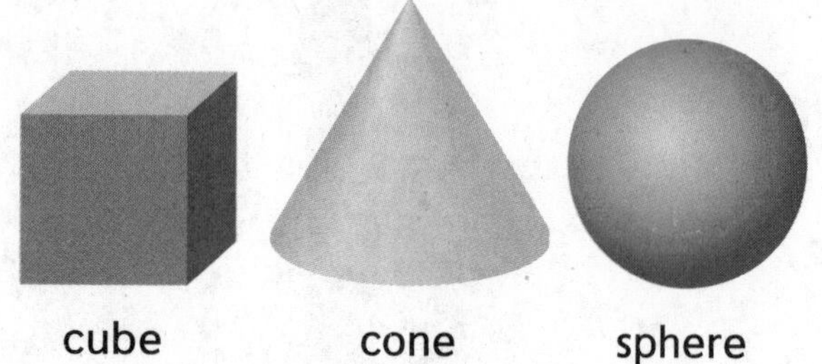


## figura tridimensional

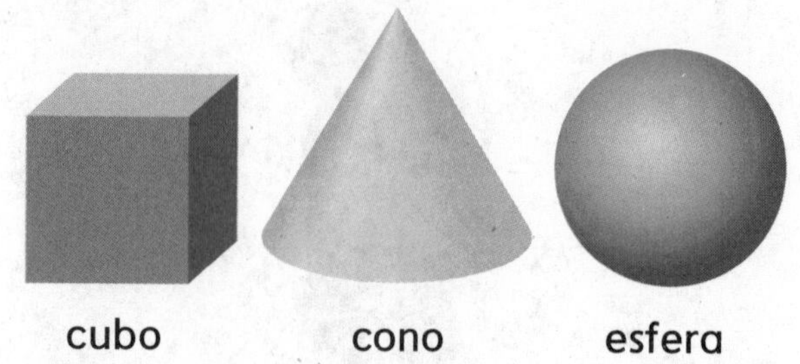


## today

yesterday
today
April
Sunday Monday Tuesday Wednesday Thursday Friday Saturday
1 2 3 4 5
6 7 8 9 10 11 12
13 14 15 16 17 18 19
20 21 22 23 24 25 26
27 28 29 30


## hoy

## tomorrow

## mañana

hoy

mañana

abril

domingo lunes martes miércoles jueves viernes sábado

1 2 3 4 5
6 7 8 9 10 11 12
13 14 15 16 17 18 19
20 21 22 23 24 25 26
27 28 29 30

| | |
|---|---|
| **triangle** 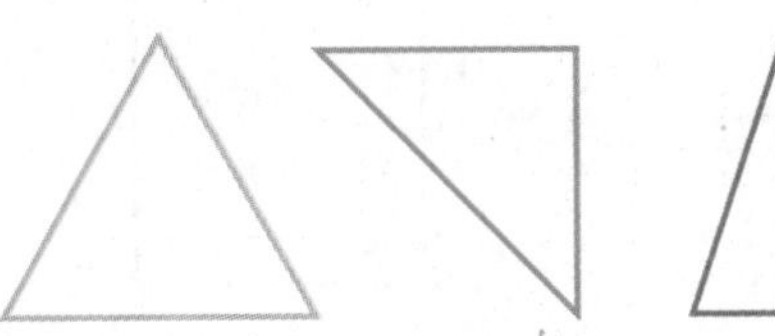 | **triángulo** 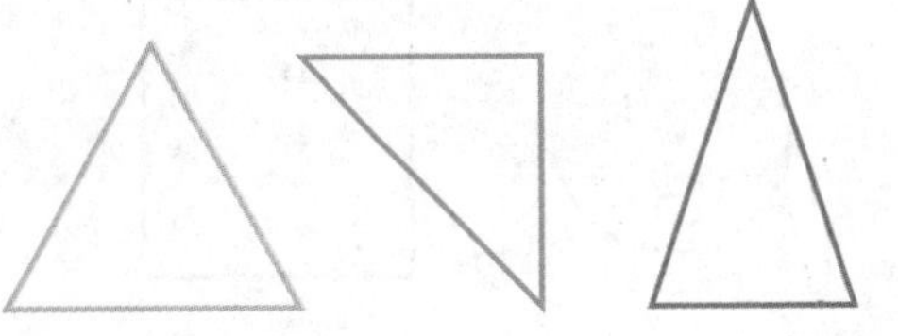 |
| **twelve**  | **doce** 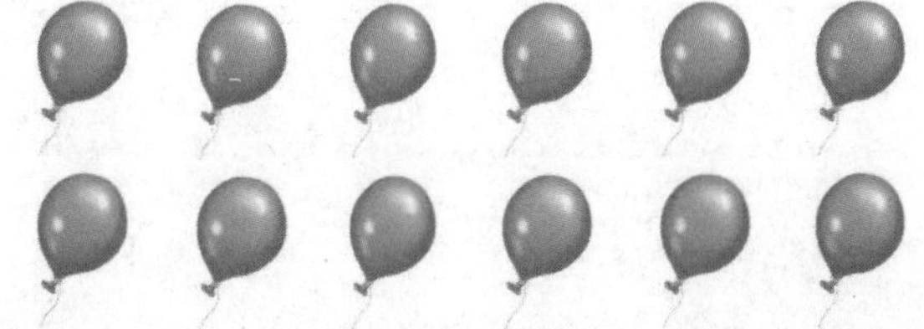 |
| **twenty**  | **veinte**  |
| **two**  | **dos**  |

## Tt

### two-dimensional shape

### figura bidimensional

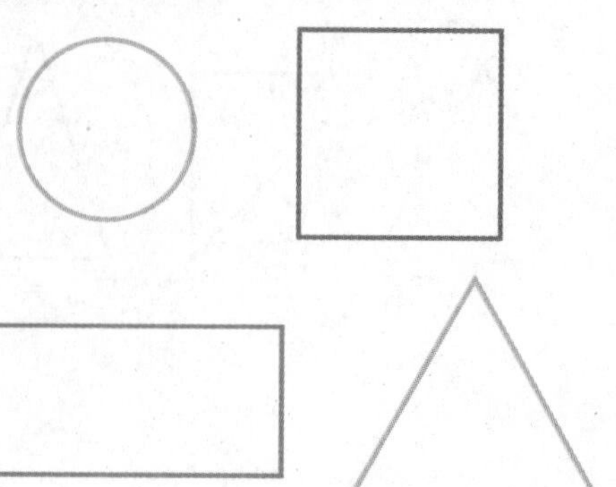

## Vv

### vertex

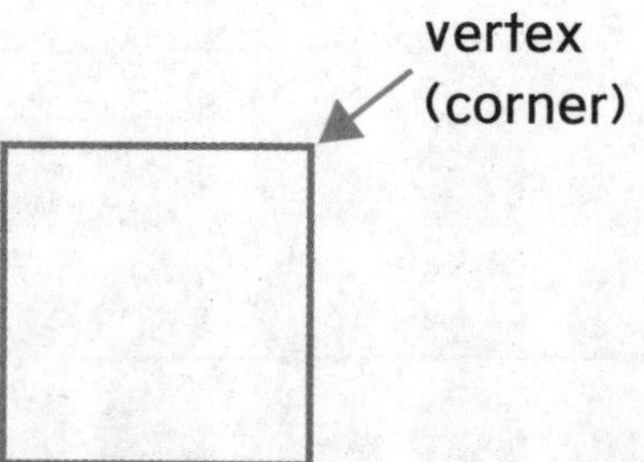


### vértice

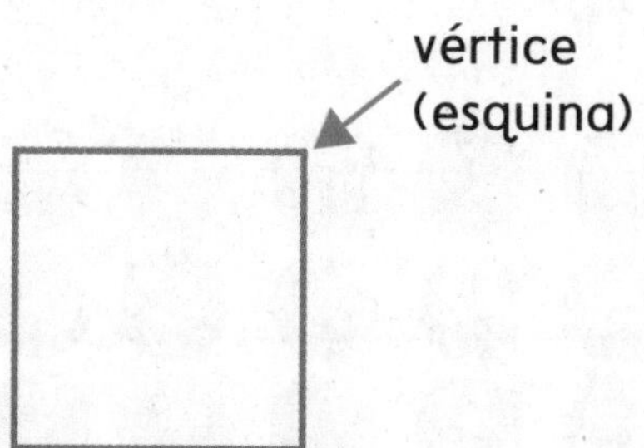

## Ww

### week

week

**April**

| Sunday | Monday | Tuesday | Wednesday | Thursday | Friday | Saturday |
|---|---|---|---|---|---|---|
| | | 1 | 2 | 3 | 4 | 5 |
| 6 | 7 | 8 | 9 | 10 | 11 | 12 |
| 13 | 14 | 15 | 16 | 17 | 18 | 19 |
| 20 | 21 | 22 | 23 | 24 | 25 | 26 |
| 27 | 28 | 29 | 30 | | | |

### semana

semana

**abril**

| domingo | lunes | martes | miércoles | jueves | viernes | sábado |
|---|---|---|---|---|---|---|
| | | 1 | 2 | 3 | 4 | 5 |
| 6 | 7 | 8 | 9 | 10 | 11 | 12 |
| 13 | 14 | 15 | 16 | 17 | 18 | 19 |
| 20 | 21 | 22 | 23 | 24 | 25 | 26 |
| 27 | 28 | 29 | 30 | | | |

## weight

heavy light

## peso

pesado liviano

# Yy

## year

**January**

| S | M | T | W | T | F | S |
|---|---|---|---|---|---|---|
| | | | | | | 1 |
| 2 | 3 | 4 | 5 | 6 | 7 | 8 |
| 9 | 10 | 11 | 12 | 13 | 14 | 15 |
| 16 | 17 | 18 | 19 | 20 | 21 | 22 |
| 23 | 24 | 25 | 26 | 27 | 28 | 29 |
| 30 | 31 | | | | | |

**February**

| S | M | T | W | T | F | S |
|---|---|---|---|---|---|---|
| | | 1 | 2 | 3 | 4 | 5 |
| 6 | 7 | 8 | 9 | 10 | 11 | 12 |
| 13 | 14 | 15 | 16 | 17 | 18 | 19 |
| 20 | 21 | 22 | 23 | 24 | 25 | 26 |
| 27 | 28 | | | | | |

**March**

| S | M | T | W | T | F | S |
|---|---|---|---|---|---|---|
| | | 1 | 2 | 3 | 4 | 5 |
| 6 | 7 | 8 | 9 | 10 | 11 | 12 |
| 13 | 14 | 15 | 16 | 17 | 18 | 19 |
| 20 | 21 | 22 | 23 | 24 | 25 | 26 |
| 27 | 28 | 29 | 30 | 31 | | |

**April**

| S | M | T | W | T | F | S |
|---|---|---|---|---|---|---|
| | | | | | 1 | 2 |
| 3 | 4 | 5 | 6 | 7 | 8 | 9 |
| 10 | 11 | 12 | 13 | 14 | 15 | 16 |
| 17 | 18 | 19 | 20 | 21 | 22 | 23 |
| 24 | 25 | 26 | 27 | 28 | 29 | 30 |

**May**

| S | M | T | W | T | F | S |
|---|---|---|---|---|---|---|
| 1 | 2 | 3 | 4 | 5 | 6 | 7 |
| 8 | 9 | 10 | 11 | 12 | 13 | 14 |
| 15 | 16 | 17 | 18 | 19 | 20 | 21 |
| 22 | 23 | 24 | 25 | 26 | 27 | 28 |
| 29 | 30 | 31 | | | | |

**June**

| S | M | T | W | T | F | S |
|---|---|---|---|---|---|---|
| | | | 1 | 2 | 3 | 4 |
| 5 | 6 | 7 | 8 | 9 | 10 | 11 |
| 12 | 13 | 14 | 15 | 16 | 17 | 18 |
| 19 | 20 | 21 | 22 | 23 | 24 | 25 |
| 26 | 27 | 28 | 29 | 30 | | |

**July**

| S | M | T | W | T | F | S |
|---|---|---|---|---|---|---|
| | | | | | 1 | 2 |
| 3 | 4 | 5 | 6 | 7 | 8 | 9 |
| 10 | 11 | 12 | 13 | 14 | 15 | 16 |
| 17 | 18 | 19 | 20 | 21 | 22 | 23 |
| 24 | 25 | 26 | 27 | 28 | 29 | 30 |
| 31 | | | | | | |

**August**

| S | M | T | W | T | F | S |
|---|---|---|---|---|---|---|
| | 1 | 2 | 3 | 4 | 5 | 6 |
| 7 | 8 | 9 | 10 | 11 | 12 | 13 |
| 14 | 15 | 16 | 17 | 18 | 19 | 20 |
| 21 | 22 | 23 | 24 | 25 | 26 | 27 |
| 28 | 29 | 30 | 31 | | | |

**September**

| S | M | T | W | T | F | S |
|---|---|---|---|---|---|---|
| | | | | 1 | 2 | 3 |
| 4 | 5 | 6 | 7 | 8 | 9 | 10 |
| 11 | 12 | 13 | 14 | 15 | 16 | 17 |
| 18 | 19 | 20 | 21 | 22 | 23 | 24 |
| 25 | 26 | 27 | 28 | 29 | 30 | |

**October**

| S | M | T | W | T | F | S |
|---|---|---|---|---|---|---|
| | | | | | | 1 |
| 2 | 3 | 4 | 5 | 6 | 7 | 8 |
| 9 | 10 | 11 | 12 | 13 | 14 | 15 |
| 16 | 17 | 18 | 19 | 20 | 21 | 22 |
| 23 | 24 | 25 | 26 | 27 | 28 | 29 |
| 30 | 31 | | | | | |

**November**

| S | M | T | W | T | F | S |
|---|---|---|---|---|---|---|
| | | 1 | 2 | 3 | 4 | 5 |
| 6 | 7 | 8 | 9 | 10 | 11 | 12 |
| 13 | 14 | 15 | 16 | 17 | 18 | 19 |
| 20 | 21 | 22 | 23 | 24 | 25 | 26 |
| 27 | 28 | 29 | 30 | | | |

**December**

| S | M | T | W | T | F | S |
|---|---|---|---|---|---|---|
| | | | | 1 | 2 | 3 |
| 4 | 5 | 6 | 7 | 8 | 9 | 10 |
| 11 | 12 | 13 | 14 | 15 | 16 | 17 |
| 18 | 19 | 20 | 21 | 22 | 23 | 24 |
| 25 | 26 | 27 | 28 | 29 | 30 | 31 |

## año

**enero**

| d | l | m | m | j | v | s |
|---|---|---|---|---|---|---|
| | | | | | | 1 |
| 2 | 3 | 4 | 5 | 6 | 7 | 8 |
| 9 | 10 | 11 | 12 | 13 | 14 | 15 |
| 16 | 17 | 18 | 19 | 20 | 21 | 22 |
| 23 | 24 | 25 | 26 | 27 | 28 | 29 |
| 30 | 31 | | | | | |

**febrero**

| d | l | m | m | j | v | s |
|---|---|---|---|---|---|---|
| | | 1 | 2 | 3 | 4 | 5 |
| 6 | 7 | 8 | 9 | 10 | 11 | 12 |
| 13 | 14 | 15 | 16 | 17 | 18 | 19 |
| 20 | 21 | 22 | 23 | 24 | 25 | 26 |
| 27 | 28 | | | | | |

**marzo**

| d | l | m | m | j | v | s |
|---|---|---|---|---|---|---|
| | | 1 | 2 | 3 | 4 | 5 |
| 6 | 7 | 8 | 9 | 10 | 11 | 12 |
| 13 | 14 | 15 | 16 | 17 | 18 | 19 |
| 20 | 21 | 22 | 23 | 24 | 25 | 26 |
| 27 | 28 | 29 | 30 | 31 | | |

**abril**

| d | l | m | m | j | v | s |
|---|---|---|---|---|---|---|
| | | | | | 1 | 2 |
| 3 | 4 | 5 | 6 | 7 | 8 | 9 |
| 10 | 11 | 12 | 13 | 14 | 15 | 16 |
| 17 | 18 | 19 | 20 | 21 | 22 | 23 |
| 24 | 25 | 26 | 27 | 28 | 29 | 30 |

**mayo**

| d | l | m | m | j | v | s |
|---|---|---|---|---|---|---|
| 1 | 2 | 3 | 4 | 5 | 6 | 7 |
| 8 | 9 | 10 | 11 | 12 | 13 | 14 |
| 15 | 16 | 17 | 18 | 19 | 20 | 21 |
| 22 | 23 | 24 | 25 | 26 | 27 | 28 |
| 29 | 30 | 31 | | | | |

**junio**

| d | l | m | m | j | v | s |
|---|---|---|---|---|---|---|
| | | | 1 | 2 | 3 | 4 |
| 5 | 6 | 7 | 8 | 9 | 10 | 11 |
| 12 | 13 | 14 | 15 | 16 | 17 | 18 |
| 19 | 20 | 21 | 22 | 23 | 24 | 25 |
| 26 | 27 | 28 | 29 | 30 | | |

**julio**

| d | l | m | m | j | v | s |
|---|---|---|---|---|---|---|
| | | | | | 1 | 2 |
| 3 | 4 | 5 | 6 | 7 | 8 | 9 |
| 10 | 11 | 12 | 13 | 14 | 15 | 16 |
| 17 | 18 | 19 | 20 | 21 | 22 | 23 |
| 24 | 25 | 26 | 27 | 28 | 29 | 30 |
| 31 | | | | | | |

**agosto**

| d | l | m | m | j | v | s |
|---|---|---|---|---|---|---|
| | 1 | 2 | 3 | 4 | 5 | 6 |
| 7 | 8 | 9 | 10 | 11 | 12 | 13 |
| 14 | 15 | 16 | 17 | 18 | 19 | 20 |
| 21 | 22 | 23 | 24 | 25 | 26 | 27 |
| 28 | 29 | 30 | 31 | | | |

**septiembre**

| d | l | m | m | j | v | s |
|---|---|---|---|---|---|---|
| | | | | 1 | 2 | 3 |
| 4 | 5 | 6 | 7 | 8 | 9 | 10 |
| 11 | 12 | 13 | 14 | 15 | 16 | 17 |
| 18 | 19 | 20 | 21 | 22 | 23 | 24 |
| 25 | 26 | 27 | 28 | 29 | 30 | |

**octubre**

| d | l | m | m | j | v | s |
|---|---|---|---|---|---|---|
| | | | | | | 1 |
| 2 | 3 | 4 | 5 | 6 | 7 | 8 |
| 9 | 10 | 11 | 12 | 13 | 14 | 15 |
| 16 | 17 | 18 | 19 | 20 | 21 | 22 |
| 23 | 24 | 25 | 26 | 27 | 28 | 29 |
| 30 | 31 | | | | | |

**noviembre**

| d | l | m | m | j | v | s |
|---|---|---|---|---|---|---|
| | | 1 | 2 | 3 | 4 | 5 |
| 6 | 7 | 8 | 9 | 10 | 11 | 12 |
| 13 | 14 | 15 | 16 | 17 | 18 | 19 |
| 20 | 21 | 22 | 23 | 24 | 25 | 26 |
| 27 | 28 | 29 | 30 | | | |

**diciembre**

| d | l | m | m | j | v | s |
|---|---|---|---|---|---|---|
| | | | | 1 | 2 | 3 |
| 4 | 5 | 6 | 7 | 8 | 9 | 10 |
| 11 | 12 | 13 | 14 | 15 | 16 | 17 |
| 18 | 19 | 20 | 21 | 22 | 23 | 24 |
| 25 | 26 | 27 | 28 | 29 | 30 | 31 |

# Yy

## yesterday

yesterday today

April

| Sunday | Monday | Tuesday | Wednesday | Thursday | Friday | Saturday |
|---|---|---|---|---|---|---|
| | | 1 | 2 | 3 | 4 | 5 |
| 6 | 7 | 8 | 9 | 10 | 11 | 12 |
| 13 | 14 | 15 | 16 | 17 | 18 | 19 |
| 20 | 21 | 22 | 23 | 24 | 25 | 26 |
| 27 | 28 | 29 | 30 | | | |

## ayer

ayer
hoy
abril
domingo lunes martes miércoles jueves viernes sábado
1 2 3 4 5
6 7 8 9 10 11 12
13 14 15 16 17 18 19
20 21 22 23 24 25 26
27 28 29 30


# Zz

## zero

## cero

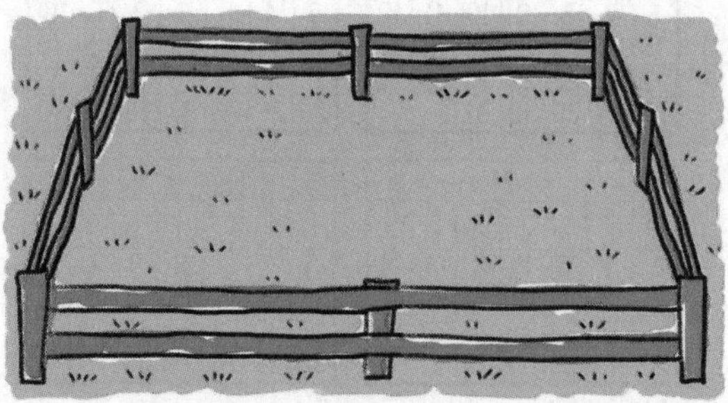

Name

# Work Mat 1: Five-Frame

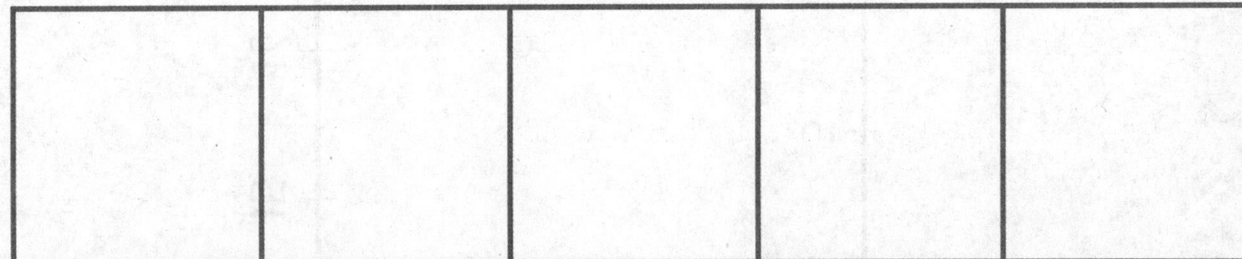

# Work Mat 2: Number Lines

0 1 2 3 4 5 6 7 8 9 10

11 12 13 14 15 16 17 18 19 20

Name

**Work Mat 3: Ten-Frame**

|   |   |
|---|---|
|   |   |
|   |   |
|   |   |
|   |   |
|   |   |

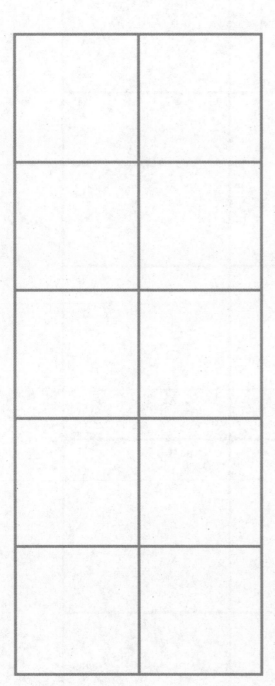

# Work Mat 4: Ten-Frames

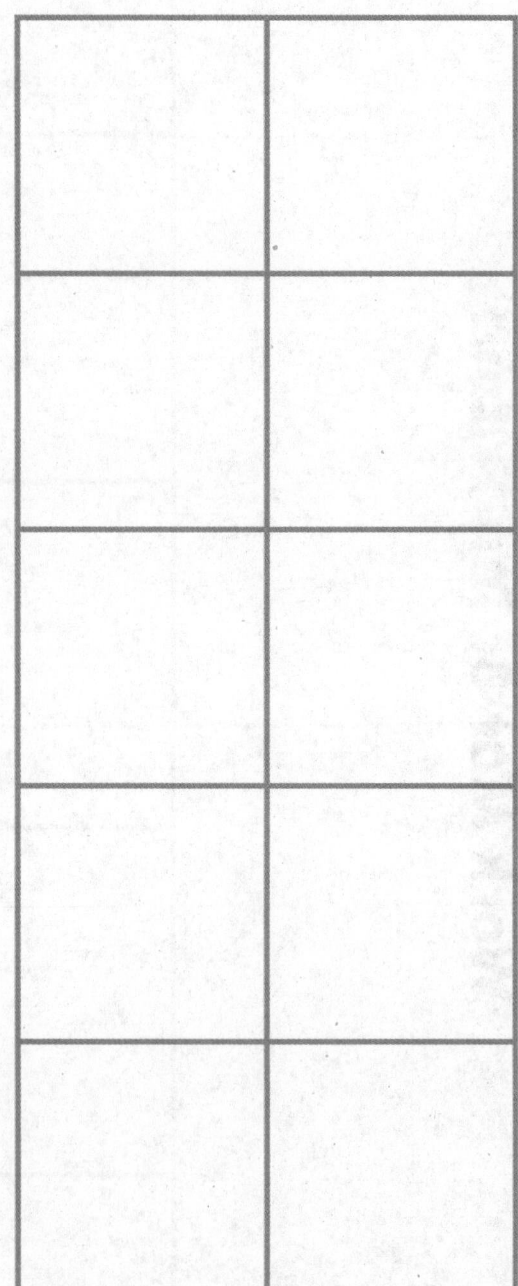

Name

Work Mat 5: Story Mat

# **Work Mat 6:** Part-Part-Whole

<table>
<tr><th>Part</th><th>Part</th></tr>
<tr><td></td><td></td></tr>
<tr><th colspan="2">Whole</th></tr>
<tr><td colspan="2"></td></tr>
</table>